ALUMINIUM ALLOYS - NEW TRENDS IN FABRICATION AND APPLICATIONS

Edited by **Zaki Ahmad**

Aluminium Alloys - New Trends in Fabrication and Applications
http://dx.doi.org/10.5772/3354
Edited by Zaki Ahmad

Contributors

Pedro Vilaça, Patiphan Juijerm, Igor Altenberger, Vaclav - Sklenicka, Jiri Dvorak, Petr Kral, Milan Svoboda, Marie Kvapilova, Wojciech Libura, Artur Rekas, Alfredo Flores, Mohamed Mazari, Mohamed Benguediab, Mokhtar Zemri, Benattou Bouchouicha, Victor Songmene, Jules Kouam, Imed Zaghbani, Nick Parson, Alexandre Maltais, Amir Farzaneh, Maysam Mohammadi, Zaki Ahmad, Nick Birbilis, Mumin SAHIN, Cenk Misirli, Paola Leo, Marek Balazinski, Patrick Hendrick

Published by InTech
Janeza Trdine 9, 51000 Rijeka, Croatia

Publishing Process Manager Iva Simcic
Typesetting InTech DTP Team
Cover InTech Design Team

Additional hard copies can be obtained from orders@intechopen.com

Aluminium Alloys - New Trends in Fabrication and Applications, Edited by Zaki Ahmad
p. cm.
ISBN 978-953-51-0861-0

For EU product safety concerns:
IN TECH d.o.o., Prolaz Marije Krucifikse Kozulić 3, 51000 Rijeka, Croatia,
info@intechopen.com or visit our website at intechopen.com.

Meet the editor

Prof. Zaki Ahmad worked at King Fahd University of Petroleum and Minerals for thirty years in rendered distinguished services in teaching and research. He is a professor Emeritus of King Fahd University and Adjunct Professor at COMSATS, Lahore, Pakistan. He obtained his PhD from LEEDS University, UK in Metallurgy. He is a chartered metallurgical engineer(C.Eng) from Engineering Council UK, fellow of the institute of Materials, Minerals and Mining(FIMMM), a member European federation of corrosion and a Fellow of The institute of Metal Finishing. He substantially contributed to the founding activities in Material Science, Corrosion Engineering, green Engineering and nanotechnology at KFUPM and at Razi Metallurgical Center Iran. Prof. Zaki Ahmad worked on international projects on aluminum with Aluminum, Ranshofen, Austria and Forschungzentrum, Geethscht, Germany and with Metallgesselscheft, Germany. He worked on international projects with Ministry of Technology, Germany. Prof. Zaki Ahmad is a founder contributor of Center of excellence in Corrosion at KFUPM, Dhahran, Saudi Arabia. He worked on the foundation and development of Nanotechnology in Saudi Arabia in 2004. He wrote over one hundred research papers in international journals and over forty papers in international research conferences.

Contents

Preface IX

Section 1 Properties and Structure of Aluminium Alloys 1

Chapter 1 **Equal-Channel Angular Pressing and Creep in Ultrafine-Grained Aluminium and Its Alloys 3**
Vaclav Sklenicka, Jiri Dvorak, Milan Svoboda, Petr Kral and Marie Kvapilova

Chapter 2 **Durability and Corrosion of Aluminium and Its Alloys: Overview, Property Space, Techniques and Developments 47**
N. L. Sukiman, X. Zhou, N. Birbilis, A.E. Hughes, J. M. C. Mol, S. J. Garcia, X. Zhou and G. E. Thompson

Chapter 3 **Influence of Structural Parameters on the Resistance on the Crack of Aluminium Alloy 99**
Mohamed Mazari, Mohamed Benguediab, Mokhtar Zemri and Benattou Bouchouicha

Chapter 4 **Effect of Micro Arc Oxidation Coatings on the Properties of Aluminium Alloys 107**
Cenk Mısırlı, Mümin Şahin and Ufuk Sözer

Section 2 Extrusion, Rolling and Machining 121

Chapter 5 **Effects of Deep Rolling and Its Modification on Fatigue Performance of Aluminium Alloy AA6110 123**
Patiphan Juijerm and Igor Altenberger

Chapter 6 **Numerical Modelling in Designing Aluminium Extrusion 137**
Wojciech Libura and Artur Rękas

Chapter 7 **Linear Friction Based Processing Technologies for Aluminum Alloys: Surfacing, Stir Welding and Stir Channeling 159**
Pedro Vilaça, João Gandra and Catarina Vidal

Chapter 8 **Dry, Semi-Dry and Wet Machining of 6061-T6 Aluminium Alloy 199**
J. Kouam, V. Songmene, M. Balazinski and P. Hendrick

Chapter 9 **Global Machinability of Al-Mg-Si Extrusions 223**
V. Songmene, J. Kouam, I. Zaghbani, N. Parson and A. Maltais

Section 3 Heat Treatment and Welding 253

Chapter 10 **Pure 7000 Alloys: Microstructure, Heat Treatments and Hot Working 255**
P. Leo and E. Cerri

Section 4 Durability, Degradation and Recycling of Aluminium Alloys 275

Chapter 11 **Mechanical and Metalurgical Properties of Friction Welded Aluminium Joints 277**
Mumin Sahin and Cenk Misirli

Chapter 12 **Elaboration of Al-Mn Alloys by Aluminothermic Reduction of Mn2O3 301**
A. Flores Valdés , J. Torres and R. Ochoa Palacios

Section 5 Application of Aluminium Alloys in Solar Power 323

Chapter 13 **Aluminium Alloys in Solar Power – Benefits and Limitations 325**
Amir Farzaneh, Maysam Mohammadi, Zaki Ahmad and Intesar Ahmad

Preface

Aluminum alloys are not only serving aerospace, automotive and renewable energy industry they are being extensively used in surface modification processes at nanoscale such as modified phosphoric acid anodizing process to create high surface activity of nanoparticles. Benign joining of ultra-fine grained aerospace aluminum alloys using nanotechnology is highly promising. Super hydrophobic surfaces have been created at a nanoscale to make the surfaces dust and water repellent. The biggest challenge lies in producing nanostructure metals at competitive costs. Severe plastic deformation (SPD) is being developed to produce nonmaterial for space applications. The focus of scientists on using aluminum alloys for direct generation of hydrogen is rapidly increasing and dramatic progress has been made in fabrication of Aluminum, Gallium and Indium alloys. It can therefore seen that the importance of aluminum has never declined and it continues to be material which has attracted the attention of scientists and engineers in all emerging technologies.

In the context of the above comments, there is ample justification for publishing this book. The chapter by Prof. Sahin Mumin describes some of the important fundamental properties related to metallurgical properties and welding. The procedure and structural details of friction stir welding and friction stir channeling has been demonstrated by Dr. Vilaça Pedro with beautiful illustrations, deep rolling ageing and and fatigue control the surface properties of auminium alloys. Dr.Ing. Juijerm Pathipham, has described the impact of the above factors comprehensively. Prof. Sklenicka Vadov has described the equal channel angular pressing in relation to producing ultra five grains materials with profuse illustrations and graphics. The readers interested in numerical modeling would find the chapter on numerical modeling very productive. Chapter on machanability by Prof. Songmene Victor focuses on auminum, magnesiun and silicon alloys. The effect of micro arc oxidation coating on structure and mechanical parameters has been shown by Prof. Sahin Mumin. Aluminum is being increasingly used in solar power due to its attributes and it is extensively used in concentrating solar power (CSP) and photovoltalic solar cells (PV). The reader interested in renewable energy would find the chapter on aluminum alloys in solar power highly interesting. The section of corrosion of PV modules has been written comprehensively in this chapter. It is a good example of international collaboration as shown by the authors from Iran, Canada, Pakistan and Saudi Arabia. InTech is to be congratulated for bringing a book on Aluminum alloys with new dimensions proliferating in venues of emerging technologies. I hope students at graduate level and all the researchers would find this book of great interest and severe topic would stimulate them in undertaking further research in areas of interest.

The spirit of my deceased father Wali Ahmed and loving mother Jameela Begum and my deceased son Intekhab Ahmed has motivated me in all my academic contributions including this book. I thank Shamsujjehan, Huma Begum, Abida Begum, Farhat Sultana for their encouragement. I thank my grandson Mr. Mishaal Ahmed for his help. I thank the director of COMSATS Dr. M Bodla, Dr. Talat Afza , Head of Academics and Research COMSATS and Dr. Assadullah Khan, Head of Chemical Department for encouragement. I thank King Fahd University of Petroleum and Minerals, Dhahran, Saudi Arabia for providing me very productive working years and environment. I thank Miss Zahra Khan and Miss Tayyeba of Chem. Eng Dept. I thank Dr Intesar Ahmed of Lahore College for Women University and Mr. Manzar Ahmed of University of South Asia for their help. Finally, I thank Allah Almighty for his countless blessings.

Zaki Ahmad
King Fahad University Of Petroleum and Minerals
Dhahran, Saudi Arabia

Section 1

Properties and Structure of Aluminium Alloys

Chapter 1

Equal-Channel Angular Pressing and Creep in Ultrafine-Grained Aluminium and Its Alloys

Vaclav Sklenicka, Jiri Dvorak, Milan Svoboda,
Petr Kral and Marie Kvapilova

Additional information is available at the end of the chapter

http://dx.doi.org/10.5772/51242

1. Introduction

Creep strength and ductility are the key creep properties of creep-resistant materials but these properties typically have opposing characteristics. Thus, materials with conventional grain sizes may be strong or ductile but there are rarely both. In this connection, recent findings of high strength and good ductility in several submicrometer metals and alloys are of special interest [1]. Reduction of the grain size of a polycrystalline material can be successfully produced through advanced synthesis processes such as the electrodeposition technique [2] and severe plastic deformation SPD [1,3-6]. Although creep is an exceptionally old area of research, above mentioned processing techniques have become available over the last two decades which provide an opportunity to expand the creep behaviour into new areas that were not feasible in earlier experiments. Creep testing of nanocrystalline (grain size $d < 100$ nm) and ultrafine-grained ($d < 1$ μm) materials is characterized by features that may be different from those documented for coarse-grained materials and thus cannot easily be compared.

Processing through the application of severe plastic deformation (SPD) is now an accepted procedure for producing bulk ultrafine-grained materials having grain sizes in the submicrometer or nanometer range. The use of SPD enhances certain material properties through the introduction of an ultrafine-grained microstructure. The ultrafine size of the grains in the bulk materials generally leads to significantly improved properties by comparison with polycrystalline materials having conventional grain sizes of the same chemical composition. Several SPD processing techniques are currently available but the most attractive technique is equal-channel angular pressing (ECAP), where the sample is pressed through a die constrained within a channel bent through an abrupt angle [4]. There are numerous reports of the processing of various pure metals and metallic alloys by ECAP and many of these reports involve a characterization of the microstructure and an investigation of the mechanical properties at ambient temperatures. There are also several reports of the tensile properties of the as-pressed materials at elevated temperatures with a special emphasis on the potential for achieving high superplastic elongations. However, the tests at elevated temperatures are invariably conducted under conditions of constant strain rate and, by contrast, only very limited reports are available describing the creep behaviour of aluminium and some aluminium alloys. Furthermore, the results for high-purity aluminium, which are the most extensive available to date, appear anomalous because under some testing conditions of stress and temperature the measured minimum or steady-state creep rates in the pressed materials with ultrafine grain sizes where slower than in the same material in a coarse-grained unpressed condition.

This chapter was initiated to provide basic information on the creep behaviour and microstructural characteristics of aluminium and some aluminium alloys. The chapter has arisen in connection with long-term research activity of the Advanced High Temperature Materials Group at the Institute of Physics of Materials, Academy of Sciences of the Czech Republic in Brno, Czech Republic. Thus, the objective of this chapter is to present an overview of some results of our current research in creep behaviour and a link between the microstructure and the creep properties of ultrafine-grained aluminium based alloys. Throughout the text, our results are compared with theoretical models and relevant experimental observations published in the literature.

2. The development of processing using equal-channel angular pressing (ECAP)

Processing by severe plastic deformation (SPD) may be defined as those metal forming procedures in which a very high strain is imposed on a bulk solid without the introduction of any significant change in the overall dimensions of the solid and leading to the production of exceptional grain refinement to that the processed bulk solids have 1000 or more grains in section [4]. Of a wide diversity of new SPD procedures, equal-channel angular pressing (ECAP) is an especially attractive processing technique. It is relatively simple procedure which can be applied to fairly large billets of many materials ranging from pure metals to precipitation-hardened alloys, intermetallics and metal-matrix composites.

2.1. Principles of ECAP

The principle of ECAP is illustrated schematically in Figure 1. For the die shown in Figure 1, the internal channel is bent through an abrupt angle, Φ, and there is an additional angle, Ψ, which represents outer arc of curvature where the two channels intersect. The sample, in the form of a rod or bar, is machined to fit within channel and the die is placed in some form of fuss so that the sample can be pressed through the die using a plunger. The nature of the imposed deformation is simple shear which occurs as the billet passes through the die. The retention of the same cross-sectional area when processing by ECAP, despite the introduction of very large strains, is the important characteristic of SPD processing and it is characteristic which distinguishes this type of processing from conventional metal-working operations such as rolling, extrusion and drawing. Since the cross-sectional area remains unchanged, the same billet may be pressed repetitively to attain exceptionally high strain.

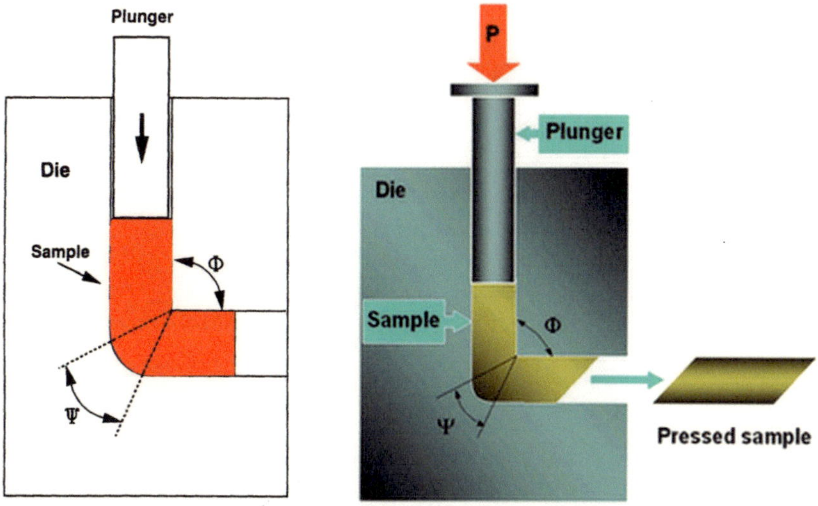

Figure 1. Principle of ECAP.

Aluminium and its alloys used in this investigation were pressed using an experimental facility for ECAP installed in the Institute of Physics of Materials, Academy of Sciences of the Czech Republic (Figure 2). The die was placed on a testing machine Zwick. ECAP was conducted mostly at room temperature with a die that had internal angle 90° between two parts of the channel and an outer arc of curvature of ~ 20°, where these two parts intersect. It can be shown from first principles that these angles lead to an imposed strain of ~ 1 in each passage of the sample. The ECAP die involved the use of billets of the length of ~ 50 – 60 mm with square cross-section of 10 mm x 10 mm. The velocity of plunger was 10 mm/min.

2.2. The processing routes in ECAP

The use of repetitive pressing provides an opportunity to invoke different slip systems on each consecutive pass by simply rotating the samples in different ways. The four different processing routes are summarized schematically in Figure 3 [7]. In route A the sample is pressed without rotation, in route B_A the sample is rotated by 90° in alternate directions between consecutive passes, in route B_C the sample is rotated by 90° in the same sense (either clockwise or counter clockwise) between each pass and in route C the sample is rotated by 180° between passes. The distinction between these routes and the difference in number of ECAP passes may lead to variations both in the macroscopic distortions of the individual grains [8] and in the capability to develop a reasonably homogeneous and equiaxed ultrafine-grained microstructure.

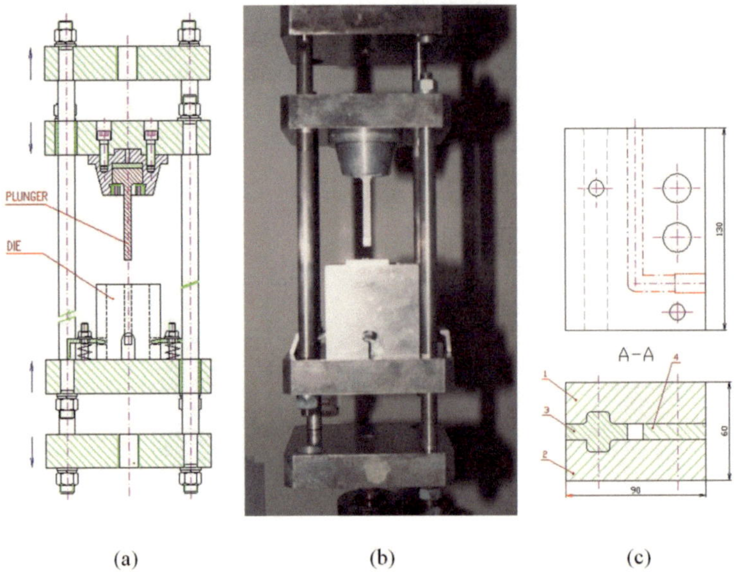

Figure 2. Adaptation of testing ZWICK machine for ECAP pressing (a, b), and (c) sketch of ECAP die design.

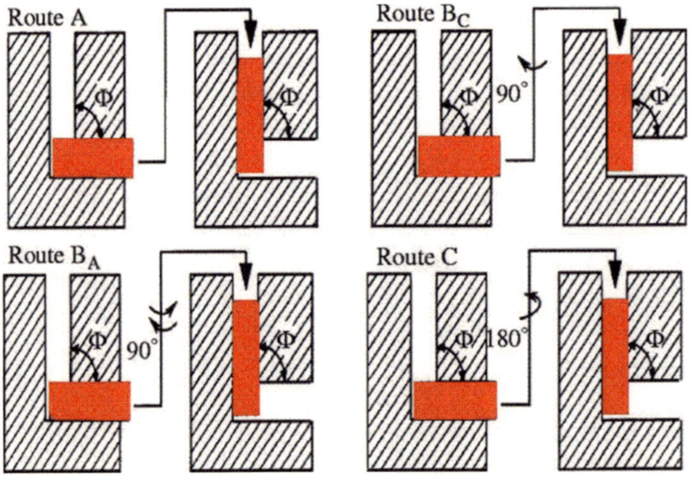

Figure 3. Schematic of four ECAP routes for repetitive pressing.

In this work the ECAP pressing was conducted in such a way that one or repetitive pressing was conducted followed either route A, B (route B_C was used only) or C. Detailed examinations of the effect of different processing routes showed that route B_C leads to the most rapid evolution into an array of high-angle grain boundaries [9,10]. The result is explained by considering the shearing patterns developed in the samples during each processing route. Thus,

http://dx.doi.org/10.5772/51242

the route B_C is most probably the optimum ECAP processing route at least for the pressing of pure aluminium and its alloys [4].

2.3. Mechanical properties and defects achieved using ECAP

During the last two decades it has been demonstrated that an ultrafine-grained structure of materials processed by ECAP may lead to significantly higher strength and hardness but to a reduction in the ductility [4]. In this connection after ECAP the mechanical properties were tested mostly at room temperature using a testing machine operating at a constant rate of 2.0 x $10^{-4}s^{-1}$ of crosshead displacement.

2.3.1. Tensile properties

Tensile tests were conducted at 293 K on pure aluminium after processing by ECAP for samples after different number of ECAP passes. In limited extent mechanical tests were performed on the samples after ECAP and static annealing at 473 K [11]. In Figure 4 the tensile data are summarized as a function of the number of passes. It is apparent from these figures that a very significant increase in yield and ultimate tensile stress occurred after the first pressing. The subsequent pressing further increased yield and ultimate stress values but to a lower rate. Further, a saturation of the level of both the parameters was attained after four passes.

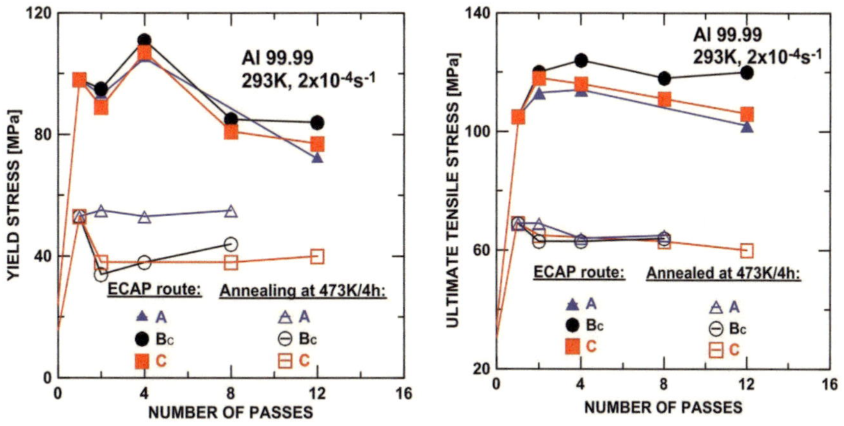

Figure 4. Influence of different ECAP routes and different number of ECAP passes on (a) yield stress, and (b) ultimate tensile stress after static annealing.

From Figure 4 it can be also noticed that static annealing at 473 K leads to a substantial decrease in the level of yield and ultimate tensile stress values due to diffusion based recovery processes for all the ECAP processed samples. No significant differences in mechanical properties among the ECAP process routes examined were found. Further, from Figure 4 is clear that although the levels of the tensile data for ECAPed Al highly decrease with the number of ECAP passes, the stress levels after 8 passes are much higher than the stress levels in the annealing state and these differences come to more than twice. This result indicates that, when compared with the tensile behaviour of the annealed state, the flow stress is considerably improved through the application of ECAP [11,12].

2.3.2. *Hardness measurements*

Figure 5a shows Vickers microhardness plotted against the number of ECAP passes for extremely high purity aluminium (99.99%) [12]. The hardness increases up to two passes to take a maximum due to the very high dislocation density. However, subsequent passes lead to a decrease in the hardness because many of the subgrain boundaries evolve into high-angle grain boundaries. Figure 5b shows Vickers microhardness plotted against different periods of time of a static annealing at 473 K for pure (99.99%) Al processed by ECAP by two different processing routes. A pronounced decrease of microhardness with an increase of annealing time can be explained by significant grain growth and softening of pressed material during an annealing exposures [11].

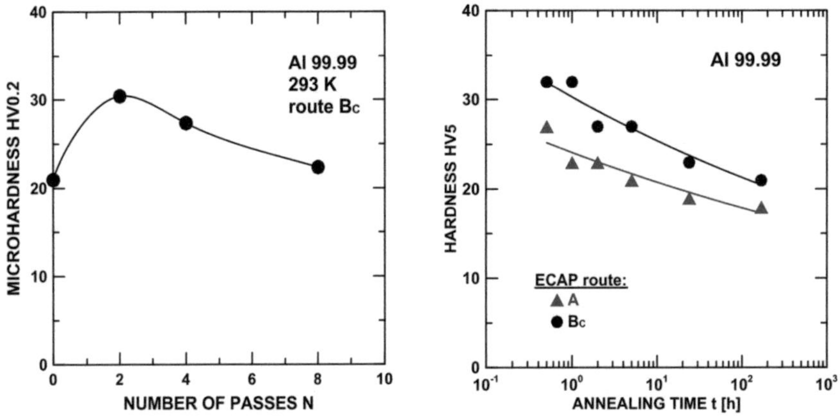

Figure 5. Hardness changes (a) with respect to number of ECAP passes, and (b) as a function of annealing time at 473 K for two different ECAP routes.

2.3.3. *Nanoporosity after ECAP processing*

It is generally recognized that the ECAP process could produce a submicrocrystalline bulk material with a relatively uniform structure and 100% density for a wide range of materials from pure metals, solid-solution alloys, commercial alloys, to metal matrix composites [1]. However, the previously performed analysis of the data on the influence of the number of passes of equal-channel angular pressing on the elastic-plastic properties and defect structure of pure aluminium demonstrated that these characteristics of mechanical properties are substantially affected by the evolution of the nanoporosity formed during equal-channel angular pressing [13-15]. Thus, to determine the total volume of nanoporosity which could be generated by ECAP, two selected samples of pure aluminium were pressed for a total of one (specimen A1) and four (specimen A4) ECAP passes, respectively, and for comparison reasons some part of these specimens were underwent by subsequent pressurization treatment by high hydrostatic pressure [16]. The samples were investigated by small-angle X-ray scattering (SAXS) and dilatometry [13].

Some differences were found in the fractional volume of the nanopores ΔV/V when compared specimen A1 to specimen A4. The values $\Delta V/V_{max}$ correspond to the as-pressed state of specimens (after ECAP only) and the values $\Delta V/V_{min}$ were evaluated for the state after ECAP and subsequent pressurization which represents a rejuvenative treatment for elimination of nanopores. The evaluated values are $\Delta V/V_{max} = 5x10^{-3}$ and $\Delta V/V_{min} = 2.5.10^{-3}$ for specimen A1 and $\Delta V/V_{max} = 7x10^{-3}$ and $\Delta V/V_{min} = 3x10^{-3}$ for specimen A4, respectively. No substantial difference in the average size of the nanopores (~ 20-30nm) was found between the specimens investigated. The values ΔV/V determined by small-angle X-ray scattering and dilatometry were about the same; e.g. the fractional volume $\Delta V/V_{min} = 3x10^{-3}$ by (SAXS) of specimen A4 agreed very well with $\Delta V/V_{min} = 2.5x10^{-3}$ as determined by dilatometry. On the basis of the aforementioned results we can conclude that ECAP deformation achieves strongly enhanced concentration of vacancy agglomerates type defects. The effect of the spectrum of the point defects and the internal stresses on elasticity and anelasticity of ECAP-ed aluminium has been reported elsewhere [17].

In recent years using a back-pressure ECAP facilities [4] has become an area of special interest. An important advantage in imposing a back-pressure may be a decrease of nanoporosity in the pressed material [18]. However, additional experiments are needed to evaluate the role of a back-pressure in elimination of nanoporosity.

3. Microstructural features of ultrafine-grained materials

Ultrafine-grained (UFG) materials processed by ECAP differ qualitatively and quantitatively from their coarse-grained (CG) counterparts in terms of their characteristic structural parameters and thus their creep behaviour cannot be easily compared with that documented for CG materials. It is important to note in this respect that UFG materials are characterized by great extension of internal interfaces; therefore, grain boundary diffusion processes have to be involved in the formation of their structure-sensitive properties, especially at elevated temperature [19].

The characteristics of the microstructures introduced by ECAP have been evaluated in numerous investigations [4]. However, most of these earlier investigations employed transmission electron microscopy (TEM) for determinations of the grain sizes produced by ECAP and the nature of any dislocation interactions occurring within grains. The application of modern imaging methods to the examination of microstructures in UFG materials processed by ECAP has permitted a more detailed investigation of a possible link between internal microstructures of UFG metals and alloys and their mechanical and/or creep behaviour [4]. Diffraction-based techniques for localized crystal orientation measurements, such as electron backscatter diffraction (EBSD), are of central importance today for characterizing fine-scale microstructural features [20-23].

The new experimental technique of EBSD considerably extended the possibilities of metallography to estimate reliably the quantitative structural characteristics of materials [23]. It enables the numerical classification of boundaries separating the regions of different orientations of their lattice structure. The magnitude of the mutual misorientation can be continuously selected and thus the regions with a misorientation less than a prescribed value as well as their boundaries can be recognized. There is a vast literature devoted to the observation by EBSD and precisely defined misorientation of boundaries and the conventional grain boundary classification based on suitably polished and etched planar surfaces as observed by optical microscopy or by boundaries observed by electron microscopy and EBSD (see e.g. [24]). As can be expected, the EBSD method is more reproducible, independent of detailed etching conditions etc., and the surface area intensities are usually higher (equivalently, the mean random profile chord is smaller). In this section a division of boundaries into true subboundaries with misorientations $\Delta < 10°$, transitional subboundaries with $10° \leq \Delta < 15°$and high-angle grain boundaries with $\Delta \geq 15°$ was made.

Such an approach is of primary importance in the examination of materials produced by severe plastic deformation (SPD), without change of shape, producing materials with ultrafine grains (e.g. [3,5]) and considerably different properties in comparison with CG materials. The reason for this difference is to a certain degree purely geometric and consists in different grain and subgrain boundary structures, which play an important role in mechanical, thermal and other properties.

This section describes the results of structural examinations of high purity aluminium and its selected precipitation-strengthened alloys processed by ECAP. The microstructure was revealed by TEM, SEM and EBSD and analyzed quantitatively by stereological methods. The various factors influencing the as-pressed microstructures including the total strain imposed in ECAP processing, the processing routes and the nature of materials are examined in detail.

3.1. Experimental materials and their microstructure after ECAP

3.1.1. Pure aluminium

The aluminium used in this investigation was an extremely coarse-grained (grain size ~ 5 mm) high purity (99.99%) Al supplied in the form of rods. The rods were cut into short billets having a length of ~ 60 mm and a cross-section 10 mm x 10 mm. ECAP was conducted at room temperature using route A, B_C and C. Full details on the processing have been described elsewhere [25-27].

TEM results have shown that one ECAP pass leads to a substantial reduction in the grain size (~ 1.4 μm), and the microstructure consists of parallel bands of grains oriented in the shearing direction. The microstructure is very inhomogeneous and the grain size varies from location to location. The inhomogeneous nature of the microstructure may reflect the coarse grain size (~ 5 mm) prior to ECAP. The grains subsequently evolve upon subsequent ECAP passes into a reasonably equiaxed and homogeneous microstructure with an average grain size of ~1 μm regardless of the particular ECAP routes. The microstructure is essentially homogeneous after four ECAP passes, although a tendency for grain elongation in the direction of the shear direction of the last pressing operation is retained. Figure 6 gives an example of the microstructure in the cross-section normal to the pressing direction after four subsequent ECAP passes performed in different routes. TEM micrographs in Figure 7 give an example of the microstructure in cross-section after four and eight subsequent ECAP passes by route B_c and C, respectively. The EBSD grain maps in Figure 8 indicate little dependence of the grain boundary disorientation distribution on the ECAPed Al processed by route B_c.

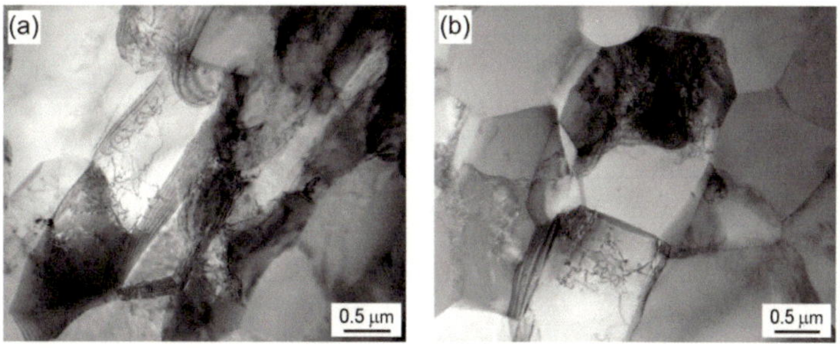

Figure 6. TEM micrographs of aluminium after four subsequent ECAP passes on route (a) A, and (b) B.

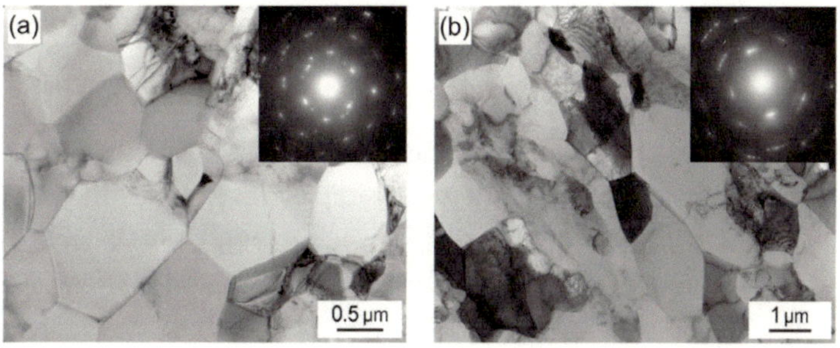

Figure 7. Typical microstructures and associated SAED patterns after passage through the die for (a) 4 pressings, route B and (b) 8 pressings, route C.

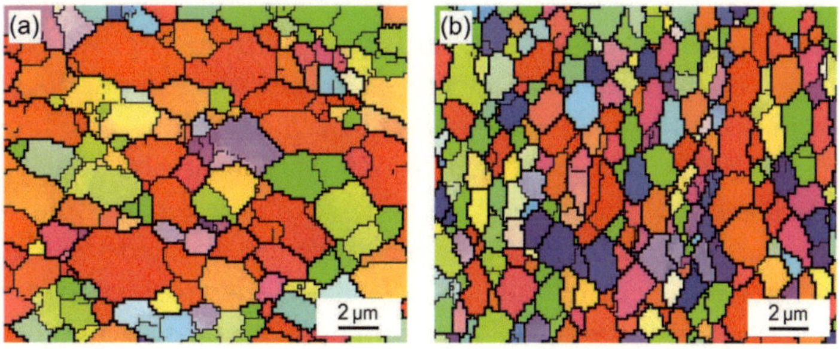

Figure 8. Grain maps for ECAPed Al after: (a) 4 passes, and (b) 8 passes by route B (EBSD).

It can be expected that the creep behaviour of the ultrafine-grained pure aluminium will critically depend on the thermal stability of the microstructure. To explore the thermal stability of ECAP processed aluminium load-less annealing was conducted at temperature of 473 K for different periods of time (i.e. at the temperature of the intended creep tests). Microscopic examination revealed that the post-ECAP annealing makes the ECAP microstructure quite unstable and a noticeable grain growth occurs at the very beginning of annealing (Table 1). Simultaneously, annealing at 473 K gives measurable change in the Vickers microhardness.

Annealing conditions	**ECAP 4 passes route A**		**ECAP 4 passes route B**	
	grain size [μm]	**microhardness HV5**	**grain size [μm]**	**microhardness HV5**
no annealing	0.9	37	0.9	38
473 K/ 0.5 h	6.6	27	4.5	32
473 K/ 1 h	7.9	23	4.8	32
473 K/ 2 h	7.3	23	4.8	27
473 K/ 5 h	7.3	21	5.3	27
473 K/ 24 h	12.2	19	5.0	23
473 K/ 168h	13.4	18	10.4	21

Table 1. Thermal stability and Vickers microhardness of the ECAP aluminium.

3.1.2. Precipitation-strengthened aluminium alloys

In evaluating the microstructure characteristics of ultrafine-grained materials processed by ECAP at elevated and high temperatures, it is very important to recognize that these ultrafine-grained microstructures are frequently unstable at these temperatures as it was just demonstrated by the above mentioned results of thermal instability of pressed pure aluminium. However, it is often feasible to retain an array of ultrafine grains even at very high temperatures by using materials containing second phases or arrays of precipitates. This was a reason why two precipitation-strengthened aluminium alloys were used in this investigation.

It has been shown that addition to aluminium alloys of even very small amounts of Sc (typically, ~ 0.2wt.%) strongly improves the microstructures of the alloys and their mechanical properties so that these alloys are suitable for use in engineering applications [28]. Scandium additions of ~ 0.2wt.%Sc to pure aluminium are sufficient to more or less retain a small grain size at elevated temperatures [29]. Further, some reports have demonstrated that it is possible to achieve high ductilities in Al-Mg-Sc alloys by using ECAP to introduce an exceptionally small grain size [30]. The creep behaviour of conventional Al-Mg alloys is extensively described in the literature. The synergy of solid-solution strengthening and precipitate strengthening has, however, not been extensively studied at elevated and high temperatures [31]. Very little information is available at present on the creep properties of ultrafine-grained Al-Sc and Al-Mg-Sc alloys [32-38]. Accordingly, the present investigation was initiated to provide a more complex information on the creep behaviour of these aluminium alloys in their ultrafine-grained states.

An Al-0.2wt.%Sc alloy was produced by diluting an Al-2.0wt.%Sc master alloy with 99.99wt.% pure aluminium. The resulting ingots were subjected to a homogenization and grain-coarsening treatment at 893 K for 12 hours and then aged in air at 623 K for 1 hour. In the as-fabricated condition, the extremely coarse grain size was measured as ~ 8 mm. The ECAP was conducted at the Institute of Physics of Materials AS CR Brno, Czech Republic, using the same die and procedure as it was reported earlier for pure aluminium (i.e. up to a total 8 ECAP passes at room temperature). The details concerning an Al-0.2wt.%Sc alloy have been reported elsewhere [33-35]. The ternary Al-Mg-Sc alloy was fabricated at the Department of Materials Science and Engineering, Faculty of Engineering, Kyushu University, Fukuoka, Japan. The alloy contained 3wt.%Mg and 0.2wt.%Sc and it was prepared from 99.99% purity Al, 99.999% purity Sc and 99.9% purity Mg. Full details on the fabrication procedure are given elsewhere [32] but, briefly, the alloy was cast, homogenized in air for 24 h at 753 K and solution treated for 1 h at 883 K. In the as-fabricated condition, the grain size was about 200 μm. Again, the ECAP was conducted using a solid die that had 90° angle between the die channels and each sample was pressed at room temperature repetitively for a total of eight passes by route B_C.

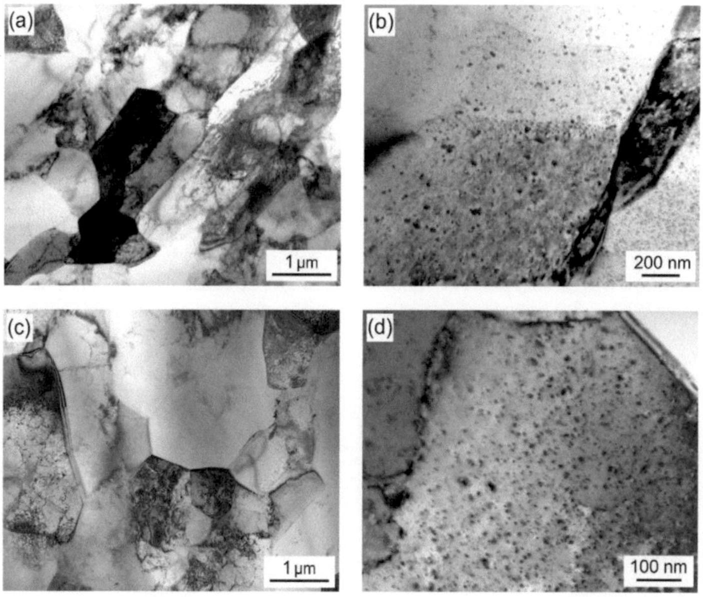

Figure 9. Microstructure in the Al-0.2wt.%Sc alloy: (a) and (b) after ECAP (B_C, 8 passes) and annealing for 1 h at 623 K, (c) and (d) after creep at 473 K.

Figures 9a,b and 10a,b show the microstructure of Al-0.2wt.%Sc and Al-3wt.%Mg-0.2wt.%Sc alloys in their as-pressed states. Experiments on Al-0.2wt.%Sc and Al-3wt.%Mg-0.2wt.%Sc alloys revealed that processing by ECAP reduced the grain size to ~ 0.4 μm and subsequent annealing at 623 K and 1 h and creep testing gave the grain sizes ~ 0.9 μm for an Al-0.2wt.%Sc alloy and ~ 1.5 μm for an Al-3wt.%Mg-0.2wt.%Sc alloy, respectively. Figures 9c,d and

10c,d give examples of the microstructure of the alloys in the longitudinal sections parallel to the pressing direction after creep exposures at 473 K. As will be shown later on no substantial difference in the relative fractions of high-angle (θ > 15°) grain boundary population after ECAP was found between the alloys investigated. These fractions were slightly increased during creep exposure up to an average value ~ 70%.

Figure 10. Microstructure in the Al-3wt.%Mg-0.2wt.%Sc alloy: (a) and (b) after ECAP (B_C, 8 passes) and annealing for 1 h at 623 K, (c) and (d) after creep at 473 K.

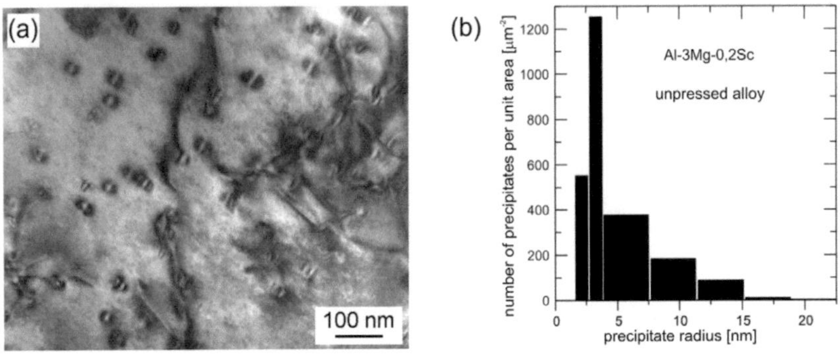

Figure 11. TEM micrograph showing the presence of coherent Al_3Sc precipitates in unpressed sample, and (b) precipitate size distribution in unpressed sample.

Figures 9b,d and 10b,d exhibit TEM micrographs, which demonstrate the presence of coherent Al_3Sc precipitates within the matrices of both alloys. Al_3Sc precipitates are indicated by a coherency strain contrast [39]. A mean size of precipitates was ~ 5 nm after creep testing of the Al-3wt.%Mg-0.2wt.%Sc alloy and a mean size of precipitates slightly large was recorded for the Al-0.2wt.%Sc alloy (~ 6 nm [35]). Figure 10d shows dislocation microstructure observed after creep exposure of ECAPed Al-3wt.%Mg-0.2wt.%Sc. The dislocation pairs present in both alloys containing the smallest precipitate radii are very frequent. For larger precipitates the dislocations are pinned efficiently by Al_3Sc precipitates as climbing becomes slower. Figure 11a exhibits TEM micrograph of an Al-3wt.%Mg-0.2 wt.%Sc alloy showing the presence of coherent Al_3Sc precipitates in an unpressed sample, and Figure 11b presents precipitate size distribution in an unpressed alloy.

3.2. Microstructure developed during creep

3.2.1. *Pure aluminium*

It can be expected that the creep behaviour of the UFG material will be influenced critically upon the subsequent thermal stability of its microstructure. To explore this effect microscopic examination of grain size change in pure aluminium during creep exposure at 473 K and 15 MPa were performed. It is important to note that each creep specimen was heated to the testing temperature in the furnace of the creep testing machine over a period of ~ 2h and then held at the testing temperature for further ~ 2h in order to reach thermal equilibrium. Consequently, the microstructure characteristics of the ECAP material at the onset of the creep testing were similar to that shown in Table 1. No substantial coarsening of grains has been observed during creep exposure at 473 K (see Table 2).

Specimen	ECAP conditions	Grain size [μm]	Time to fracture [h]
A4	route A, 4 passes	6.4	79
A8	route A, 8 passes	7.0	26
A12	route A, 12 passes	6.7	17
B4	route B, 4 passes	8.7	62
B8	route B, 8 passes	7.2	60
B12	route B, 12 passes	8.8	39

Table 2. Grain size of the ECAP material after creep at 473 K and 15 MPa.

TEM observations were used also to established details of microstructure evolution during creep. The micrographs in Figure 12a,b illustrate a dislocation substructure inside the grains. The dislocation lines were wavy and occasionally tangled with each other. It is know that large grains in UFG materials contain dislocations while grains smaller than a certain size are dislocation free [3,6]. EBSD measurements were taken to determine the grain boundary misorientation and the value of relative fraction of a high-angle grain boundary ($\theta > 15°$) population (for details see 3.3.2.).

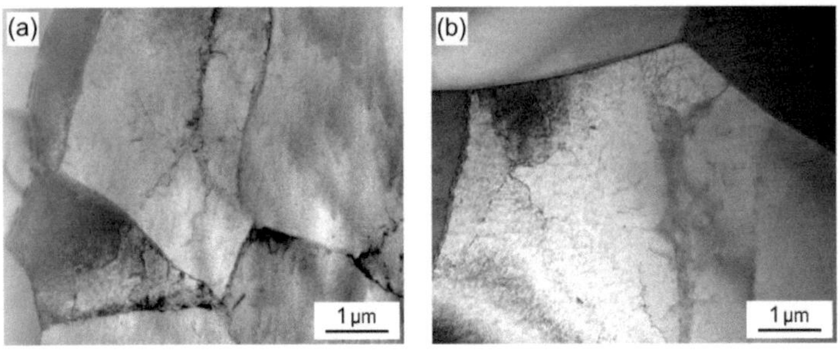

Figure 12. TEM micrographs from the longitudinal section of an aluminium processed by ECAP route B_c (a) after 1 ECAP pass and creep, (b) after 8 ECAP passes and creep. Creep at 473 K and 15 MPa.

3.2.2. Precipitation-strengthened alloys

For comparison reasons, some results of microstructural changes in Al-0.2wt.%Sc and Al-3wt. %Mg-0.2wt.%Sc alloys during creep were presented earlier in 3.1.2. It was found that creep exposures of an Al-0.2wt.%Sc alloy at 473 K and 20 MPa caused the changes in (sub)grain sizes initially resulting from ECAP pressing. Figure 13a shows the microstructure after 8 ECAP passes and subsequent creep exposure. TEM analysis revealed that the average (sub)grain size increases from ~ 0.4 um to ~ 1.3 μm after creep exposure. The (sub)grain growth was effected by presence of coherent Al_3Sc precipitates (Figure 13b) which to some extent pinned the boundaries against their migration and restricted the movement of dislocation.

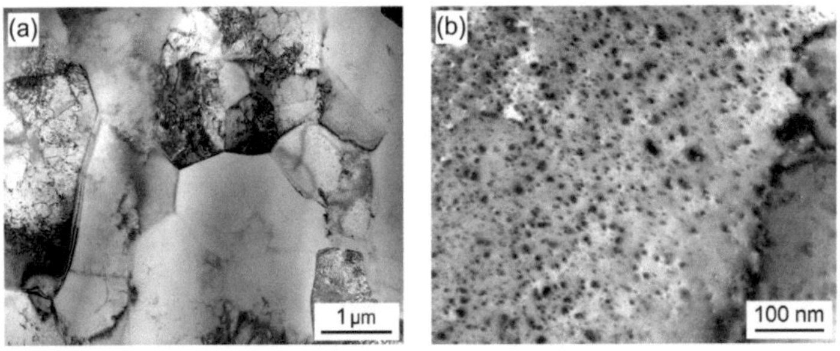

Figure 13. Microstructure of Al-0.2wt.%Sc alloy after 8 ECAP passes and subsequent creep exposure at 473 K and 20 MPa. (a) microstructure, and (b) precipitates Al_3Sc.

The EBSD data indicate that the number of high angle-boundaries (θ>15°) measured in the specimens after ECAP and subsequent creep exposure is strongly dependent on the number of ECAP passes. The number of high-angle grain boundaries is increasing with increasing number of ECAP passes from approximately 2% in the specimen after 1 ECAP pass and subsequent creep to ~ 70% in the specimen after 8 ECAP passes and subsequent creep. It was reported [4] that the grain boundary sliding can occur in UFG materials at elevated temperatures. Thus we can suppose that changes in the number of high-angle grain boundaries in the microstructure of ECAPed materials during creep tests can affect their creep behaviour by increasing the contribution of grain boundary sliding to the total creep strain [27].

The EBSD analyses were performed on the several places of the gauge length of creep specimen after ECAP and subsequent creep revealed scatter in the number of high angle grain boundaries (HAGBs). In the Figure 14 the minimal and maximal measured values of the number of HAGBs are plotted. The inspection of Figure 14 shows that the scatter in HAGBs can be particularly expected after creep tests in the specimens with lower number of ECAP passes. The heterogeneous distribution of HAGB can probably influence the homogeneity of grain boundary sliding. In the areas with the higher number of HAGBs the grain boundary sliding will be more intensive than in the surrounding areas [8].

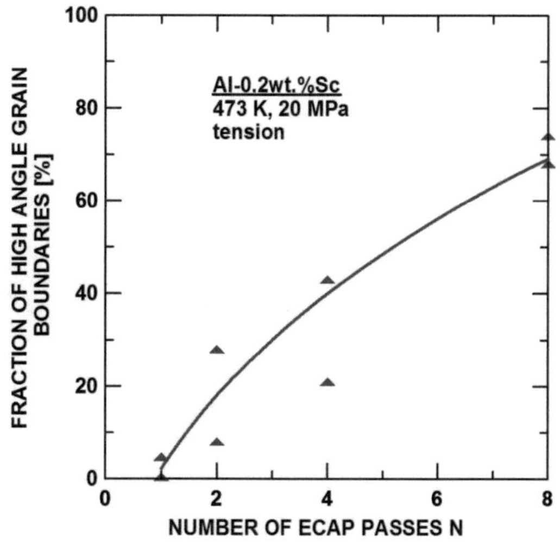

Figure 14. Fraction of high angle grain boundaries as a function of the number of ECAP passes in the Al-0.2wt.%Sc alloy.

The investigation of the unetched surfaces of the specimens after 2-8 ECAP passes and after creep exposure revealed the appearance of mesoscopic shear bands [14,15,35, 40-42] lying near to the shear plane of the last ECAP pass (Figure 15). On the surface of specimens the mesoscopic shear bands were particularly observed near the fracture region and their frequency decreased rapidly with increasing distance from the fracture. On the specimen surface after 8 ECAP passes the mesoscopic shear bands already covered almost the whole gauge length. It was found that the width of the bands decreases with increasing number of ECAP passing and after 8 ECAP passes the average width of the bands was ~ 35 μm as it is shown in Figure 15. The analyses of microstructure on the interfaces of the bands found that in the vicinity of these interfaces high heterogeneity in the distribution of HAGBs can be observed (Figure 15). The formation of the mesoscopic shear band can be related to inhomogeneity of microstructure of ECAPed alloy after creep exposure. Examination by EBSD revealed that the microstructure of mesoscopic shear bands is created by high-angle grain boundaries (Figure 15 and 16).

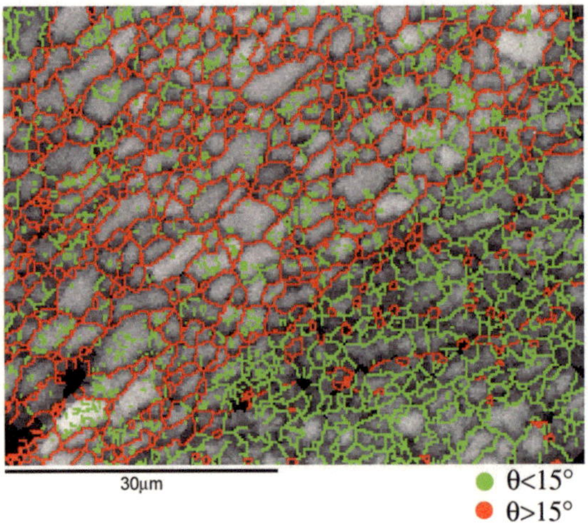

Figure 15. The heterogeneous distribution of HAGBs (red coloured) in the sample of an Al-0.2wt.%Sc alloy after 4 ECAP passes and creep at 473 K and 20 MPa.

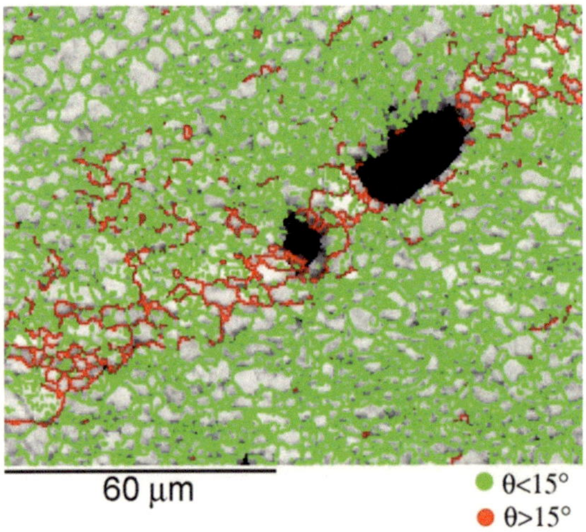

Figure 16. Appearance of the microstructure in the Al-0.2wt%Sc alloy after 2 ECAP passes and subsequent creep at 473 K. Tensile axis is horizontal, SEM.

3.3. Unique features of microstructure in ultrafine-grained materials

The processing technique used to obtain the UFG microstructure should strongly influence the creep properties of the material. This is primarily due to difference in microstructure as distribution of grains, subgrains, dislocation density and boundary character. The grain boundary character is usually quantified using the misorientation angle θ across grain boundaries, with high and low angle grain boundaries defined as $\theta \geq 15°$ an $2° < \theta < 15°$, respectively. Electron back scatter diffractions (EBSD) mapping has been used to quantitatively characterize boundaries in UFG materials [10,20,21,23,24]. Although the boundary spacing saturates after the first few ECAP passes, the fraction of high angle boundaries continues to increase with increasing ECAP passes [27,34,43]. In addition to grain size determination, there are a number of important microstructural parameters evaluated from EBSD but not available from conventional methods of grain characterization in particulars parameters relating to the grain orientations and boundary characters [44 - 46]. The following text describes representative results of quantitative characterization of UFG microstructure.

3.3.1. *Stereological estimates of UFG microstructure characteristics*

At each examined specimen there were made three mutually perpendicular planar metallographic sections denoted as XY, XZ (longitudinal sections) and YZ (transverse section), where X, Y and Z are the axes of the Cartesian coordinate system with X along the last pressing direction and Z perpendicular to the bottom of the channel. The technique of automated EBSD in the scanning electron microscope was used for quantitative metallography. Four ranges of the boundary misorientation Δ were selected; 2° ≤ Δ, 5° ≤ Δ, 10° ≤ Δ and 15° ≤ Δ. Then standard intercept counting [45] resulting in the mean number N_L of profile chords per unit length of the examined test lines was carried out. In each specimen, six systematically selected directions of the test lines in each section were examined. The mean boundary areas unit volume were then estimated by the stereological relation $S_V = [2N_L]$. Another important feature of the grain boundary structure is its inhomogeneity. The dispersion of grain profile areas can be qualified by the coefficient of variation CV_a of the grain profile areas in a plane $CV_a = \frac{\sqrt{V}}{\bar{x}}$, where V is grain profile areas variation and $\bar{x}$ is the mean value of the grain profile area [22,47,48]. The coefficient of variation CV_a of the profile areas is perhaps the best stereometric characteristic to evaluate homogeneity of microstructure and nowadays it is relatively easily attainable by a computer image analysis [49].

3.3.2. *Inhomogeneity of UFG microstructure*

There are numerous reports of the processing of various pure metals and metallic alloys by ECAP and many of these reports involve a detailed characterization of the microstructure. These results are summarized in recent reviews [3,4]. However, information seldom is reported on the percentage of high angle grain boundaries (HAGB´s), an important parameter in the comparison of plasticity of different processing routes and materials [50]. It can be expected that samples with different distributions of misorientation across the grain boundaries will deform differently. Further, to provide information on the optimum microstructure of UFG materials we need to use an additional quantitative microstructural parameter other than just the average grain size critical for the creep behaviour and properties [51]. Such parameter could be a coefficient of profile CV_a as a measure of homogeneity of materials microstructure [48].

Hence, the grain and subgrain structure of the creep specimens was revealed by means of EBSD and characterized by the coefficient of variation CV_a of the profile areas. Four ranges of the boundary misorientation Δ between adjacent pixels were selected for examination using

EBSD, which correspond namely to subboundaries, transitive and high angle grain boundaries within 2° ≤ Δ and 5° ≤ Δ, transitive and high angle grain boundaries for 10° ≤ Δ, and mostly grains with HAGB´s for Δ ≥ 15°. Selected examples of images of XZ sections produced by EBSD of an Al-0.2wt.%Sc are shown in Figure 17.

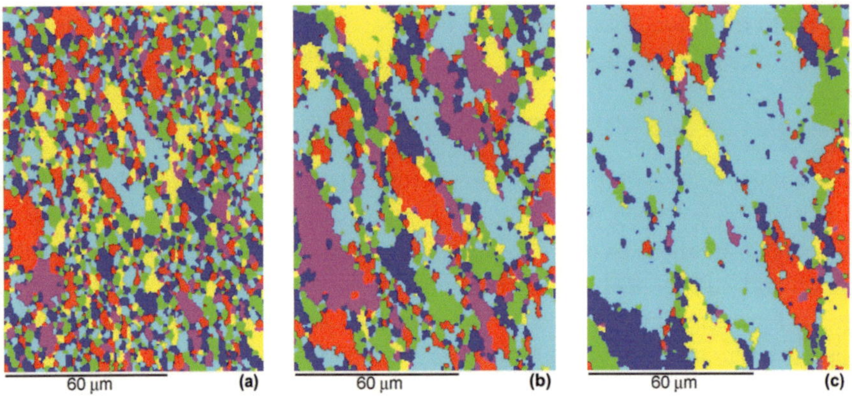

Figure 17. Selected examples of EBSD grain maps of an Al-0.2wt.%Sc alloys after ECAP constructed for different grain misorientations: (a) subboundaries Δ ≥ 2°, (b) transitive subboundaries and high angle boundaries Δ ≥ 10°, and, (c) high-angle grain boundaries Δ ≥ 15°.

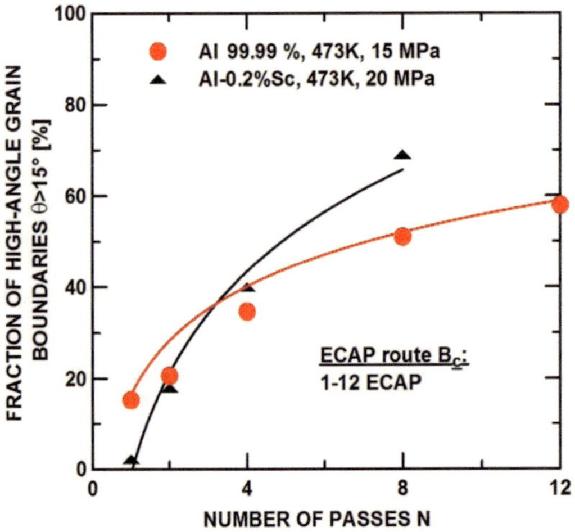

Figure 18. The fraction of high-angle grain boundaries in the crept samples as a function of the number of ECAP passes.

It was generally observed that with the increasing number of ECAP passes N, a considerable amount of subgrain boundaries was gradually transformed to HAGB´s as shown in Figure 18. At the same time, the local homogeneity of structure as characterized by the values of CV_a with the increasing number N is gradually improved as demonstrates by Figure 19. Figure 19 also shows the microscopic appearance of the specimens of an Al-0.2wt.%Sc alloy crept under the same loading conditions (473 K, 20 MPa) but processed by different numbers of ECAP passes N. The values of CV_a as high as 10 at N = 2, $1 \leq CV_a < 2$ at N = 4, and $0.55 \leq CV_a \leq 1$ at N = 8 were found. Extremely high value CV_a at N = 2 demonstrates very high inhomogeneity of a mixture of subgrain and grain structures. The value of CV_a in very homogeneous grain systems should not exceed the value of 1 [23]. It should be noted that extremely high values of the coefficient of variation CV_a is a natural consequence of the short as well as long-range inhomogeneity of microstructure.

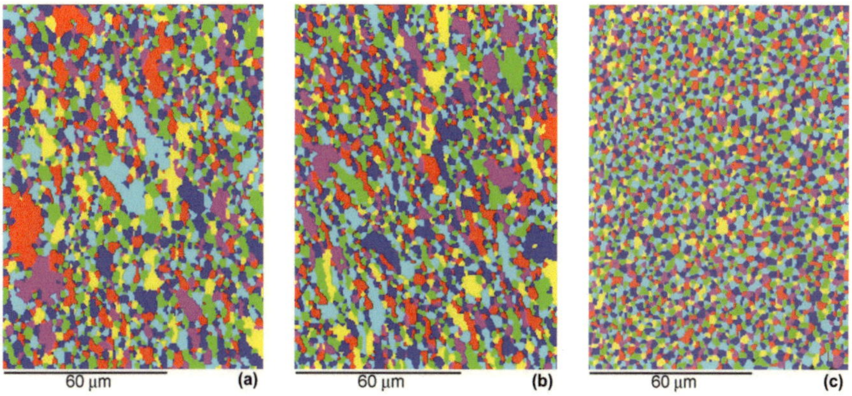

Figure 19. Grain maps of an Al-0.2wt.%Sc alloy processed by different number N of ECAP passes and crept at 473 K and 20 MPa, and corresponding parameter CV_a: (a) 2 passes, $CV_a >> 2$, (b) 4 passes, $CV_a < 2$ and, (c) 8 passes, $0.55 \leq CV_a < 1$.

The substantial grain coarsening especially in the case of pure metals came up during the creep exposures depending on stress and temperature thus manifesting the thermal instability of ultrafine-grained microstructure [4,52,53]. It is clear that in pure aluminium the grains grow rapidly at elevated temperatures because there are no precipitates within the crystalline lattice to restrict the movement of the grain boundaries by a "pinning effect". By contrast, submicrometer grains may be retained to relatively high temperatures in materials containing a distribution of fine precipitates as in the case of an Al-0.2%Sc alloys containing Al_3Sc precipitates [29]. The quantitative characterization of the inhomogeneity of the boundary structure is shown in Figure 20. It should be stressed that the values of the coefficient of profile area CV_a as a measure of structure homogeneity strongly depend on the chosen ranges of misorientation Δ in EBSD analysis. Whereas the fraction of the subboundaries (low-angle grain boundaries) are dominating for $\Delta \geq 2°$ (Figure 21a), the fractions of high-angle grain boundaries ($\theta \geq 15°$) confirm their high share for the range of $\Delta \geq 15°$ (Figure 21b). Detailed inspection of Figure 20b shows a strong dependence of CV_a after creep on the number of ECAP passes. Substantial decrease of the values CV_a 2° for the precipitation-strengthened Al-0.2Sc alloy with increasing number of passes N for $\Delta \geq 2°$ may be connected with the more rapid evolution boundaries having misorientation angles $\theta > 15°$ (Figure 20a).

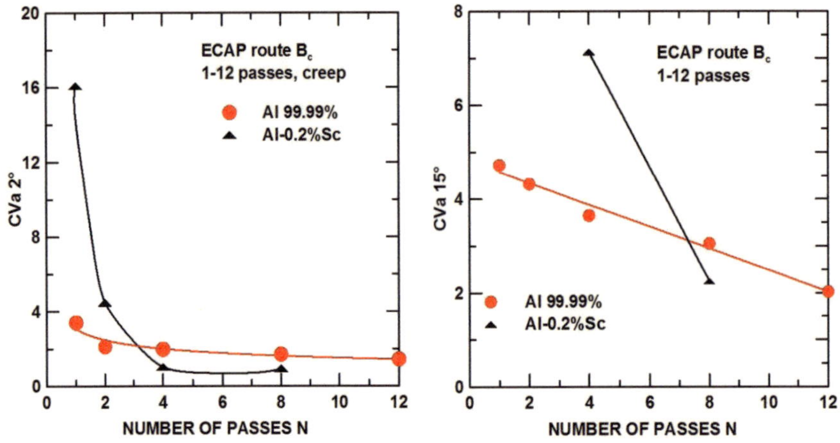

Figure 20. Coefficient of profile area CVa as a measure of homogeneity: $0.55 \leq CVa < 1$ (homogeneous system), and CVa >> 2 (multimodal grain size distribution). The chosen ranges of misorientation Δ in EBSD analysis: (a) Δ ≥ 2°, (b) Δ ≥ 15°.

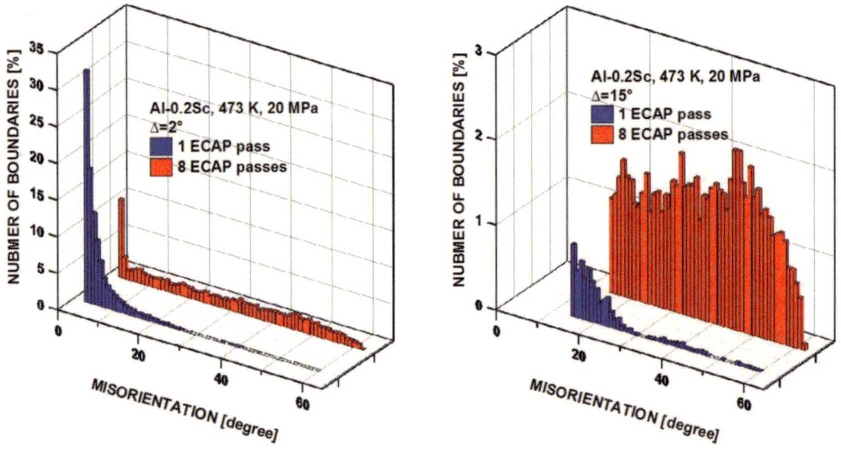

Figure 21. Distribution of boundaries with different misorientation θ for an Al-0.2wt.%Sc alloy analysed in Figure 8: (a) EBSD analysis for Δ ≥ 2°, (b) Δ ≥ 15°.

4. Creep behaviour of UFG aluminium and its alloys

The mechanical properties of bulk ultrafine-grained (UFG) materials at elevated and/or high temperatures are a new and important area of research [4]. However, there have been only a few investigations on the creep behaviour of bulk UFG materials processed by equal-channel angular pressing (ECAP) [43,54]. By comparison with the unpressed (coarse-grained) state, processing by ECAP may lead to considerable changes in the creep properties in bulk

UFG materials including a decrease and/or an increase [8] in the minimum creep rate and the creep life. It is important to note that these trends may be noticeably dependent on the number of ECAP passes. Many investigations concerned with the identification of creep mechanisms have been undertaking using coarse-grained pure aluminium [55] and commercial aluminium alloys [56]. However, it is logical to expect that the mechanism of hardening/softening observed in the aluminium processed by ECAP may be different from that observed in the coarse-grained material. Consequently, it cannot be excluded that creep in the ECAP aluminium and its alloys is controlled by different creep mechanism(s) than that in the coarse material. Thus, neither phenomenological nor microscopic aspects of the creep behaviour of materials processed by ECAP have been understand sufficiently as yet.

This section reports a series of creep experiments that were conducted on specimens of pure aluminium and its Al-0.2wt.%Sc and Al-2wt.%Mg-0.2wt.%Sc alloys processed by equal-channel angular pressing. For comparison purposes, some creep tests were performed also on the unpressed materials. Creep tests were performed both in tension and compression.

4.1. Effect of processing route on creep behaviour

As discussed in more detail in Section 2.2. four distinct pressing routes have been identified (Figure 3). The ECAP processing was conducted by one or repetitive passes following either route A, B (route B_C was used only) or C. Figure 22 contains a complete record of the main creep parameters for the ECAP specimens of Al after creep testing in tension at 473 K and 15 MPa (each point represents the average results of two to three individual creep tests at the same loading conditions). Inspection of Figure 22 shows there are not very significant differences in creep properties of specimens prepared by the various ECAP processing routes. All three processing routes produce a significant increase in the minimum creep rate through the first four passes and a slight increase during subsequent pressing (Figure 22a). By contrast, the time to fracture (creep life) dramatically drops through four passes and then there is no significant differences among the number of following passes – Figure 22b.

Recently, attention has been given to effectiveness of the various ECAP routes in producing grain refinement in aluminium [10]. It has been demonstrated that ECAP is capable of producing refined structures with large fractions of high-angle boundaries [10] although the mechanisms involved in the formation of fine grains and high-angle boundaries in the deformation microstructure remain to be clarified. In this work microstructural investigation all routes examined indicated little differences in the grain size produced via the various ECAP routes. With increasing number of ECAP passes this difference decreases. Further, there was little apparent dependence of the misorientation on the various process route for an ECAP die having an internal angle equals to 90°. The misorientation data confirmed that repetitive pressing results in a progressive increase in the fraction of high-angle grain boundaries (Figure 18).

In related work, Sklenicka et al. [11,57] carried out an extensive creep testing on pure aluminium processed by various ECAP routes. It was found that processing route had a little apparent effect on the creep behaviour of a pressed aluminium. However, the effect of grain growth during creep may tend to obscure the effect of different processing routes and the creep experiments are probably not a sufficiently refined procedure for picking up these rather small differences in the creep behaviour.

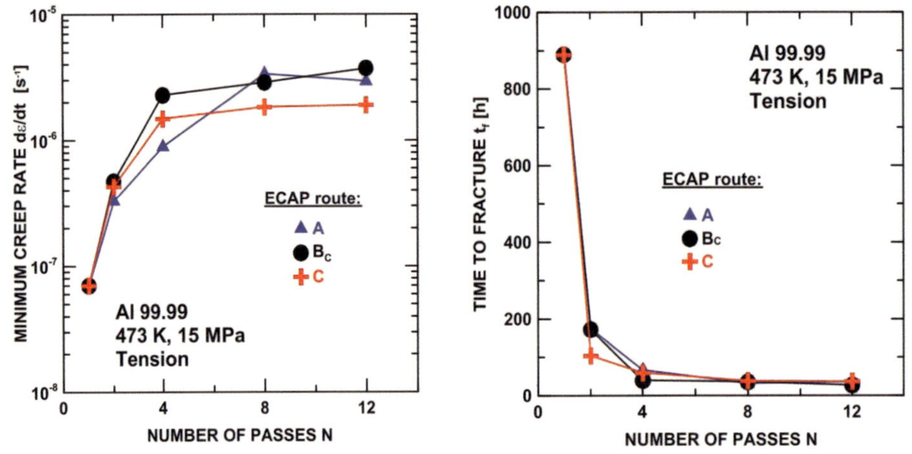

Figure 22. Influence of different ECAP routes and different number of ECAP passes on (a) creep rate, and (b) time to fracture.

4.2. Creep behaviour of pure aluminium

Creep tests were performed on the as-pressed specimens both in tension and compression in the temperature interval from 423 to 523 K under an applied stress range between 10 and 25 MPa. The subsequent ECAP passes were performed by route B_C (see part 2.2.) up to 12 passes.

4.2.1. Creep behaviour

Representative creep curves are shown in Figures 23 and 24. All of these plots were obtained at temperature of 473 K (~ 0.5 T_m) under an initial applied uniaxial tensile or compression stress of 15 MPa. The creep tests in tension were run up to the final fracture of the creep specimens, whereas the creep tests in compression were interrupted at a true strain of about ~ 0.35.

Standard ε vs t creep curves in Figures 23a and 24a can be easily replotted in the form of the instantaneous strain rate dε/dt versus strain as shown in Figures 23b and 24b. As demonstrated by figures, significant differences were found in the creep behaviour of the ECAP material when compared to its coarse-grained counterpart. First, the ECAP materials exhibits markedly longer creep life (Figure 23a) or markedly longer duration of creep exposure to obtain a strain of ~ 0.35 (Figure 24b) than coarse grained aluminium. Second, the minimum creep rate for the ECAP material is about one to two orders of magnitude less than that of coarse-grained material. Third, the shapes of tensile creep curves for the ECAP material after high number of pressing differ considerably from the tests conducted at small number of the ECAP passes by the extent of individual stages of creep.

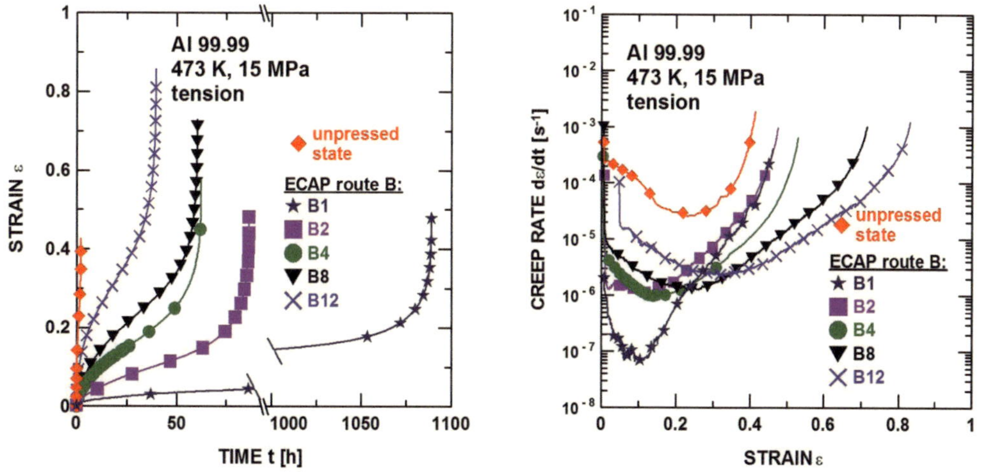

Figure 23. Standard creep and creep rate versus strain curves for unpressed state and various number of ECAP passes via route B_C (creep in tension up to fracture).

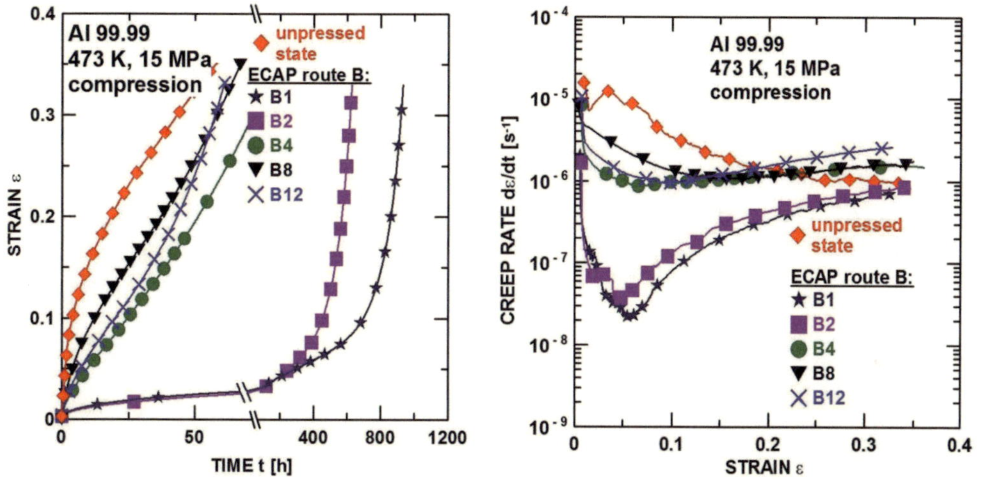

Figure 24. Standard creep and creep rate versus strain curves for unpressed state and various number ECAP passes via route B_C (creep in compression up to strain ~ 0.35).

The difference in the minimum creep rate for the ECAP material and unpressed state consistently decreases with increasing number of ECAP passes (Figures 23b and 24b). An additional difference is illustrated by Figure 25a, which shows the variation of the minimum creep rate with the applied stress for the ECAP specimens after 8 passes. The results demonstrate that at high stresses the minimum compressive creep rate of the ECAP material may be up to one order of magnitude lower than that of the unpressed material, although this difference decreases with decreasing applied stress and becomes negligible at 10 MPa. The observed values of the stress exponent n = $(\partial \ln\dot{\varepsilon} / \partial \ln\sigma)_T$ are ~ 6.6 for the unpressed material, ~ 4.8 (creep in compression) and ~ 5.7 (creep in tension) for the ECAP Al, respectively.

To determine the apparent activation energy for creep Q_c, the minimum creep rate was measured in the temperature interval from 423 to 523 K and at two tensile applied stresses 15 and 20 MPa, respectively. The activation energy for creep Q_c is defined as

$$Q_c = \left[\frac{\partial \ln\dot{\varepsilon}_{min}}{\partial(-1/kT)} \right]_\sigma \tag{1}$$

Thus, the activation energy Q_C can be derived from the slope of log dε/dt versus 1/T plots shown in Figure 25b. A value of the apparent activation energy, Q_C, was determined by the least square methods. The Q_C is stress dependent and equals to 129.7 ± 16 and 110.9 ± 9 kJ/mol for stresses 20 and 15 MPa, respectively.

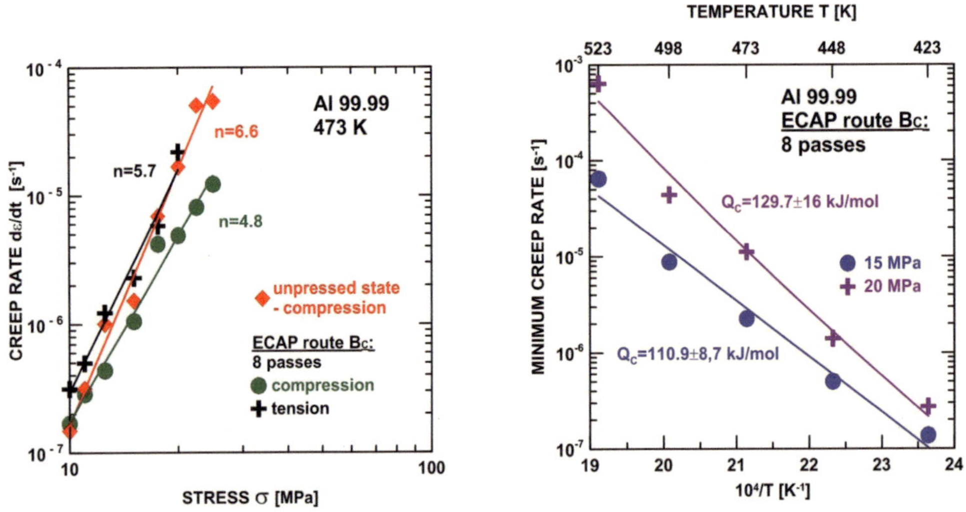

Figure 25. Dependence of minimum creep rate for unpressed state and 8 ECAP passes on: (a) applied stress, (b) testing temperature at two levels of stress.

4.2.2. Grain boundary sliding (GBS)

The total creep strain ε generally consists of following contributions: the strains caused by dislocation glide and nonconservative motion of dislocations, grain boundary sliding (GBS) [58-60], stress directed diffusion of vacancies and by intergranular void nucleation and growth, respectively. However, it should be noted that not all of the above processes operating are independent of each other, as frequently assumed. A possible explanation for the occurrence of intensive GBS in UFG materials is that diffusion is more rapid in ECAP processed materials with highly non-equilibrium grain boundaries [4, 19, 61,62]. Accordingly, it appears that GBS is easier in these UFG materials.

The amount of grain boundary sliding (GBS) was determined by measuring the surface offsets produced at the intersections of grain boundaries with marker lines transverse to the stress axis [27,58]. Figure 26 shows one clear example of the occurrence of grain boundary sliding in creep of the ECAP aluminium. Longitudinal displacements of the marker lines, u, due to GBS, together with the fraction of boundaries, κ_S, with observable GBS, were measured using SEM.

Grain boundary sliding was measured on the surfaces of the tensile specimens crept up to a predetermined strain $\varepsilon \approx 0.15$. Scanning electron microscopy made it possible to detect GBS characterized by $u \geq 0.1\mu m$. However, GBS was not observed at all grain boundaries; that is why the relative frequency of sliding boundaries κ_s was determined. Then the strain component ε_{gb} due to GBS is expressed as [27,58]:

$$\varepsilon_{gb} = (1 + \varepsilon)\, \bar{u} . \kappa_s / \overline{L} \tag{2}$$

where the mean grain size $\overline{L}$ was determined by the linear intercept method and the overall contribution of GBS to the total creep strain in the specimen, γ, was estimated as $\gamma = \varepsilon_{gb}/\varepsilon$. The results of GBS measurements are summarized in Table 3. It is evident that the fraction of boundaries κ_s increases as the number of ECAP passes increases. This result supports the idea that GBS is connected with microstructural changes of grain boundaries [27]. It is to note that in the best case (12 passes) the contribution of GBS to creep strain is only 33%.

No of Passes	$\bar{u}$[μm]	κ_s	$\bar{L}$ [μm]	$\varepsilon_{gb}.10^2$	$\varepsilon_{gb}/\varepsilon.10^2$
1	0.51	0.80	14.9	3.15	21.0
2	0.48	0.83	12.7	3.60	24.0
4	0.55	0.93	12.2	4.80	32.0
8	0.49	0.91	10.8	4.70	31.0
12	0.52	0.92	11.0	5.00	33.0

Table 3. Summary of GBS measurements ($\varepsilon \cong 0.15$).

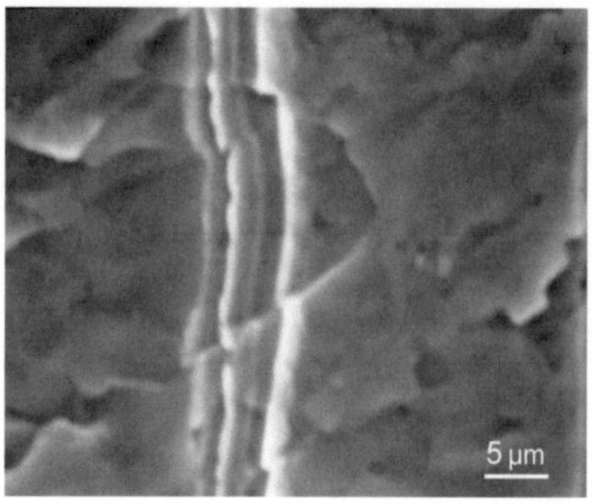

Figure 26. Example of grain boundary sliding in the ECAPed aluminium (route B_c, 8 passes) after creep testing at 473 K and 15 MPa. Tensile stress axis is horizontal.

4.2.3. *Creep deformation mechanisms*

The mechanisms controlling the creep properties of pure metals have been usually identified from the dependence of the minimum and/or steady-state creep rate $\dot{\varepsilon}_m$ on stress σ, absolute temperature T and grain size d, using a power-law expression of the form

$$\dot{\varepsilon}_m = A\sigma^n (1/d)^p \exp(-Q_c/RT) \tag{3}$$

where Q_c is the activation energy for creep. With this approach, the fact that n, p and Q_c are themselves functions of stress, temperature and grain size is conventionally explained by assuming that different mechanisms, each associated with different values of n, p and Q_c, control the creep characteristics in different stress/temperature regimes. In turn, the dominant mechanisms under specific test conditions are then generally determined by comparing experimentally determined values of n, p and Q_c with the values predicted theoretically for different creep mechanisms.

In Figure 27, the minimum creep rates $\dot{\varepsilon}_{min}$ are plotted against applied stress σ for pure aluminium using the data presented in Figure 25a. Experimental points are shown for both the unprocessed (coarse-grained) and for the UFG aluminium after 8 ECAP passes (only the results of tensile creep testing are used).

The broken lines in Figure 27 denote the model predictions of the theoretical creep rates according models for various creep deformation mechanisms, namely for superplastic flow, Nabarro-Herring [59, 63,64] and Coble [59,65] diffusion creep and power-law creep by dislocation climb and glide processes. It should be stressed that the phenomenon of dislocation diffusion is not well understood on the fundamental level at present. The theoretical creep rates were calculated from equation (3) using the material data presented in Table 4.

Al (473K) d_{ECAP} = 1μm d_{CREEP} = 12μm [27]								
number of curve	**creep mechanism**	**n**	**p**	**A**	**D [$m^2 s^{-1}$]**		**Q [$kJ\ mol^{-1}$]**	**source**
(1), (1*)	superplastic flow	2	2	10	D_{GB}	5.9×10^{-14}	86	[66]
(2), (2*)	Nabarro – Herring creep	1	2	28	D_L	2.72×10^{-20}	143.4	[66]
(3), (3*)	Coble creep	1	3	62	D_{GB}	5.9×10^{-14}	86	[66] [67]
(4)	dislocation climb and glide	5	0	10^3	D^*	1.9×10^{-14}	124	[66]

D^* is the effective diffusion coefficient which incorporates contribution from both lattice and grain boundary diffusion, D_{gb} and D_L are the grain boundary and lattice diffusion coefficients, respectively.

Table 4. Creep mechanisms and material data.

It should be noted that grain growth occurs easily at the elevated temperatures used in creep experiments of pure metals. Indeed, the occurrence of significant grain growth in creep tests conducted on high-purity aluminium processed by ECAP at room temperature was observed [27,36]. Accordingly, two sets of the predicted theoretical rates were calculated in this analysis for both states of materials using the measured grain size after ECAP processing (d_{ECAP}) and after subsequent creep exposure (d_{CREEP}) [27]. The theoretical rates using d_{ECAP} will be marked by simple number (e.g. 1) while a numbering with asterisk (e.g. 1*) will be used for the rates corresponding to d_{CREEP} in Figure 27.

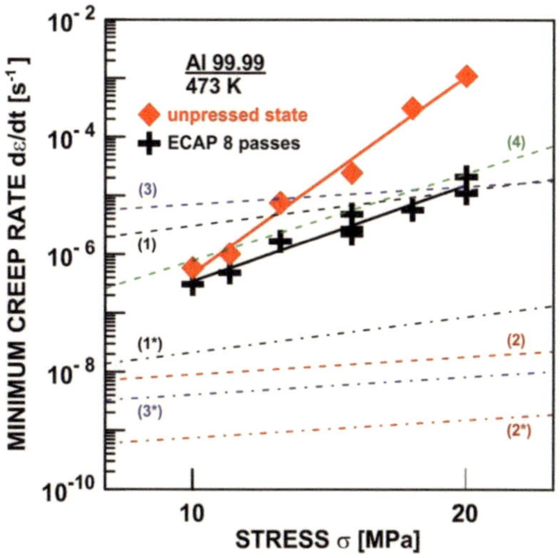

Figure 27. Experimentally determined and theoretically predicted the stress dependences of the minimum creep rates for various creep mechanisms in aluminium.

Figure 27 demonstrates that at high applied stresses the experimentally determined minimum creep rate of the ECAP aluminium may be up to two orders of magnitude lower than that of the unpressed material, although this difference decreases with decreasing applied stress and becomes nearly negligible at 10 MPa. The predictions show that under the creep loading conditions investigated Nabarro-Herring and Coble diffusion creep and superplastic flow are too slow to account for the creep deformation considering a significant grain growth in the pressed materials. Also shown in Figure 27 are the predicted theoretical creep rates for ultrafine-grained states (d_{ECAP}) after ECAP which are within two to five orders of magnitude faster than that for the creep of coarsened materials. However, such predictions are not correct a priori due to thermal instability of microstructure of the pressed materials. Inspection of Figure 27 shows that for the pressed material there is an excellent agreement between the experimental datum points and the predicted creep behaviour based on dislocation climb and glide. Further, for $n \geq 4$ creep is known to occur by diffusion-controlled movement of dislocations within grains and/or along grain boundaries (grain boundary sliding). The high value of n_{CG} for the unpressed aluminium (Figure 27) could represent a regime leading into a power-law breakdown (PLB) region at rapid strain rates and/or high stress levels.

The analysis of Figure 27 indicates that creep in pure aluminium after ECAP occurs by the same mechanism as in conventional coarse-grained materials with intragranular dislocation glide and climb as the dominant rate-controlling flow process. Therefore, the activation energy for creep Q_C should be the same as the value of the activation enthalpy of lattice self-diffusion ΔH_{SD} ($\Delta H_{SD} \cong$ 127 – 143 kJ/mol in aluminium [59]). The values obtained for Q_C (Figure 25b) are somewhat lower than that of ΔH_{SD}. Supposing that grain boundary sliding is controlled by grain boundary diffusion, which is assumed to be about 0.7 times that for lattice self-diffusion, the presented results give support to the assumption that GBS may be increasingly important in creep of the ECAP aluminium at low applied stresses.

It can be concluded that the creep resistance of high purity aluminium is increased considerably already after one ECAP pass. However, successive ECAP pressing leads to a noticeable decrease in the creep resistance. The results of microstructure investigations indicate that an inhomogeneity and thermal instability of the ECAP microstructure may strongly influence the creep behaviour of the pressed material [48].

4.3. Creep behaviour of aluminium alloys

The combination of solid-solution strengthening and precipitation strengthening in creep of aluminium alloys at elevated temperatures has not been extensively studied. Numerous reports dealt with the creep behaviour of Al-Mg solid solution [68-70]. Most precipitation strengthened aluminium alloys currently being used are limited to relatively low temperature usage, because of the dissolution and/or rapid coarsening of their precipitates. An exception represents Al-Sc alloys containing low volume fractions of very fine coherent cuboidal Al_3Sc precipitates. The dislocation creep behaviour of coarse-grained binary Al-Sc alloys at 573 K and the precipitation strengthening effect of the Al_3Sc phase were investigated by Fuller et al. [71] and Seidman et al. [72]. Recently, the effect of Mg addition on the creep behaviour of an Al-Sc alloy was reported by Marquis et al. [31]. It was found that the creep strength of an Al-3wt.% Mg-0.2wt.%Sc alloy, containing Mg in solid solution and Al_3Sc as nanosize precipitates, is significantly improved compared to binary Al-Sc alloys.

4.3.1. Creep behaviour

As it was reported in Section 4.2. the processing by ECAP of a coarse-grained high purity aluminium provided a potential for marked improvement in the creep properties. Accordingly, Section 4.3. reports on a systematic study of the creep behaviour of the ECAP processed aluminium alloys containing low volume fraction of Al_3Sc precipitates to elucidating the effect of ECAP on their creep resistance.

Figure 28a shows standard strain ε versus time t curves for the as-received (unpressed) Al-0.2wt.%Sc and Al-3wt.%Mg-0.2wt.%Sc alloys and those for the same alloys processed by ECAP through 8 passes at 473 K and 50 MPa (an exception is 80 MPa for Al-3wt.%Mg-0.2wt. %Sc alloy in the as-received state). The creep tests in compression were interrupted at a true strain of about ~ 0.35. These standard creep curves were replotted in the form of the instantaneous creep rate $d\varepsilon/dt$ versus time t as shown in Figure 28b. It is clear that no-one of the creep curves exhibits a well-defined steady state. In fact this stage is reduced to an inflection point of the $d\varepsilon/dt$ versus t curve. Supposing that the instantaneous creep rate $d\varepsilon/dt$ at given stress and temperature is a certain measure of the "softness" of the microstructure, then the dε/dt-t plots reveal the time evolution of this "softness". However, the dε/dt-ε plots may give additional information, since they reflect the effect of the plastic creep strain on the instantaneous "softness" of the microstructure (Figure 28c).

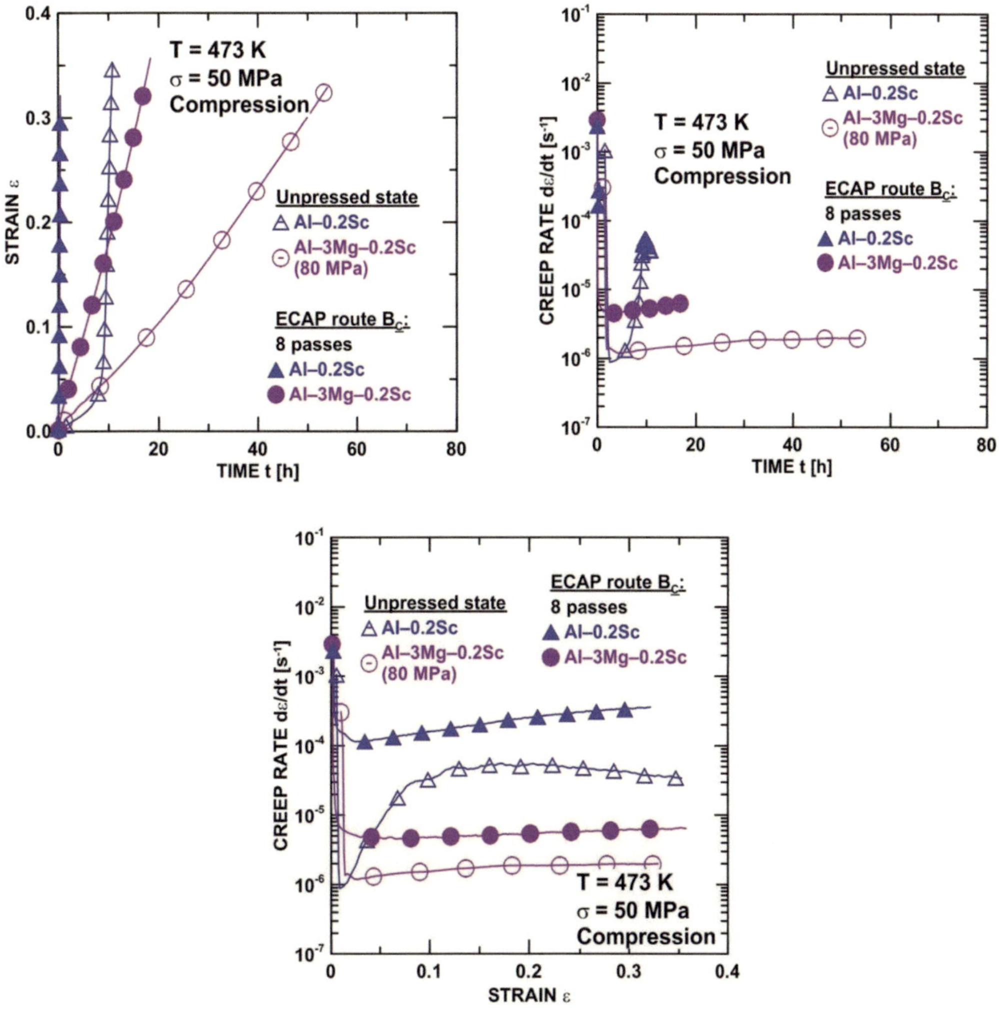

Figure 28. Creep curves for specimens after ECAP processing through 8 passes and for unpressed specimens: (a) standard creep curve, (b) creep rate vs. time, (c) creep rate vs. strain.

The differences in the minimum creep rates for pure Al, Al-0.2wt.%Sc and Al-3wt.%Mg-0.2wt.%Sc in the as-received and as-pressed conditions are illustrated most readily in Figure 29 showing the variation of the minimum creep rate with applied stress. The results demonstrate that for pure aluminium at high stresses the minimum creep rate of ECAP material may be up to one order of magnitude lower than that of the unpressed material, although this difference decreases with decreasing applied stress so that, at 10 MPa is negligible. By contrast, when tests of Al-Sc and Al-Mg-Sc alloys are performed at the same stress, the creep rates in the as-pressed alloys are faster than in the unpressed alloys by more than two and/or three orders of magnitude on the strain rate scale. The stress dependence of the minimum creep rate for the as-pressed Al-Mg-Sc alloy at lower stresses (σ < 20 MPa) is different in trend, which is clearly demonstrated by the characteristic curvature on the plot in Figure 29.

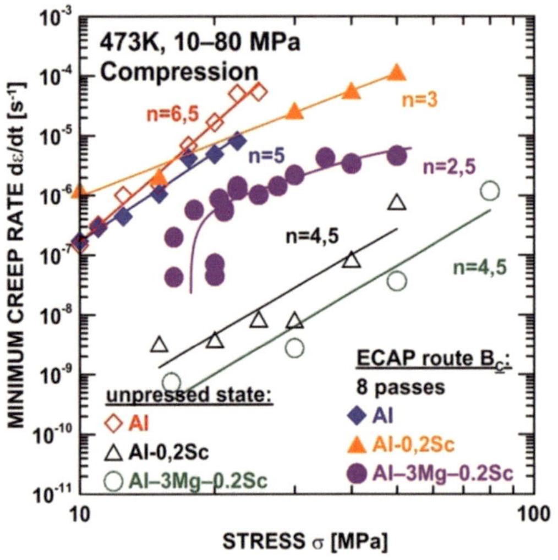

Figure 29. Stress dependences of minimum creep rate for pure aluminium and its alloys in the unpressed and ECAPed conditions.

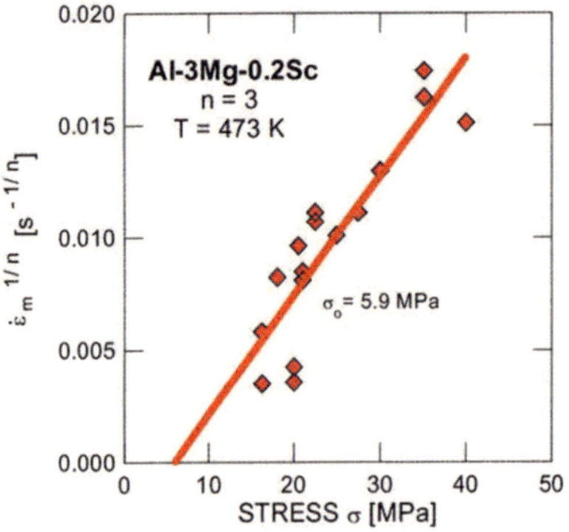

Figure 30. The linear extrapolation procedure for determining the threshold stress.

Such behaviour is generally associated with the presence of a threshold stress marking a lower limit stress below which no measurable strain rate can be achieved [73,74]. The threshold stress value determined from linear plot [75] of $\dot{\varepsilon}^{1/n}$ versus σ is about 6 MPa for the alloy studied. The validity of this approach is illustrated in Figure 30, where the data fall on straight line with the slope of 3, consistent with the assumed value of ~ 3 (Figure 29) for the stress exponent of the minimum creep rate $n = (\partial \ln \dot{\varepsilon}_{min} / \partial \ln \sigma)_T$.

Recently the tensile creep experiments on the same binary Al-0.2%Sc alloy were reported [38]. The stress dependences of the minimum creep rates and the times to fracture for this alloy after 1 and 8 ECAP passes are shown in Figure 31. As demonstrates by Figure 31, the pressed alloy after 8 ECAP passes exhibits the very similar value of n as the results obtained by compression tests (Figure 29).

4.3.2. Creep deformation mechanisms

The results from this investigation on precipitation-strengthened aluminium alloys do not confirm a general validity of the conclusion of our earlier results that processing by ECAP of a coarse-grained aluminium gave a potential for an improvement in the creep resistance [26,27]. By contrast, the Al-0.2wt.%Sc and Al-3wt.%Mg-0.2wt.%Sc alloys exhibited faster creep rate than their coarse-grained counterparts when creep tested under the same loading conditions.

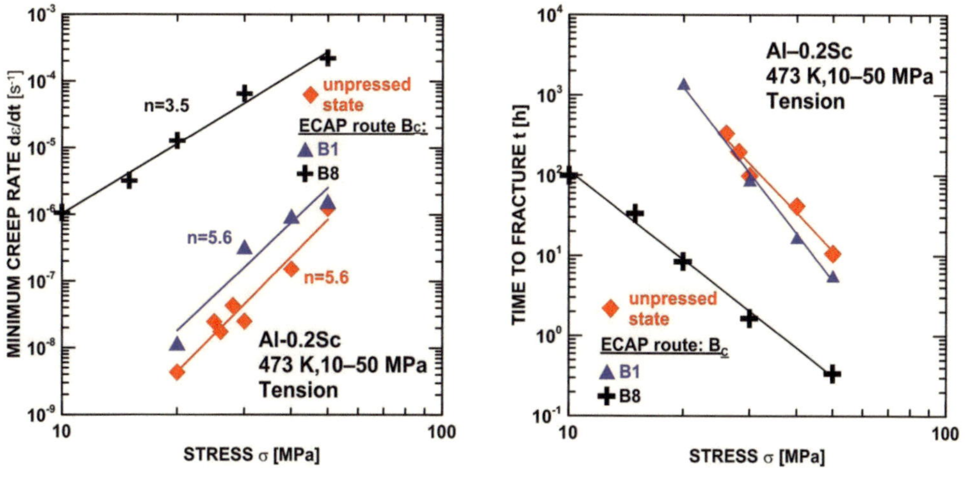

Figure 31. Stress dependence of (a) the minimum creep rates, and (b) times to fracture for Al-0.2Sc alloy.

The observed values of the stress exponents $n = (\partial \ln \dot{\varepsilon} / \partial \ln \sigma)_T$ are ~ 4.5 for the unpressed alloys and ~ 3 for the ECAPed alloys, respectively (see Figure 29). The mechanism which most probably plays the dominant role in the power-law creep (n ~ 4.5) of coarse-grained Al-0.2wt.%Sc and Al-3wt.%Mg-Sc alloys is the dislocation climb-bypass mechanism in the presence of elastic interactions between dislocations and coherent precipitates. The lower value of the stress exponent (n ~ 3) found for the ECAPed alloys may reflect the synergetic effect of more intensive grain boundary sliding in the creep of ultrafine-grained materials [27]. Thus, an important contribution of grain boundary sliding to the total creep strain in the ultrafine-grained Al-0.2Sc and Al-3Mg-0.2Sc alloys may explain the observed detrimental effect of ECAP on their creep resistance. In order to examine rigorously the differences in creep behaviour of the alloys investigated more quantitative results of microstructural analysis are needed.

The detailed analysis was undertaken by Kawasaki et al. [36] to examine the flow characteristics of the ultrafined-grained Al-3Mg-0.2Sc and Al-0.2Sc alloys [33-35,43]. The theoretical predictions of the minimum creep rates have shown that Nabarro-Herring creep is too slow to account for the creep deformation in an ultrafine-grained Al-3wt.%Mg-0.2wt.%Sc alloy [33] but there is good agreement, to within an order of magnitude, with the predictions of superplastic flow except only at the lower stresses where the points deviate from linearity and there is evidence for the presence of a threshold stress (see Figure 29).This threshold stress probably arises from the presence of coherent Al_3Sc precipitates.

For an Al-0.2wt.%Sc alloy crept in compression at 473 K [33] predictions for Nabarro-Herring diffusional creep were again too slow but there was excellent agreement between the experimental datum points and the predicted behaviour in superplastic flow. The slightly higher stress exponent after ECAP may reflect an inhibition in GBS at the lowest strain rates due to the presence of intergranular Al_3Sc precipitates. Finally, the results on an Al-0.2wt. %Sc alloy tested in creep under tensile conditions at 473 K [43] again shown that the predicted behaviour for Nabarro-Herring creep is too slow but there is reasonable agreement with the model for superplastic flow.

Thus, the results from both sets of experiments exhibit a general consistency with the predicted behaviour for conventional superplasticity. The theoretical predictions provide a clear demonstration that conventional creep mechanisms, already developed for coarse-grained materials, may be used to explain the flow characteristics of materials with ultrafine grain size. Furthemore, at least for the aluminium alloys examined in this Chapter, it is not necessary to involve any new and different creep deformation mechanisms.

4.4. Creep ductility

Creep ductility is very important for various shaping and forming technological operations at elevated and high temperatures and especially for avoiding catastrophic failure in load-bearing parts of high temperature components. Creep strength and ductility are the key creep properties of creep-resistant materials but these properties typically have opposing characteristics. Thus, these materials may be strong or ductile but they are rarely both. In this connection, recent findings of high strength and good ductility in several bulk ultrafine-grained (UFG) metals produced by severe plastic deformation (SPD) are of special interest [4].

Typically, creep ductility in tension which can be characterized by the strain to fracture ε_f, is given by

$$\varepsilon_f(\sigma, T, S) = \int_0^{t_f(\sigma,T,S)} \dot{\varepsilon}(\sigma, T, S, t)dt \tag{4}$$

where σ is the applied stress, T is the absolute temperature, S is some parameter that characterizes the structure of specimen and t_f is the time to fracture. According to the first mean-value theorem of a definite integral eq. (4) can be expressed as

$$\varepsilon_f(\sigma, T, S) = t_f(\sigma, T, S)\dot{\varepsilon}(\sigma, T, S) \tag{5}$$

where $\dot{\varepsilon}(\sigma, T, S)$ is some strain rate in the interval from 0 to t_f.

A strong influence of the microstructure on creep behaviour has been observed in various UFG materials [22,23,38,49,54,76]. By contrast, no report is available describing the link between microstructure and creep ductility in UFG materials processed by ECAP at elevated and high temperatures.

To investigate a course of creep strain during creep exposure and for mutual comparison of the fracture strains Figures 32a and 33a can be replotted in the form of the instantaneous strain rate dε/dt versus creep exposure time t normalized to the time to fracture t_f as shown in Figures 32b and 33b. The figures demonstrate that very significant strain contributions to the fracture strains are generated during the last tenth of creep life, however, an accumulation of the creep strains during the course of creep exposure slightly differ.

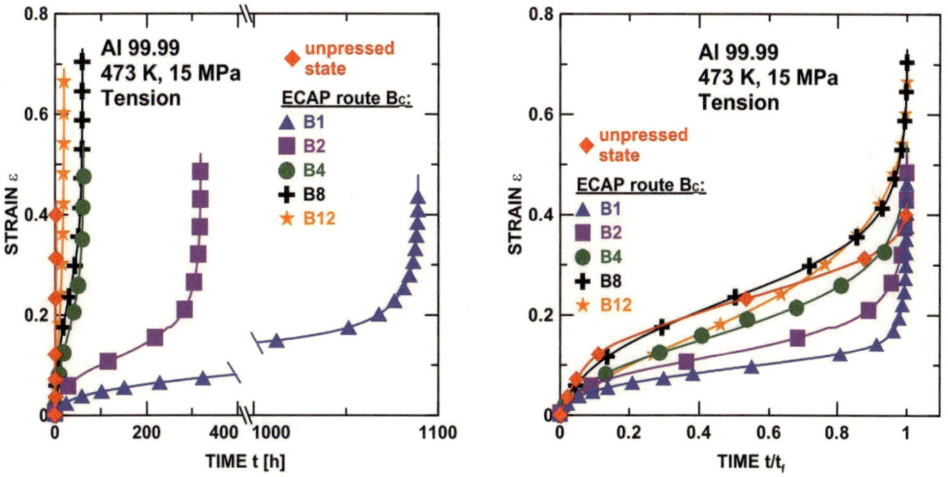

Figure 32. Creep curves of pure aluminium for unpressed state and various number of ECAP passes (creep in tension up to fracture): (a) standard creep curves, (b) creep strain ε vs. time t/t_f.

A question naturally arises about an approach to the problem of creep ductility enhancement. In this connection different approaches may be considered [4]. It has been suggested that some ductility enhancement may be associated with an increase in the fraction of high-angle grain boundaries with increasing number ECAP passes and with a consequent change in the controlling deformation mechanisms due to the increasing tendency for the occurrence of grain boundary sliding. A possible explanation for the occurrence of intensive grain boundary sliding in UFG materials is that diffusion is more rapid in ECAP processed materials with highly non-equilibrium grain boundaries [77,78]. Accordingly, it appears that grain boundary sliding is easier in these UFG materials during creeping at ambient temperatures leading to the possibility of increased ductility. Figure 34a shows the measured strain to fracture as a function of the fraction of HAGBs when testing under creep loading conditions presented in Figures 32 and 33.

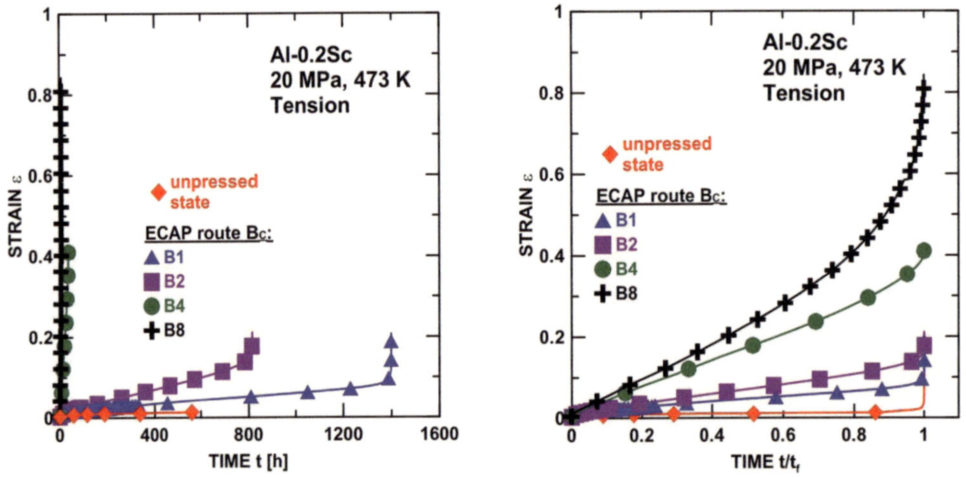

Figure 33. Creep curve of an Al-0.2wt.%Sc alloy for unpressed state and various number of ECAP passes (creep in tension up to fracture): (a) standard creep curves, (b) creep strain ε vs. time t/t$_f$.

It is apparent from Figure 34a that specimens with higher fraction of high-angle grain boundaries exhibit higher ductility. Further approach to the problem of ductility enhancement at room temperature in ultrafine-grained materials was suggested through the introduction of a bimodal or multi-modal grain size distribution [79-81]. In this ductility strategy, the ultrafine-grained matrix in the bimodal microstructure provides the high strength, while the relatively large grains of the order of micrometers contribute to the ductility. The existence of large grains may also release stress concentrations, thereby delay the early fracture of the specimens, and allow further plastic deformation to take place in the ultrafine-grained matrix. Furthermore, the investigation of copper showed that bimodal structures may increase the ductility not only during tensile testing but also during cyclic deformation [82].

Unfortunately, the results of this work have not advocated the advantages of using material with a bimodal and/or multimodal grain size distribution for obtaining at the same time good creep strength and ductility at elevated or high temperatures. As it follows by inspection of Figures 19 and 20 typical feature of the boundary structure of UFG materials is its inhomogeneity; especially at low values of ECAP passes N completely different structures are observed. The dispersion of observed grain profile areas attains quite enormous values of the coefficient of variation CV_a (between 6 to 16) in the whole range of N ≤ 12. In such circumstances, a great variability of times to fracture and creep fracture strain is a natural consequence of short as well as long-range inhomogeneity of microstructure of tested specimens. A very pronounced improvement of creep ductility and strong evidence for a significant role of microstructure homogeneity are documented in Figure 34b. These data indicate possible relationship between the ECAP microstructure and the creep ductility of the pressed materials [48]. The better is homogeneity of pressed material the better is its creep ductility. Thus, using a non-uniform grain structure by mixing ultrafine – sized and larger grains to derive enhanced ductility [83] seems not to be beneficial at elevated and/or high temperatures.

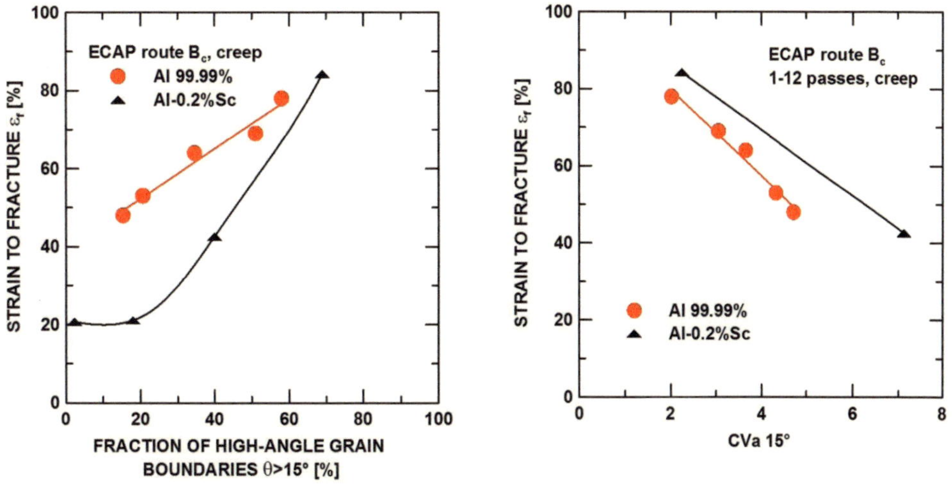

Figure 34. Dependence of the strain to fracture ε_f on (a) the fraction of high-angle grain boundaries θ > 15°, and (b) on the coefficient CVa for EBSD range Δ ≥ 15°. Crept at 473 K and 15 MPa (Al 99.99%) or 20 MPa (Al – 0.2% Sc) – see Figures 32 and 33.

Finally, due to a frequent use of miniaturized tensile specimens in research on UFG materials the specimen dimension and/or geometry effects could be considered for ductile behaviour of these materials. Thus, the thickness effect is mainly caused by the necking geometry and/or fracture modes, and the gauge length effect originates from the strain definition [84]. Therefore, no existence of a standardized protocol of creep specimen and a great variety of specimen sizes and geometries have been used by different authors, primarily depending on SPD techniques used and on the availability of material. Thus, the results creep ductile behaviour measured using such different specimens are hardly comparable.

5. Conclusions

High purity (4N) aluminium, a ternary Al-3wt.%Mg-0.2wt.%Sc and a binary Al-0.2wt.%Sc alloys were processed by equal-channel angular pressing (ECAP) through 1-12 passes and examined by TEM, SEM and EBSD microscopy. The microstructural investigations reveal that from 1 to 4 ECAP passes the microstructure evolves from elongated subgrains to an essentially equiaxed array of ultrafine grains (d < 1μm) and from 4 to 12 passes there is no substantial change in average grain size. The boundary misorientation angle and the fraction of high-angle grain boundaries (HAGBs) increase rapidly up to 4 ECAP passes and at a slower rate from 4 to 12 passes.

It was found, that the creep resistance of pure aluminium is considerably increased in comparison with the coarse-grained unpressed state already after first ECAP pass. However, successive ECAP pressing lead to a noticeable decrease in the creep properties. This softening may be related to the decrease of the spacing of HAGBs at approximately constant subgrain size with increasing number of ECAP passes, resulting in the fraction of low-angle boundaries decreasing considerably. This indicates that HAGBs have lower strengthening effect under creep than low-angle ones. The softening by HAGBs may be explained in terms of the indirect effect which grain boundaries exert on the creep resistance by influencing the evolution of the dislocation microstructure in modifying the rates of generation and annihilation of dislocations. Further, a progressively increasing contribution of grain boundary sliding along HAGBs to overall creep strain can be expected as a consequence of accompanying transformation of low angle boundaries towards an equilibrium state when the number of passes increases. The results demonstrate that creep occurs in pure aluminium after processing by ECAP by the same mechanism as in conventional coarse-grained materials with intragranular dislocation glide and climb as the dominant rate-controlling deformation processes. Therefore, the higher creep resistance of pressed aluminium cannot be a consequence of any significant change in the rate controlling process.

The present results concerning Al-3wt.%Mg-0.2wt.%Sc and Al-0.2wt.%Sc alloys do not confirm a general validity of the conclusion that processing by ECAP of a coarse-grained material gives a potential for an improvement in the creep resistance. By contrast, the minimum creep rates for the pressed alloys are more rapid than the rate attained in the coarse-grained state. The creep mechanism which most probably plays the dominant role in creep of the coarse-grained alloys is the dislocation climb-bypass mechanism in the presence of elastic interaction between dislocations and coherent Al_3Sc precipitates. Strong support for making use of the same conventional creep mechanism in interpreting the creep characteristic of pressed alloy was presented. Although it seems acceptable to associate these faster creep rates with the smaller grain size after ECAP due to more intensive grain boundary sliding, there are probable further reasons strongly influencing an intragranular deformation mechanism which seems to be the rate-controlling process. First, the pressed alloys contain a high dislocation density because of the intense straining imposed during pressing. Second, an important strain contribution during creep of pressed alloys is produced by intensive mesoscale sliding of groups of grains along shear bands. Finally, the occurrence of nanopores formed during ECAP pressing at matrix/precipitate interfaces and resulting in the decohesion at coherent particle surface could have a strong effect on the creep behaviour of pressed alloys. It is important to note that the experimental results are in reasonable agreement with a theoretical model for superplastic flow that was developed earlier for coarse-grained alloys. This conclusion is important because of the many results now available demonstrating the occurrence of superplastic elongations in materials processed by ECAP and because of the new possibilities for achieving exceptional superplastic elongations in materials with ultrafine grain sizes.

Acknowledgements

The authors acknowledge the financial support for this work by the Czech Science Foundation under the Grant No. P108/11/2260. This work was realized in CEITEC – Central European Institute of Technology with research infrastructure supported by the project CZ. 1.05/1.1.00/02.0068 financed from European Regional Development Fund. We would like to cordially thank late Dr. Ivan Saxl of Mathematical Institute of the Academy of Sciences of the Czech Republic for his useful advices concerning the correct applications and interpretations of strereological methods for quantitative characterization of microstructure.

Author details

Vaclav Sklenicka[1,2*], Jiri Dvorak[1,2], Milan Svoboda[1,2], Petr Kral[1] and Marie Kvapilova[1]

*Address all correspondence to: V. Sklenicka, sklen@ipm.cz

1 Institute of Physics of Materials Academy of Sciences of the Czech Republic, Brno, Czech Republic

2 CEITEC-IPM, Institute of Physics of, Materials, Academy of Sciences of the Czech Republic, Brno, Czech Republic

References

[1] Valiev, R. Z. (2011). Producing Bulk Nanostructured Metals and Alloys by Severe Plastic Deformation (SPD). *In: Whang S H, editor. Nanostructured Metals and Alloys: Processing, Microstructure, Mechanical Properties and Applications*, Cambridge, Woodhead Publishing Ltd., 3-39.

[2] Erb, U., Palumbo, G., & McCrea, J. L. (2011). The Processing of Bulk Nanocrystalline Metals and Alloys by Electrodeposition. *In: Whang S H, editor. Nanostructured Metals and Alloys: Processing, Microstructure, Mechanical Properties and Applications*, Cambridge, Woodhead Publishing Ltd., 118-177.

[3] Valiev, R. Z., Islamgaliev, R. K., & Alexandrov, I. V. (2000). Bulk Nanostructured Materials from Severe Plastic Deformation. *Progr Mater Sci*, 45, 103-189.

[4] Valiev, R. Z., & Langdon, T. G. (2006). Principles of Equal-Channel Angular Pressing as a Processing Tool for Grain Refinement. *Progr Mater Sci*, 51, 881-981.

[5] Zhu, Y. T., Valiev, R. Z., Langdon, T. G., Tsuji, N., & Lu, K. (2010). Processing of Nanostructured Metals and Alloys via Plastic Deformation. *MRS Bulletin*, 35, 977-981.

[6] Valiev, R. Z., & Langdon, T. G. (2011). Achieving Exceptional Grain Refinement Through Severe Plastic Deformation: New Approaches for Improving the Processing Technology. *Metall Mater Trans*, 42, 2942-2951.

[7] Nakashima, K., Horita, Z., Nemoto, M., & Langdon, T. G. (2000). Development of a Multi-Pass Facility for Equal-Channel Angular Pressing to High Total Strains. *Mater Sci Eng A*, 281, 82-87.

[8] Furukawa, M., Iwahashi, Y., Horita, Z., Nemoto, M., & Langdon, T. G. (1998). The Shearing Characteristics Associated with Equal-Channel Angular Pressing. *Mater Sci Eng A*, 257, 328-332.

[9] Iwahashi, Y., Horita, Z., Nemoto, M., & Langdon, T. G. (1998). The Process of Grain Refinement in Equal-Channel Angular Pressing. *Acta Mater*, 46, 3317-3331.

[10] Mc Nelley, T. R., Swisher, D. L., Horita, Z., & Langdon, T. G. (2002). Influence of Processing Route on Microstructure and Grain Boundary Development During Equal-Channel Angular Pressing of Pure Aluminium. *In: Zhu Y T et al., Ultrafine Grained Materials II*, Warrendale, TMS, 15-24.

[11] Sklenicka, V., Dvorak, J., Svoboda, M., Kral, P., & Vlach, B. (2005). Effect of Processing Route on Microstructure and Mechanical Behaviour of Ultrafine-Grained Metals Processed by Severe Plastic Deformation. *Mater Sci Forum*, 482, 83-88.

[12] Dvorak, J., Sklenicka, V., & Horita, Z. (2008). Microstructural Evolution and Mechanical Properties of High Purity Aluminium Processed by Equal-Channel Angular Pressing. *Mater Trans*, 49, 15-19.

[13] Betekhtin, V. I., Kadomtsev, A. G., Sklenicka, V., & Saxl, I. (2007). Nanoporosity of Fine-Crystalline Aluminium and an Aluminium-Based Alloy. *Phys Solid State*, 49, 1787-1790.

[14] Betekhtin, V. I., Kadomtsev, A. G., Kral, P., Dvorak, J., Svoboda, M., Saxl, I., & Sklenicka, V. (2007). Significance of Microdefects Induced by ECAP in Aluminium, Al-0.2%Sc Alloy and Copper. *Mater Sci Forum*, 567-568, 93-96.

[15] Betekhtin, V. I., Sklenicka, V., Saxl, I., Kardashev, B. K., Kadomtsev, A. G., & Narykova, M. V. (2010). Influence of the Number of Passes under Equal-Channel Angular Pressing on the Elastic-Plastic Properties, Durability, and Defect Structure of the Al-0.2wt%Sc Alloy. *Phys Solid State*, 52, 1517-1523.

[16] Betekhtin, V. I., Kadomtsev, A. G., Sklenicka, V., & Narykova, M. V. (2011). Effect of Hydrostatic Pressure on Defect Structure and Durability of Ultrafine-Grained Aluminium. *Tech Phys Letters*, 37, 977-979.

[17] Betekhtin, V. I., Kadomtsev, A. G., & Kardashev, B. K. (2006). Elasticity and Anelasticity of Microcrystalline Aluminum Samples Having Various Deformation and Thermal History. *Phys Solid State*, 48, 1506-1512.

[18] Betekhtin, V. I., Tabachnikova, E. D., Kadomtsev, A. G., Narykova, M. V., & Lapovok, M. V. (2011). Effect of Counterpressure During Equal-Channel Angular Pressing on Nanoporosity Formation in Ultrafine-Grained Copper. *Tech Phys Letters*, 37, 767-768.

[19] Kolobov, Yu. R., & Ratochka, J. V. (2005). Grain Boundary Diffusion and Plasticity/ Superplasticity of Polycrystalline and Nanostructured Metals and Alloys. *Mater Sci Eng A*, 410- 411, 468-471.

[20] Humphreys, F. J. (2001). Grain and Subgrain Characterization by Electron Backscatter Diffraction. *J Mater Sci*, 36, 3833-3854.

[21] Ilucova, l., Saxl, I., Svoboda, M., Sklenicka, V., & Kral, P. (2007). Structure of ECAP Aluminium after Different Number of Passes. *Image Anal Stereol*, 26, 37-43.

[22] Saxl, I., Sklenicka, V., Ilucova, L., Svoboda, M., Dvorak, J., & Kral, P. (2009). The Link Between Microstructure and Creep in Aluminium Processed by Equal-Channel Angular Pressing. *Mater Sci Eng A*, 503, 82-85.

[23] Saxl, I., Kalouskova, A., Ilucova, L., & Sklenicka, V. (2009). Grain and Subgrain Boundaries in Ultrafine-Grained Materials. *Mater Characterization*, 60, 1163-1167.

[24] Gao, N., Wang, S., Ubhi, H. S., & Starink, M. A. (2005). A Comparison of Grain Size Determination by Light Microscopy and EBSD Analysis. *J Mater Sci*, 40, 4971-4974.

[25] Sklenicka, V., Dvorak, J., & Svoboda, M. (2004). Creep Behaviour of Pure Aluminium Processed by Equal-Channel Angular Pressing. *In: Zehetbauer M J, Valiev R Z, editors. Nanomaterials by Severe Plastic Deformation*, Weinham, Wiley VCH, 200-206.

[26] Sklenicka, V., Dvorak, J., & Svoboda, M. (2004). Creep in Ultrafine-Grained Aluminium. *Mater Sci Eng A*, 387-389, 696-701.

[27] Sklenicka, V., Dvorak, J., Kral, P., Stonawska, Z., & Svoboda, M. (2005). Creep Processes in Pure Aluminium Processed by Equal-Channel Angu-lar Pressing. *Mater Sci Eng A*, 410-411, 408-412.

[28] Venkateswarlu, K., Rajinikanth, V., Ray, A. K., Xu, C., & Langdon, T. G. (2010). The Characteristics of Aluminum-Scandium Alloys Processed by ECAP. *Mater Sci Eng A*, 527, 1448-1452.

[29] Royset, J., & Ryum, N. (2005). Kinetics and Mechanisms of Precipitation in an Al-0.2wt.%Sc Alloy. *Mater Sci Eng A*, 396, 409-422.

[30] Furukawa, M., Utsunomiya, A., Matsubara, K., Horita, Z., & Langdon, T. G. (2001). Influence of Magnesium on Grain Refinement and Ductility in a Dilute Al-Sc Alloy. *Acta Mater*, 49, 3829-3838.

[31] Marquis, E., Seidman, D. N., & Dunand, D. C. (2003). Effect of Mg Addition on the Creep and Yield Behaviour of an Al-Sc Alloy. *Acta Mater*, 51, 4751-4760.

[32] Komura, S., Horita, Z., Furukawa, M., Nemoto, M., & Langdon, T. G. (2001). An Evaluation of the Flow Behavior During High Strain Rate Superplasticity in an Al-Mg-Sc Alloy. *Metall Mater Trans A*, 32, 707-716.

[33] Sklenicka, V., Dvorak, J., Svoboda, M., Kral, P., Kvapilova, M., & Horita, Z. (2006). Compressive Creep in an Al-3%Mg-0.2%Sc Alloy Processed by Equal-Channel Angular Pressing. *In: Zhu Y, Langdon T G, Horita Z, Zehetbauer M J, Semiatin S L, Lowe T C, editors. Ultrafine Grained Materials IV*, Warrendale, TMS, 459-464.

[34] Sklenicka, V., Dvorak, J., Kvapilova, M., Svoboda, M., Kral, P., Saxl, I., & Horita, Z. (2007). Effect of Equal-Channel Angular Pressing (ECAP) on Creep in Aluminium Alloys. *Mater Sci Forum*, 539-543, 2904-2909.

[35] Kral, P., Dvorak, J., & Sklenicka, V. (2008). Microstructural Evolution and Creep of an Al-0.2wt.%Sc Alloy After Equal-Channel Angular Pressing. *Mater Sci Forum*, 584-586, 846-851.

[36] Kawasaki, M., Sklenicka, V., & Langdon, T. G. (2010). An Evaluation of Creep Behavior in Ultrafine-Grained Aluminum Alloys Processed by ECAP. *J Mater Sci*, 45, 271-274.

[37] Kawasaki, M., Sklenicka, V., & Langdon, T. G. (2011). Creep Behavior of Metals Processed by Equal-Channel Angular Pressing. *Kovove Mater*, 49, 75-83.

[38] Marquis, E. A., & Seidman, D. N. (2001). Nanoscale Structural Evolution of Al_3Sc Precipitates in Al(Sc) Alloys. *Acta Mater*, 49, 1909-1919.

[39] Sklenicka, V., Kral, P., Dvorak, J., Kvapilova, M., Kawasaki, M., & Langdon, T. G. (2011). Effect of Equal-Channel Angular Pressing on the Creep Resistance of Precipitation-Strengthened Alloys. *Mater Sci Forum*, 667-669, 897-902.

[40] Hahn, H., Mondal, P., & Podmanabhan, K. (1997). Plastic Deformation of Nanocrystalline Materials. *NanoStruc Mater*, 9, 603-607.

[41] Vinogradov, A., Hashimoto, S., Patlan, V., & Kitagawa, K. (2001). Atomic Force Microscopic Study on Surface Morphology of Ultra-Fine Grained Materials After Tensile Testing. *Mater Sci Eng A*, 319-321, 862-866.

[42] Huang, Y., & Langdon, T. G. (2003). Using Atomic Force Microscopy to Evaluate the Development of Mesoscopic Shear Planes in Materials Processed by Severe Plastic Deformation. *Mater Sci Eng A*, 358, 114-121.

[43] Sklenicka, V., Dvorak, J., Kral, P., Svoboda, M., & Saxl, I. (2009). Some Factors Affecting the Creep Behaviour of Metallic Materials Processed by Equal-Channel Angular Pressing. *Int J Mat Res*, 100, 762-766.

[44] Prochazka, J., Ponizil, P., & Saxl, I. (2008). Grain Size Estimation in Anisotropic Materials. *Mater Sci Forum*, 567- 568, 285-288.

[45] Saxl, I., & Sklenicka, V. (2009). Interpretation of Profile and Intercept Counts in Microstructure Characterization. *Mater Sci Forum*, 604-605, 403-410.

[46] Kral, P., Dvorak, J., Kvapilova, M., Benes, V., Ponizil, P., Sedivy, O., & Sklenicka, V. (2011). Quantitative Characterization of Microstructure in Copper Processed by Equal-Channel Angular Pressing. *Mater Sci Forum*, 667-660, 235-240.

[47] Saxl, I., Sulleiova, K., & Ponizil, P. (2001). Simulating Grain Size Estimation. *Kovove Mater*, 39, 396-409.

[48] Sklenicka, V., Kral, P., Ilucova, L., Saxl, I., Dvorak, J., & Svoboda, M. (2006). Inhomogeneity of Microstructure and Creep of ECAP Aluminium. *Mater Sci Forum*, 503-504, 245-250.

[49] Sklenicka, V., Dvorak, J., Kral, P., Svoboda, M., Kvapilova, M., & Langdon, T. G. (2012). Creep Ductility of Ultrafine-Grained Metallic Materials. *Submited to Mat Sci Eng A*.

[50] Kapoor, R., Kumar, N., Mishra, R. S., Huskamp, C. S., & Sankaran, K. K. (2010). Influence of Fraction of High Angle Boundaries on the Mechanical Behavior of an Ultrafine Grained Al-Mg Alloy. *Mater Sci Eng A*, 527, 5246-5254.

[51] Phaniraj, M. P., Prasad, M. J. N. V., & Chokshi, A. H. (2007). Grain-Size Distribution Effects in Plastic Flow and Failure. *Mater Sci Eng A*, 463, 231-237.

[52] Hasegawa, H., Komura, S., Utsunomiya, A., Horita, Z., Furukawa, M., Nemoto, M., & Langdon, T. G. (1999). Thermal Stability of Ultrafine-Grained Aluminium in the Presence of Mg and Zr Additions. *Mater Sci Eng A*, 265, 188-196.

[53] Mughrabi, H., Höppel, H. W., Kautz, M., & Valiev, R. Z. (2003). Annealing Treatments to Enhance Thermal and Mechanical Stability of Ultrafine-Grained Metals Produced by Severe Plastic Deformation. *Z Metallkde*, 94, 1079-1083.

[54] Kawasaki, M., Beyerline, I. J., Vogel, S. C., & Langdon, T. G. (2008). Characterization of Creep Properties and Creep Textures in Pure Aluminum Processed by Equal-Channel Angular Pressing. *Acta Mater*, 56, 2307-2317.

[55] Wilshire, B., & Palmer, C. (2002). Deformation Processes During Creep of Pure Aluminium. *In: Mishra R S, Earthman J C, Raj S V, editors. Creep Deformation. Fundamentals and Applications*, Warrendale, TMS, 51-60.

[56] Wilshire, B., & Scharning, P. J. (2008). Creep and Creep Fracture of Commercial Aluminium Alloys. *J Mater Sci*, 43, 3992-4000.

[57] Sklenicka, V., Dvorak, J., & Svoboda, M. (2004). Influence of Processing Route on Creep of Ultrafine Grained Aluminium Prepared by ECAP. *In: Zhu Y T, Langdon T G, Valiev R Z, Semiatin S L, Shin D H, Lowe T C, editors. Ultrafine Grained Materials III*, Warrendale, TMS, 647-652.

[58] Sklenicka, V., & Cadek, J. (1970). Determination of Strain Component Caused by Grain Boundary Sliding. *Z Metallkde*, 61, 575-580.

[59] Cadek, J. (1988). *Creep in Metallic Materials*, Amsterdam, Elsevier Science Publ., 372.

[60] Langdon, T. G. (2006). Grain Boundary Sliding Revisited. Developments in Sliding over Four Decades. *J Mater Sci*, 41, 597-609.

[61] Zhilyaev, A., & Pshenichnyuk, A. (2011). *Superplasticity and Grain Boundaries in Ultrafine-Grained Materials*, Cambridge, Woodhead Publishing, 312.

[62] Divinski, S. V., Reglitz, G., Rosner, H., Estrin, Y., & Wilde, G. (2011). Ultra-Fast Diffusion Channels in Pure Ni Severly Deformed by Equal-Channel Angular Pressing. *Acta Mater*, 59, 1974-1985.

[63] Nabarro, F. R. N. (1948). Deformation of Crystals by the Motions of Single Ions. *In: Rep. Conf. Strength Solids*, The Physical Society London, 75.

[64] Herring, C. (1950). Diffusional Viscosity of a Polycrystalline Materials. *J Appl Phys*, 21, 437.

[65] Coble, R. L. (1963). A Model for Boundary Diffusion Controlled Creep in Polycrystalline Materials. *J Appl Phys*, 31, 1679.

[66] Mohamed, F. A., & Langdon, T. G. (1974). Deformation Mechanism Maps Based on Grain-Size. *Metall Trans*, 5, 2239-2345.

[67] Zeng, X. H., Li, Y. J., & Blum, W. (2004). On Coble Creep in Ultrafine-Grained Cu. *Phys Stat Sol (a)*, 201(14), R114-117.

[68] Yavari, P., Mohamed, F. A., & Langdon, T. G. (1981). Creep and Substructure Formation in an Al-5Percent Mg Solid Solution Alloy. *Acta Metall*, 29, 1495-1507.

[69] Yavari, P., & Langdon, T. G. (1982). An Examination of the Breakdown in Creep by Viscous Glide in Solid Solution Alloys at High Stress Levels. *Acta Metall*, 30, 2181-2196.

[70] Mc Nelley, T. R., Michel, D. J., & Salama, A. (1989). The Mg-Concentration Dependence of the Strength of AlMg Alloys During Glide-Controlled Deformation. *Scripta Met*, 23, 1657-1662.

[71] Fuller, C. B., Seidman, D. N., & Dunand, D. C. (1999). Creep Properties of Coarse-Grained Al(Sc) Alloys at 300°C. *Scripta Mater*, 40, 691-696.

[72] Seidman, D. N., Marquis, E. A., & Dunand, D. C. (2002). Precipitation Strengthening at Ambient and Elevated Temperatures of Heat-Treatable Al(Sc) Alloys. *Acta Mater*, 50, 4021-4035.

[73] Gibeling, J. C., & Nix, W. D. (1980). The Description of Elevated Temperature Deformation in Terms of Threshold Stresses and Back Stresses: A Review. *Mater Sci Eng*, 45, 123-135.

[74] Marquis, E. A., & Dunand, D. C. (2002). Model for Creep Threshold Stress in Precipitation-Strengthened Alloys with Coherent Particles. *Scripta Mat*, 47, 503-508.

[75] Cadek, J., Zhu, S. J., & Milicka, K. (1998). Threshold Creep Behaviour of Aluminium Dispersion Strengthened by Fine Alumina Particles. *Mater Sci Eng A*, 252, 1-5.

[76] Blum, W., Eisenlohr, P., & Sklenicka, V. (2009). Creep Behavior of Bulk Nanostructured Materials- time dependent deformation and deformation kinetics. *In: Zehetbauer M J, Zhu T Z, editors. Bulk Nanostructured Materials*, Weinhaim, Wiley-VCH Verlag GmBH & Co., 519-538.

[77] Kolobov, Y. R., Grabovetskaya, G. P., Ivanov, M. B., Zhilyaev, A. P., & Valiev, R. Z. (2001). Grain Boundary Diffusion Characteristics of Nanostructured Nickel. *Scripta Mater*, 44, 873-878.

[78] Wang, Z. B., Lu, K., Wilde, G., & Divinski, S. V. (2011). Effect of Grain Growth on Interface Diffusion in Nanostructured Cu. *Scripta Mater*, 64, 1055-1058.

[79] Wang, Y., Chen, M., Zhou, F., & Ma, E. (2002). High Tensile Ductility in a Nanostructured Metal. *Nature*, 419, 912-914.

[80] Kawasaki, M., Horita, Z., & Langdon, T. G. (2009). Microstructural Evolution in High Purity Aluminum Processed by ECAP. *Mater Sci Eng A*, 524, 143-150.

[81] Llorca-Isern, N., Grosdidier, T., & Cabrera, J. M. (2010). Enhancing Ductility of ECAP Pressed Metals. *Mater Sci Forum*, 654-656, 1219-1222.

[82] Korn, M., Lapovok, R., Böhner, A., Höppel, H. W., & Mughrabi, H. (2011). Bimodal Grain Size Distributions in UFG Materials Produced by SPD- Their Evolution and Effect on the Fatigue and Monotonic Strength Properties. *Kovove Mater*, 49, 51-53.

[83] Ma, E. (2003). Instabilities and Ductility of Nanocrystalline and Ultrafine-Grained Metals. *Scripta Mater*, 49, 663-668.

[84] Zhao, Y. H., Guo, Y. Z., Wie, Q., Dangelewitz, A. M., Xu, C., Zhu, Y. T., Langdon, T. G., Zhou, Y. Z., & Lavernia, E. J. (2008). Influence of Specimen Dimensions on the Tensile Behaviour of Ultrafine-Grained Cu. *Scripta Mater*, 59, 627-630.

Chapter 2

Durability and Corrosion of Aluminium and Its Alloys: Overview, Property Space, Techniques and Developments

N. L. Sukiman, X. Zhou, N. Birbilis, A.E. Hughes,
J. M. C. Mol, S. J. Garcia, X. Zhou and G. E. Thompson

Additional information is available at the end of the chapter

http://dx.doi.org/10.5772/53752

1. Introduction

Aluminium (Al) is an important structural engineering material, its usage ranking only behind ferrous alloys (Birbilis, Muster et al. 2011). The growth in usage and production of Al continues to increase (Davis 1999). The extensive use of Al lies in its strength:density ratio, toughness, and to some degree, its corrosion resistance. From a corrosion perspective, which is most relevant to this chapter, Al has been a successful metal used in a number of applications from commodity roles, to structural components of aircraft. A number of Al alloys can be satisfactorily deployed in environmental/atmospheric conditions in their conventional form, leaving the corrosion protection industry to focus on market needs in more demanding applications (such as those which require coating systems, for example, the aerospace industry).

Relatively pure aluminium presents good corrosion resistance due to the formation of a barrier oxide film that is bonded strongly to its surface (passive layer) and, that if damaged, reforms immediately in most environments; i.e. re-passivation (Davis 1999). This protective oxide layer is especially stable in near-neutral solutions of most non-halide salts leading to excellent pitting resistance. Nevertheless, in open air solutions containing halide ions, with Cl^- being the most common, aluminium is susceptible to pitting corrosion. This process occurs, because in the presence of oxygen, the metal is readily polarised to its pitting potential, and because chlorides contribute to the formation of soluble chlorinated aluminium (hydr)oxide which interferes with the formation of a stable oxide on the aluminium surface.

Aluminium and its alloys readily oxidises, including when Al is present in either in solid solution or in intermetallic (IM) particles. Industrial alloy surfaces however, tend to be as heterogeneous as their underlying microstructures. The surface of a wrought or cast alloy is likely to contain not only aluminium oxide alone, but may for example contain a fragment of a mixed Al-Mg oxide for alloys rich in Mg (Harvey, Hughes et al. 2008)). This is primarily because of the heat of segregation of Mg is high and it has a favorable free energy for oxide formation. If however an Al surface is mechanically undisturbed - then the surface oxide is relatively protective. Though, most real surfaces have some sort of mechanical finishing which results in the formation of a near surface deformed layer (NSDL) and shingling. Shingling occurs where the alloy matrix is spread across the surface including IM particles in abrasion and milling (Scholes, Furman et al. 2006; Muster, Hughes et al. 2009). This is because the IM particles are harder than the surrounding matrix and less susceptible to deformation (Zhou, Liu et al. 2011). Even on polished surfaces, the matrix and the IM particles rapidly form different oxide structures (Juffs, Hughes et al. 2001; Juffs, Hughes et al. 2002). This is almost certainly due to different chemical environments and different electrochemical reactions over the IM particles compared to the matrix. Furthermore, the morphology and the oxide are not continuous from the IM particles to the matrix and this represents a potential defect site in the context of corrosion. For the purposes of descriptions herein, IM particles can be classified into three main types; i) precipitates, ii) constituent particles and iii) dispersoids. Precipitates are typically in the shape of needles, laths, plates or spherical with the size ranging from Angstroms to fractions of a micrometer. They are formed by nucleation and growth from a supersaturated solid solution during low temperature aging and may be concentrated along the grain boundaries. Constituent particles however, are relatively large with irregular shape and the size can be up to 10 micrometers. This type of particle forms during solidification of the alloy and is not fully dissolved by subsequent thermomechanical processing (including solution heat treatments). They can be found in colonies of several IM crystals or different compound types. On the contrary, dispersoids are small particles with size ranging from 0.05 to 0.5 micrometers. They are thermally stable intermetallics of a fine size that are functional for controlling grain size and recrystallisation behavior. Dispersoids form by low level additions of highly insoluble elements such as Cr, Mn or Zr.

This chapter will aim to cover some of the important aspects related to the corrosion of Al-alloys, bearing in mind the role of alloy chemistry. In addition, some of the topical aspects related to techniques and ongoing developments in the general field of Al-alloy corrosion are presented. An attempt has been made to give the reader an overview of the key technical aspects, but unfortunately for comprehensive insight into the topic overall, the size of this chapter alone cannot be a replacement to dedicated monographs on the specific topics at hand; nor the ever-evolving journal literature that represents the state of the art. To aid in the transfer of information, this chapter has been divided into a number of sections to treat the widely varying topics independently.

http://dx.doi.org/10.5772/53752

1.1. The general performance of the Al-alloy classes

The corrosion potential of an aluminium alloy in a given environment is primarily determined by the composition of the aluminium rich solid solution, which constitutes the predominant volume fraction and area fraction of the alloy microstructure (Davis 1999). While the potential is not affected significantly by second phase IM particles of microscopic size, these particles frequently have corrosion potentials (when measured in isolation) differing from that of the solid solution matrix resulting in local (micro-) galvanic cells, when IMs are polarised to the corrosion potential of the alloy. The result is that local currents on the alloy surface differ, establishing anodes and cathodes. Since most of the commercial aluminium alloys contain additions of more than one type of alloying element, the effects of multiple elements on solution potential are approximately additive. The amounts retained in solid solution, particularly for more highly alloyed compositions, depend on production and thermal processing so that the heat treatment and other processing variables influence the final electrode potential of the product.

By measuring the potentials of grain boundaries and grain bodies separately, the difference in potential responsible for local types of corrosion such as intergranular corrosion, exfoliation, and stress corrosion cracking (SCC) can be quantified (Guillaumin and Mankowski 1998; Zhang and Frankel 2003). By measuring the corrosion potential of IMs (Buchheit 1995), and indeed by measurement of the polarisation response of IMs, even more significant insights into localised corrosion can be gained (Birbilis and Buchheit 2005). Such specialist topics are not dealt with in their entirety herein, however an abridged written summary of the performance of the key Al-alloy classes (as outlined by the Aluminium Association (Hatch 1984)) is provided below.

1.1.1. 1xxx series alloys

Corrosion resistance of aluminium increases with increasing metal purity, however the use of the >99.8% grades is usually confined to those applications where very high corrosion resistance or ductility is required. In regards to such specialist applications however, the actual number of applications are very few. Consequently 1xxx series alloys are not commonly used or sold (but do serve as important feedstock to secondary alloy producers or production). In the instance where general-purpose alloys for lightly stressed applications are required, such alloys are approximately 99% pure aluminium and offer adequate corrosion resistance in near neutral environments. 1xxx is also sometimes used in cladding for example AA1230 is used as clad on AA2024 (Hatch 1984)

1.1.2. 2xxx series alloys

Copper is one of the most common alloying additions - since it has appreciable solubility and can impart significant strengthening by promotion of age hardening (in fact, the Al-Cu system was the classical/original age hardening system (Hatch 1984)). These alloys were the foundation of the modern aerospace construction industry and, for example AA2024 (Al-4.4Cu-1.5Mg-0.8Mn), which is still used in many applications to this day, can achieve strengths in excess of 500MPa depending on temper (Polmear 2006).

1.1.3. 3xxx series alloys

The 3xxx series alloys are a commodity product that is nominally available in the form of thin sheet (for beverage can usage). The key alloying element, Manganese, has a relatively low solubility in aluminium but can improve corrosion resistance when remaining in solid solution. Additions of manganese of up to about 1% form the basis of the non-heat treatable wrought alloys with good corrosion resistance, moderate strength (i.e. AA3003 tensile strength ~110MPa) and exceptionally high formability (Polmear 2006).

1.1.4. 5xxx series alloys

Magnesium has significant solubility in aluminium and imparts substantial solid solution strengthening (which can also contribute to enhanced work hardening characteristics) (Davis 1999; Polmear 2006). The 5xxx series alloys (containing <~6% Mg) do not age harden. Whilst it is possible for β-phase (Mg_2Al_3) to precipitate in systems with above ~3%Mg, the β-phase is not a strengthening precipitate and actually weakens the alloy (by depleting the solute of Mg). Nominally, the corrosion resistance of 5xxx series alloys is good and their mechanical properties make them ideally suited for structural use in aggressive conditions (such as marine vessels). Fully work-hardened AA5456 (Al-4.7Mg-0.7Mn-0.12Cr) has a tensile strength of >380MPa. One corrosion issue with fully work-hardened 5xxx series alloys is that the heavy dislocation density (and supersaturation of the solid solution with Mg) can permit the sensitization of the microstructure by precipitation of deleterious β-phase (Mg_2Al_3) during sustained high temperature exposure (i.e. in service) (Baer, Windisch et al. 2000; Searles, Gouma et al. 2002; Davenport, Yuan et al. 2006; Goswami, Spanos et al. 2010).

1.1.5. 6xxx series alloys

Silicon additions alone can lower the melting point of aluminium whilst simultaneously increasing fluidity (which is why the vast majority of cast Al products contain various amounts of Si). These alloys are increasing in importance in automotive applications for engine and drive train components – however are yet to realise the majority of market share. Heat-treatable Al-Mg-Si are predominantly structural materials (strengths >300MPa are possible), all of which have an appreciable resistance to corrosion, immunity to SCC and are weldable. To date, 6xxx series alloys are mainly used in extruded form, although increasing amounts of sheet are being produced (Birbilis, Muster et al. 2011). Magnesium and silicon additions are made in balanced amounts to form quasi-binary Al-Mg_2Si alloys, or excess silicon additions are made beyond the level required to form Mg_2Si. Alloys containing magnesium and silicon in excess of 1.4% develop higher strength upon aging.

1.1.6. 7xxx series alloys

The Al-Zn-Mg alloy system provides a range of commercial compositions, primarily where strength is the key requirement (and this can be achieved without relatively high cost or complex alloying). Al-Zn-Mg-Cu alloys have traditionally offered the greatest potential for age hardening and as early as 1917 a tensile strength of 580MPa was achieved, however, such alloys were not suitable for commercial use until their high susceptibility to stress corrosion cracking could be moderated (Song, Dietzel et al. 2004; Birbilis, Cavanaugh et al. 2006; Lin, Liao et al. 2006; Lynch, Knight et al. 2009). Aerospace needs led to the introduction of a range of high strength aerospace alloys of which AA7075 (Al-5.6Zn-2.5Mg-1.6Cu-0.4Si-0.5Fe-0.3Mn-0.2Cr-0.2Ti) is perhaps the most well-known, and which is now essentially wholly superseded by AA7150 (or the 7x50 family). The high strength 7xxx series alloys derive their strength from the precipitation of η-phase ($MgZn_2$) and its precursor forms. The heat treatment of the 7xxx series alloys is complex, involving a range of heat treatments that have been developed to balance strength and stress corrosion cracking performance - including secondary (or more) heat treatments that can include retrogression and re-aging (Fleck, Calleros et al. 2000; Ferrer, Koul et al. 2003; Zieliński, Chrzanowski et al. 2004; Marlaud, Deschamps et al. 2010).

1.1.7. 8xxx series alloys

Nominally reserved for the sundry alloys, 8xxx series alloys include a number of Lithium (Li) containing alloys. Li is soluble in aluminium to ~ 4 wt% (corresponding to ~ 16 at%). As these alloys of high specific strength and stiffness readily respond to heat treatment, research and development has intensified due to their potential for widespread usage in aerospace applications (Lavernia and Grant 1987; Dorward and Pritchett 1988; Giummarra, Thomas et al. 2007). Based on the impressive lightweight of such alloys, present day aircraft are comprised of some portion of Al-Li based alloys (modern generations of which actually include low Li levels and hence are nowadays designated as 2xxx alloys (Ambat and Dwarakadasa 1992; Garrard 1994; Semenov 2001; Giummarra, Thomas et al. 2007). First generation Li-containing alloys displayed some of the highest corrosion rates of all aluminium alloys, where susceptibility to intergranular corrosion was challenging. Modern Al-Cu-Li seem to have overcome this challenge; however it is also important to recognise that production requires specialised melting and casting, not presently available in most commercial facilities.

2. Corrosion of aluminium and its alloys in aqueous environment

2.1. Environmental corrosion of aluminium

Corrosion in aluminium alloys is generally of a local nature, because of the separation of anodic and cathodic reactions and solution resistance limiting the galvanic cell size. The basic anodic reaction is metal dissolution:

$Al \rightarrow Al^{3+} + 3e^-$

While the cathodic reactions are oxygen reduction:

$O_2 + 2H_2O + 4e^- \rightarrow 4OH^-$

or hydrogen reduction in acidified solution such as in a pit environment as a result of aluminium ion hydrolysis:

$$2H^{+} + 2e \rightarrow H_2$$

It is the interaction between local cathodes and anodes and the alloy matrix that leads to nearly all forms of corrosion in aluminium alloys. These include pitting corrosion, selective dissolution, trenching, intermetallic particle etchout, intergranular attack and exfoliation corrosion. Surface and subsurface grain etchout is also influenced by grain energy which is derived from grain defect density. Grain etchout, has a significant role in exfoliation corrosion since the volume of hydrated aluminium oxide generated during dissolution is larger than the original volume of the grain.

The general consensus for Al and its alloys is that they are resistant towards corrosion in mildly aggressive aqueous environments. The protective oxide layer represents the thermodynamic stability of Al alloys in corrosive environment - acting as a physical barrier as well as capable to repair itself in oxidizing environments if damaged. While the passive layer breakdown mechanism by chloride ions is still in debate (Sato 1990; McCafferty 2010) due to the complexity of the process (Szklarska-Smialowska 2002), the general consensus is that localized attack starts by adsorption of aggressive anions and formation of soluble transitional complexes with the cations at the oxide surface. Thermodynamic principles to explain and predict the passivity phenomenon that controls the corrosion behavior of Al are summarised by Pourbaix-type analysis. This results in a plot of potential vs. pH based on the electrochemical reaction of the species involved, the representation known as a Pourbaix diagram (Pourbaix 1974) as shown in Figure 1.

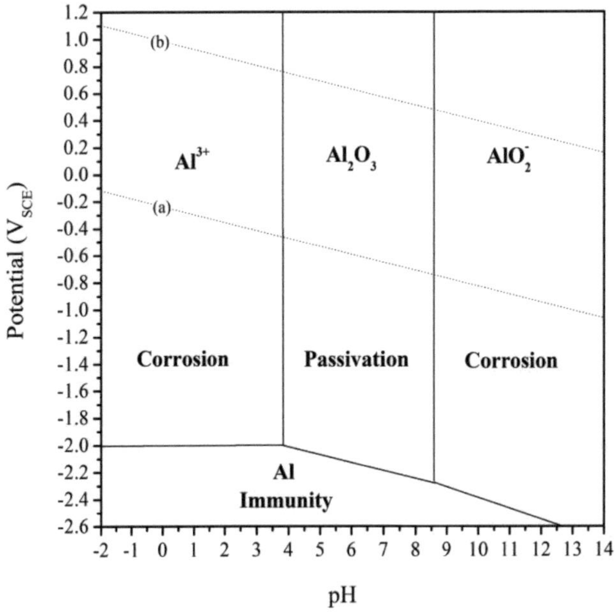

Figure 1. E-pH diagram for pure Al at 25°C in aqueous solution (adapted from Pourbaix 1974). The lines (a) and (b) correspond to water stability and its decomposed product.

It is seen that Al is nominally passive in the pH range of ~4 to 9 due to the presence of an Al_2O_3 film. In environments that deviate from the near neutral range, the continuity of this film can be disrupted in which the film becomes soluble, facilitating the relatively rapid of dissolution the alloy. In the acidic range, Al is oxidised by forming Al^{3+}, whilst AlO^{2-} occurs in alkaline range.

The E-pH diagram gives an impression that corrosion prediction is a straightforward process, however in actual engineering applications, there are several variables that weren't considered by Pourbaix. These include (i) the presence of alloying elements in most engineering metals (ii) the presence of substances in the electrolyte such as chloride (albeit that this has been addressed in more modern computations), (iii) the operating temperature of the alloy, (iv) the mode of corrosion, and (v) the rate of reaction. Taking these factors into account is nominally done on a case by case (i.e. alloy by alloy) basis, and a revised version of an E-pH diagram for 5xxx series alloys in 0.5M sodium chloride is given in Figure 2.

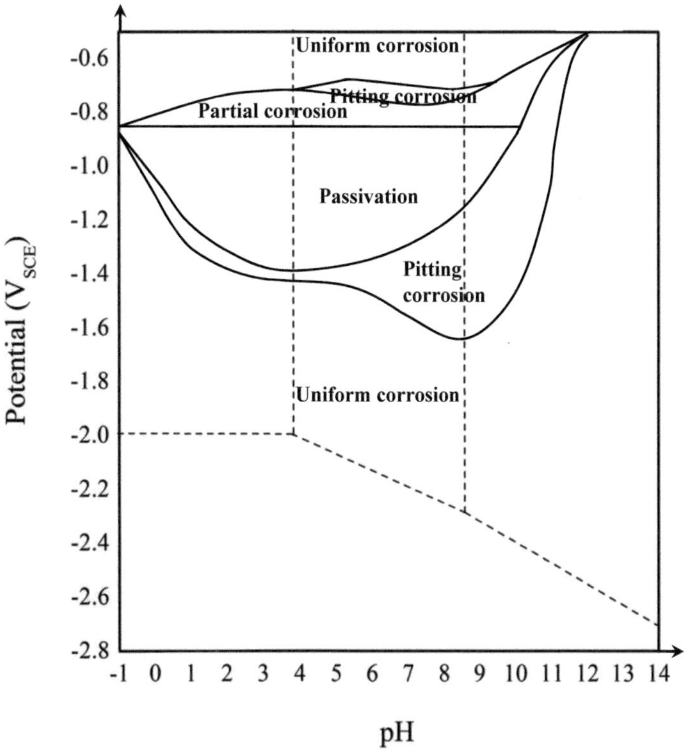

Figure 2. Mode of corrosion based on experimental data for AA5086 in the presence of 0.5M sodium chloride (adapted from Gimenez, Rameau et al. 1981)

Figure 2 indicates windows where localized attack is highly possible in the supposed passive region (Gimenez, Rameau et al. 1981). It is also seen that localised attack is possible across the whole range of pH depending on the specific potential. One should therefore not rely solely on the Pourbaix diagram as a direct index to actual corrosion rates, with rates needing to be independently measured for a given alloy-electrolyte combination (Ambat and Dwarakadasa 1992). Finally, whilst not to be discussed in detail here, it is prudent to indicate that effectively all Al-alloys do not attain practical/empirical immunity as evidence in Figure 1. Cathodic polarisation tends to contribute to alloy deterioration by two modes. Firstly, the accumulation of hydroxyl ions at the Al-surface will cause chemical dissolution of the Al. Secondly, Al is a very strong hydride former, and hydrogen from the cathodic reaction at such negative potentials will serve combine with Al to form hydrides (Perrault 1979).

2.2. Kinetic stability of aluminium alloys

Kinetics represents the rate of reaction during corrosion. When exposed to an aqueous environment, metals stabilise to a value of electrochemical potential that is characteristic of the material and its composition for a given electrolyte. This potential is the potential at which anodic and cathodic reactions upon the metal surface are equal, and the value of this potential is thus significantly influenced by factors that can alter the relative rates of anodic or cathodic reaction efficiency upon the metal surface (i.e. alloying, precipitate state, etc.).

Most typically, the potentiodynamic polarisation test is used to characterise the corrosion performance of an alloy (as far as determination of mechanistic aspects from an instantaneous test). This method gives vital kinetic information such as current density over a range of potentials, pitting potential (if it exists), corrosion potential, the passive current density and potentially more information in reverse scans, etc. Thus factors affecting corrosion as discussed in the previous sections can be evaluated with much higher confidence. For example, Figure 3 shows a polarisation curve of Pure Al compare to AA2024-T3 (Al-4.3Cu-1.5Mg-0.6Mn) in 0.1M NaCl.

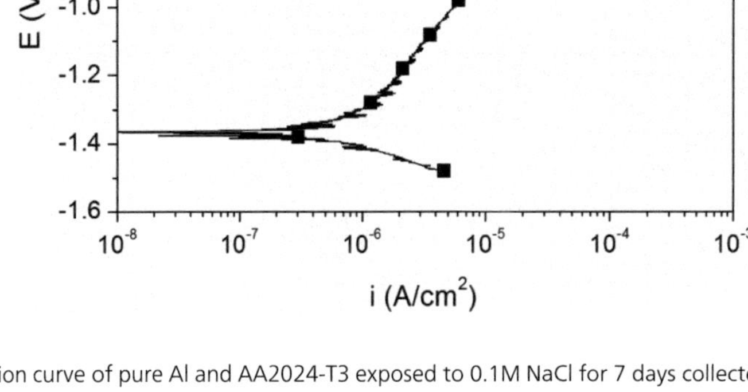

Figure 3. Polarisation curve of pure Al and AA2024-T3 exposed to 0.1M NaCl for 7 days collected at 1mV/s^{-1} (adapted from (Sukiman, Birbilis et al. 2010))

Comparison of alloy behavior and dominant reactions can be made in a quantitative manner. The anodic branch of the polarisation curve gives information related to the anodic/dissolution reaction while the cathodic branch represents the reduction reaction (nominally oxygen reduction, but at lower potentials or in strong acids, water reduction). The addition of more noble alloying elements typically increases the corrosion potential to more noble values (Davis 1999) and this is dramatically observed in Figure 3. This ennoblement however does not correlate to the rate of corrosion (as judged by Figure 3), whereby we see the pure Al versus AA2024-T3 has a difference in potential of ~0.5V. In addition, the main practical threat for Al alloys is localised attack such as pitting, so in that vein, a more noble value of pitting potential does not necessarily signify a better corrosion resistance (Frankel 1998; Birbilis and Buchheit 2005). Rather generally, the electrochemical reactions upon Al-alloys are heavily influenced by the chemistry and microstructure of the alloy – which we attempt to discuss in the following section.

Moving beyond potentiodynamic polarisation towards a true measure of kinetic stability in the E-pH domain (similar in concept to Pourbaix diagrams however which give 'speed' and not just thermodynamic likelihood) there are tests which can be done in this regard. In order to develop an improved understanding of overall kinetic stability of a metal over the potential-pH space, methods including the staircase potentio-electrochemical impedance spectroscopy (SPEIS) can be used to establish so-called kinetic stability diagrams, as previously demonstrated for pure Al (Zhou, Birbilis et al. 2010) and depicted in Figure 4. The specifics of SPEIS will be introduced in section 5.3.

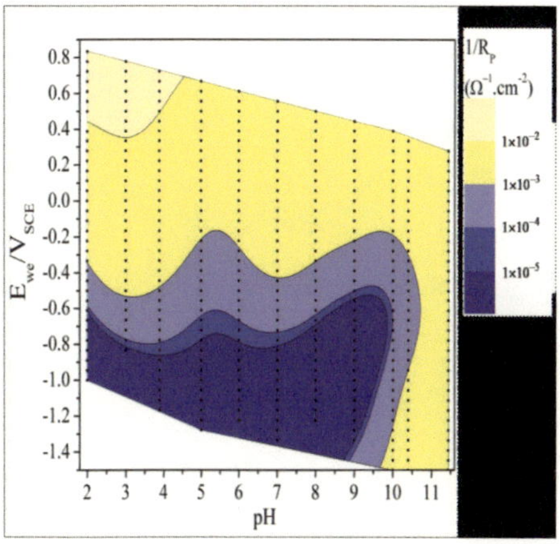

(a)

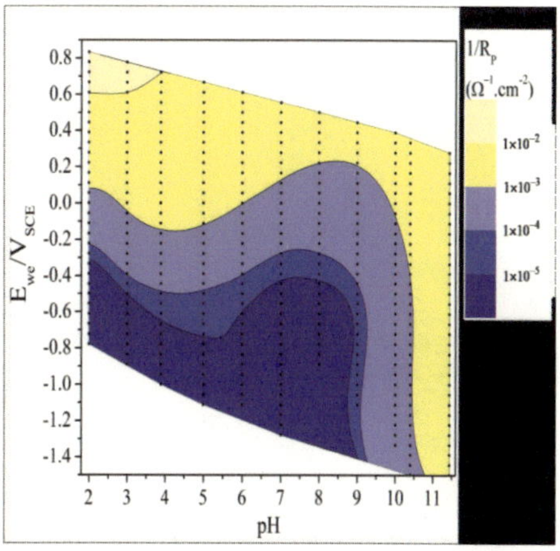

(b)

Figure 4. Contour plots of 1/RP for the E-pH space and their data for (a) A portion of a 99.9999% Al ingot single crystal and (b) a polycrystalline specimen from the identical ingot in (a) (adapted from (Zhou, Birbilis et al. 2010)

In Figure 4, presentation of the reciprocal of polarisation resistance ($1/R_P$) is the metric of reaction 'speed', as it is proportional to the reaction rate at a given condition. The influence of potential and pH is presented not only for pure Al in the sample which was single crystal, but also for a polycrystalline sample. As a result, one is also able to assess the effect of grain structure from Figure 4. Most importantly however, Figure 4 reveals that the rate of reaction stays in a low range in areas that extend beyond that of the Pourbaix diagram in terms of certain E-pH combinations – indicating that although Al may be in a thermodynamically unstable region, the reaction rate can be maintained to be low enough to make it still be useful in an engineering context. Similarly, there are regions of high potential where pure aluminium may be in a thermodynamically stable region, but unusable – owing to transpassive dissolution. Finally, in terms of microstructure effects, it is seen that the exact same material can have a different kinetic response based on structural factors alone. Such differences are not detectible or predicted from thermodynamic analysis in any way, and this highlights the importance of approaches which provide kinetic information to meet demands of engineering applications.

2.3. The property space and corrosion property profile of aluminium alloys

As technologies continue to advance with more challenging applications and environments, a general understanding of durability limits across the class of Al-alloys is essential. Durability needed in a broad sense is the ability to withstand an environment while maintaining mechanical integrity. This indicates a requirement to understand the property space for Al and its alloys. Figure 5 shows a trend that is in line with the notion that increases in hardness (used here as a proxy to yield strength) show increases in corrosion rate.

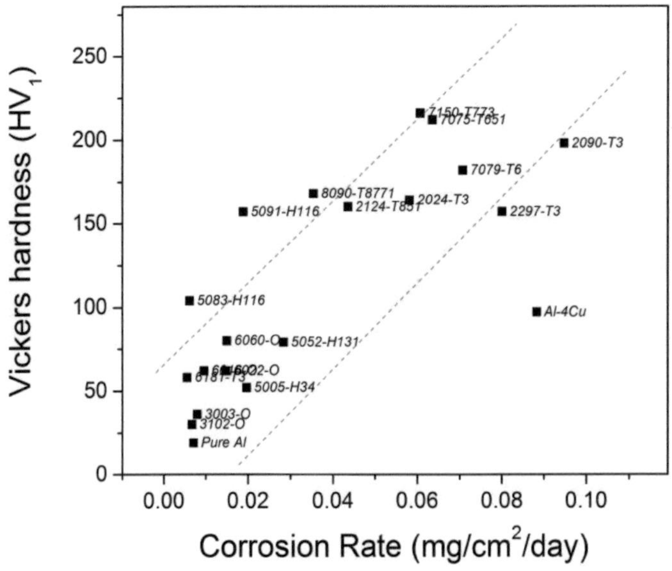

Figure 5. Corrosion rate as determined from weight loss data for commercial Al alloys collected after 14 days exposure in quiescent 0.1M NaCl presented against alloy hardness.

http://dx.doi.org/10.5772/53752

From Figure 5, it can be argued that the data is divided into two main groups, one at each end of the corrosion rate spectrum. High hardness/strength aluminium alloys were found to populate the high corrosion rate space. This is in contrast to the medium to low hardness/strength alloys that revealed considerably lower corrosion rates. The alloys that show the highest corrosion rates are the 'precipitation hardenable' family. Besides the high number density of precipitate particles in such alloys, they also contain an appreciable population of constituent type particles (Chen, Gao et al. 1996; Wei 2001; Andreatta, Terryn et al. 2004; Ilevbare, Schneider et al. 2004; Birbilis, Cavanaugh et al. 2006; Boag, Hughes et al. 2009; Hughes, MacRae et al. 2010; Xu, Birbilis et al. 2011). These particles are industrially necessary, since the complex chemistry of precipitation hardenable alloys (that can contain up to 10 alloying elements) have significant alloying additions added via alloy rich master alloys. It is also observed that the alloys that show the highest corrosion rates also contain appreciable amount of copper.

The plot in Figure 5 allows one to identify a region of property space (at the top left corner) that has potential for future alloys, with ongoing efforts aiming to reach such space (in addition to controlling ductility). Efforts that regard in are underway, focusing on corrosion rate (Carroll, Gouma et al. 2000; Norova, Ganiev et al. 2003; Rosalbino, Angelini et al. 2003; Cavanaugh, Birbilis et al. 2007; Lucente and Scully 2007; Fang, Chen et al. 2009; Graver, Pedersen et al. 2009; Ralston, Birbilis et al. 2010; Tan and Allen 2010; Xu, Birbilis et al. 2011; Brunner, Birbilis et al. 2012) and strength (Poole, Seter et al. 2000; Pedersen and Arnberg 2001; Fuller, Krause et al. 2002; Raviprasad, Hutchinson et al. 2003; Lee, Shin et al. 2004; Oliveira Jr, de Barros et al. 2004; Zhao, Liao et al. 2004; Kim, Kim et al. 2005; Teixeira, Bourgeois et al. 2007; McKenzie and Lapovok 2010; Wang, Zhang et al. 2010; Puga, Costa et al. 2011; Zhong, Feng et al. 2011; Westermann, Hopperstad et al. 2012). However, such studies are done independently of both properties thus the symbiotic effect can't be readily evaluated to date.

3. Corrosion of aluminium and its alloys in aqueous environment

3.1. The role of chemistry on corrosion

Alloying elements are added to aluminium for various reasons, with improving mechanical properties the principal reason. These elements introduce heterogeneity into the microstructure, which is the main cause of localised corrosion that initiates in the form of pitting. Each alloying element has a different effect on the corrosion of Al, and in this section we *briefly* discuss the role of alloying elements on corrosion of Al.

3.1.1. Influence of magnesium

Mg is one of the major elements added to Al to improve mechanical properties by solid solution strengthening – and can be found in 5xxx alloys, as well as 2xxx, 3xxx, 6xxx and 7xxx commercial alloys. Mg can stabilize GP zones, has a high solubility in Al and decreases the alloy density. Muller and Galvele showed that Mg when present in solid solution does not have a significant effect on the pitting corrosion of Al which can be understood on the basis of standard potentials of Al and Mg (Muller and Galvele 1977). Moreover, Mg decreases the rate of the cathodic reaction when present in solid solution, increasing corrosion resistance, which may appear counterintuitive, but is rather obvious (as Mg has a very low exchange current density and hence retards the cathodic reaction). In contrast, excess amounts of Mg in the alloy or a long term exposure to elevated temperature will cause the precipitation of either Al_8Mg_2 or Al_3Mg_2 (Searles, Gouma et al. 2002; Davenport, Yuan et al. 2006; Oguocha, Adigun et al. 2008; Jain, Lim et al. 2012). These phases form typically along grain boundaries (Baer, Windisch et al. 2000; Goswami, Spanos et al. 2010) and are known to be anodic with respect to Al matrix therefore prone to localized corrosion (Vetrano, Williford et al. 1997; Aballe, Bethencourt et al. 2001; Jones, Baer et al. 2001; Brunner, May et al. 2010). Mg in 2xxx, 6xxx and 7xxx alloys however, forms precipitates with other alloying elements to strengthen the alloy where role of Mg mainly depends upon the other alloying additions (Ringer, Hono et al. 1996; Buchheit, Grant et al. 1997; Campestrini, van Westing et al. 2000; Guillaumin and Mankowski 2000; Eckermann, Suter et al. 2008).

3.1.2. Influence of silicon

The addition of Si in conjunction with Mg, which is typical in 6xxx series Al alloys, allows Mg_2Si particles to precipitate. There is vast literature on the chemical composition of the Mg-Si phase and its role on mechanical properties (Hirth, Marshall et al. 2001; Usta, Glicksman et al. 2004; Stelling, Irretier et al. 2006; Eckermann, Suter et al. 2008; Zeng, Wei et al. 2011). This particle is beneficial in terms of increasing strength but renders the alloy prone to localised corrosion (Eckermann, Suter et al. 2008). The electrochemical behavior of Mg_2Si was investigated recently and it was shown that Mg_2Si remains more 'anodic' (i.e.. less noble) than the matrix in Al-alloys. As a consequence of this, Mg_2Si remains anodic and undergoes selective dissolution in the Al-matrix. Some 6xxx series alloys contain excess Si. Excess amount of Si however increases the cathodic reaction rate (Eckermann, Suter et al. 2008) and are unfavorable since Si tends to be present along the grain boundary and this may lead to intergranular corrosion and stress corrosion cracking (Guillaumin and Mankowski 2000; Larsen, Walmsley et al. 2008; Zeng, Wei et al. 2011).

3.1.3. Influence of copper

The presence of Cu is viewed as detrimental to corrosion due to the formation of cathodic particles capable of sustaining the cathodic reaction locally and efficiently, such as Al_2Cu and $AlCu_2Mg$. In some cases where low Cu content is used, the impact of Cu is minimal, however given that corrosion is not the principal alloy design criteria in most instance, Cu is common in many (most) Al-alloys. The 2xxx series alloys are Cu rich, however Cu is added to other alloy classes such as the 6xxx series where it can increase strength when present in trace amounts, and also enhance precipitation hardening. The same is true in 7xxx alloys, with most modern aerospace alloys having appreciable amounts of Cu that can increase strength by modifying precipitation and minimising SCC via incorporation into precipitates (such as $Mg(Zn,Cu)_2$).

In general however, there is still some debate on the precise role of Cu, which also depends on the temper condition. Muller and Buchheit found that Cu in the form of solid solution decreases pitting susceptibility through the ennoblement of pitting potential. While Muller and Galvele reported an increase in pitting potential for solid solution content of Cu up to 5 wt%. In the case of Al-Cu-Mg alloys which contain *S* phase (Al_2CuMg), large differences in solution potential between Cu (highly noble) and Mg exist, with significant focus on corrosion of *S* phase (Buchheit, Grant et al. 1997; Guillaumin and Mankowski 1998; Buchheit, Montes et al. 1999; Ilevbare, Schneider et al. 2004; Boag, Hughes et al. 2011) revealing dealloying and selective dissolution that leads to preferential dissolution of Mg and Al with Cu remnant being redistributed at or near the site of the Al_2CuMg. A range of other particles associated with Cu have been reported such as Al_7Cu_2Fe. However recent microprobe studies of a number of batches of AA2024-T351 indicate five common compositions across modern alloys which do not have the same composition as older alloy stock indicating that this is still an active area of research (Hughes, Glenn et al. 2012). In general, Cu, or Cu containing particles are capable of supporting high oxygen reductions rates and hence undesirable from corrosion perspective (Mazurkiewicz and Piotrowski 1983; Scully, Knight et al. 1993; Buchheit 2000; Birbilis, Cavanaugh et al. 2006).

3.1.4. Influence of zinc

In high strength commercial aluminium alloys such 7xxx series alloys, Zn is added to stimulate precipitation hardening. Alloys containing high levels of Zn such as the modern aerospace alloys 7050 and 7150 are amongst the highest strengths of Al-alloys owing to the high number density of precipitates such as $MgZn_2$ which is evenly distributed throughout the Al matrix (Ringer, Hono et al. 1996; Andreatta, Terryn et al. 2004; Sha and Cerezo 2004; Birbilis and Buchheit 2005; Polmear 2006) in 5xxx alloys. The addition of Zn to Al-Mg alloys was reported to improve resistance against SCC (Unocic, Kobe et al. 2006) where a small amount of Zn added into AA5083 alloy was found to reduce the corrosion - reporting that Zn can promote the formation of Al-Mg-Zn (τ phase) instead of Al_3Mg_2 (β phase) the latter of which is responsible for stress corrosion cracking (Carroll, Gouma et al. 2000; Carroll, Gouma et al. 2001).

3.1.5. Influence of iron

Iron is typically present as an impurity in all commercial Al alloys due to the production process of Al alloys. Unless specifically required for specialist applications, it is simply too expensive to remove all iron (even in aluminium destined for aerospace applications). Despite having a small fraction of the composition, iron is detrimental to corrosion due to its low solubility and hence ability to form constituent particles which are cathodic to the Al-matrix such as Al_3Fe (Nisancioglu 1990). Additionally, iron is capable of sustaining cathodic reactions more efficiently than Al (Galvele 1976; Szklarska-Smialowska 1999). In more complex alloys, Fe can also combine with other alloying elements such as Mn or Cu (in the latter case forming Al_7Cu_2Fe), which are also a major issue for corrosion (Birbilis, Cavanaugh et al. 2006) since the combination of Fe and Cu provides even greater cathodic efficiency for such particles. Corrosion associated with such noble cathodic constituents/intermetallics leads to an increase in local pH of the solution further enhancing anodic dissolution of the Al matrix adjacent to say, Al_3Fe (Seri 1994; Park, Paik et al. 1999; Birbilis and Buchheit 2005; Ambat, Davenport et al. 2006).

3.1.6. Influence of manganese

The addition of Mn is effective in reducing the pitting susceptibility of Al alloys particularly in the context of modifying Fe containing intermetallic particles (Nisancioglu 1990) (where Mn can substitute for Fe, rendering the resulting constituent particle somewhat less noble) The presence of Mn has been noted as reducing the concentration of Fe and reducing the degree of resultant corrosion (Koroleva, Thompson et al. 1999); owing to the formation of Al_6MnFe has a similar electrochemical potential with that of the Al matrix. However, it has also been noted that an excess of Mn can lead to an increase in the cathodic activity when beyond the solubility limit (solubility of Mn in Al is 1.25 wt%) – with constituents such as Al_6Mn readily forming (Liu and Cheng 2010). Generally however, the presence of Mn constituent particles are not as detrimental as particles rich in Fe or Cu (Birbilis and Buchheit 2005; Cavanaugh, Birbilis et al. 2007), which is evidenced by the reliable corrosion performance of 3xxx commercial Al alloy (Zamin 1981; Seri and Tagashira 1986; Tahani, Chaieb et al. 2003; Liu and Cheng 2011).

3.1.7. Influence of lithium

The addition of Li in Al alloys is efficient at significantly reducing alloy density whilst increasing strength – making it an obvious contender in the range of transport, namely aero, applications. Al-Li alloys are a rather specialised field that spans the past five decades, with descriptions originally in the 8xxx series compositional space (with principally Li rich compositions). Such so-called 1st generation Al-Li alloys were only used in specialised applications owing to their susceptibility to cracking. The cracking issue was later managed via new alloy compositions and thermomechanical processing (2nd generation Al-Li alloys), however until relatively recently Al-Li alloys were not so popular owing to relatively high corrosion rates and localised forms of corrosion propagation. Most recently, the 3rd generation of Al-Li alloys has gained significant attention and growing usage in commodity aerospace applications. These 3rd generation alloys are actually 2xxx series alloys, with less Li than Cu. These new 2xxx series alloys will be a significant alloy of the future, whilst still further research is required (from a corrosion perspective) to fully understand the performance, particularly as a function of thermomechanical treatment. Some abridged information regarding the role of Li upon corrosion is included here. In Li rich alloys, the formation of strengthening phase, Al_3Li which is dispersed homogeneously throughout the matrix, is responsible for the increase in strength (Lavernia and Grant 1987). It is however detrimental to corrosion as Al_3Li initially form along the grain boundaries. As Li is an active (i.e. less noble) element, this will localise dissolution to Li rich regions and therefore susceptibility to attack such as intergranular corrosion is high (Martin 1988). When Cu is also added in conjunction with Li (in alloys such as AA2090) the precipitation of phases such as T_1 (Al_2CuLi) occurs. There are two modes of attack associated with T_1, one of which T_1 at the precipitate free zone is dissolved forming small pits, while the other is when T_1 undergo selective dissolution along with dissolution of the adjacent Al matrix leaving larger pits (Buis and Schijve 1992; Buchheit, Wall et al. 1995).

3.1.8. *Influence of activating elements (i.e. Pb, Sn)*

Lead (Pb) and tin (Sn) are usually present in low levels as trace elements (Gundersen, Aytaç et al. 2004; Premendra, Terryn et al. 2009). When present in trace amounts, their influence is minimal or negligible. When (by say, recycling or contamination) the levels rise above the solubility limits, the presence of Pb leads to segregation that results in Pb-rich film at the metal - oxide interface when the alloy is heat treated at elevated temperature (Sævik, Yu et al. 2005) causing the Al oxide film to destabilise particularly in the presence of chloride. The disruption of Al oxide film leads to an increase in anodic reaction rate which not only increases the pitting susceptibility, but can activate the entire surface. This process is called anodic activation, and has been well documented for a number of years by studies from the group of Nisancioglu (Keuong, Nordlien et al. 2003; Gundersen, Aytaç et al. 2004; Yu, Saevlk et al. 2004; Yu, Sævik et al. 2005; Walmsley, Sævik et al. 2007; Jia, Graver et al. 2008; Graver, Pedersen et al. 2009; Anawati, Graver et al. 2010; Graver, van Helvoort et al. 2010; Anawati, Diplas et al. 2011). There have been some efforts to reduce the activation effect of Pb by addition of more noble alloying elements such as Cu in the hope that the addition of Cu may alter the surface potential - hence reducing the activation (Anawati, Diplas et al. 2011). A similar result was observed for the addition of Mg (Jia, Graver et al. 2008), however, such methods are not viable on the basis that the Pb interfering with the oxide is an effect in addition to any changes in the alloy potential. The presence of Sn along with Pb however reduces the dissolution rate when annealed at the maximum temperature of 600°C for an hour at which Sn is found to dissolve in the aluminium solid solution diluting the Sn concentration on the surface (Graver, Pedersen et al. 2009).

3.1.9. *Influence of other element, including Zr, Cr, Sc, Ti, W and Sr*

These elements are typically added independently in small amounts (i.e. <0.2 wt%) for the purpose of grain refinement, to reduce recrystallisation as well as minimising the effect of intermetallic compounds (Vetrano, Henager Jr et al. 1998). Elements such as Zr and Ti are able to form intermetallics at high temperatures in the Al melt, and persist as finely dispersed particles of Al_3Zr and Al_3Ti within the solidified matrix, which, based on their fine size (i.e. <<1 μm), have a minimal impact on corrosion (Scully, Knight et al. 1993). Similarly with scandium (Sc) additions above the solubility limit the formation of Al_3Sc will occur, contributing to the strength and significantly reducing recrystallisation during thermomechanical processing (Cavanaugh, Birbilis et al. 2007). In general, and neglecting Al_3Fe, such dispersoids are based on the Al_3X system where X is Zr, Ti, Sc, Er, etc, and taking the form of fine insoluble dispersoids which are functional in grain inoculation and refinement. As such, there are specific alloying additions of Ti and Zr to high strength alloys such as AA7075 (Senkov, Bhat et al. 2005; Zou, Pan et al. 2007; He, Zhang et al. 2010).

Whilst not studied in detail, it has been posited that the ability to suppress recrystallisation leads to lower corrosion rates by avoiding the formation of high angle grain boundaries (Fang, Chen et al. 2009). Furthermore, there are also complex second and third order interactions between sparingly soluble elements that extend beyond the predictions of simple phase diagrams. An example is that the addition of Sr will impact intermetallics such as Al_5FeSi, making them smaller in both their size and volume fraction (Ashtari, Tezuka et al. 2003; Eidhed 2008). Such an effect has a role in corrosion by minimising the number of intermetallic sites. The purpose however of this section, is not to describe the metallurgy, as that has been done classically as far back as Mondolfo (Mondolfo 1971; Mondolfo and Barlock 1975), but to emphasise the microstructures direct impact on corrosion.

In common alloys prepared by conventional casting technologies, transition metals (TMs) such as W, Mo and Cr are not employed owing to their very low solubility limits. However it is important to note that when prepared in sputtered or thin film forms, such Al-TM alloys display the lowest corrosion rates of all the Al-alloys. Shaw successfully produced the alloys by sputter deposition and found that these elements increased the pitting potential and passivity of the alloys as well as inability to form second phase upon heat treatment (Shaw, Fritz et al. 1990; Shaw, Davis et al. 1991). The work of Frankel also showed promising results (Frankel, Russak et al. 1989; Frankel, Davenport et al. 1992; Tailleart, Gauthier et al. 2009).

3.2. The role of microstructure on corrosion

In order to understand the corrosion performance of Al alloys, and following on from the previous section, an appreciation of the microstructure is vital. Alloying elements and thermomechanical processing play an important role in dictating the type of microstructure produced. For homogeneous alloys, such as pure Al or 5xxx series alloys, corrosion susceptibility is low due to lack of pre-existing microstructural attack sites. The main concern however is regarding heterogeneous alloys, particularly the higher strength Al alloys such as the 2xxx, 7xxx and heat-treatable 6xxx series, where microstructural heterogeneity is a necessity. The most common features of a microstructure are the intermetallic particles which are classified into precipitates (forming from nucleation and growth, nominally 1nm to 300nm in diameter), constituent particles (from insoluble or impurity elements, unable to redissolve, nominally a few microns, to a few tens of microns, in size) and dispersoids (nominally << 1 micron in size) (Polmear 2006). Each of these features consists of different electrochemical characteristics (including their native electrochemical potential and the currents they can sustain at a given potential characteristic of the alloy which they populate) and act as the sites which dictate the severity of corrosion attack. Work on categorising such intermetallics in relation to corrosion is plentiful and now has been going on for several decades (Mazurkiewicz and Piotrowski 1983; Scully, Knight et al. 1993; Buchheit, Grant et al. 1997; Birbilis and Buchheit 2005; Eckermann, Suter et al. 2008; Goswami, Spanos et al. 2010; Boag, Hughes et al. 2011; Hughes, Boag et al. 2011). We make the distinction in this chapter that whilst such intermetallics are responsible for corrosion initiation steps, including pitting, we will not cover corrosion propagation in detail (i.e. stress corrosion, or intergranular corrosion) since they would require a dedicated chapter.

The knowledge of intermetallic chemistry and the electrochemistry allows a prediction of the mode of corrosion and the propensity of the attack (Cavanaugh, Buchheit et al. 2009). Theoretically, a more active particle (i.e. less noble) will become a local anode and consequently corrode while the more noble particles become cathodes (Szklarska-Smialowska 1999). This is not always a true reflection of kinetics however, as reported by Birbilis where the capability of the element to sustain the cathodic reduction process can not be the deduced by relative nobility as well (Birbilis and Buchheit 2005). As corrosion occurs upon Al alloys, particularly pitting and early damage accumulation, two types of corrosion mode are identified. In one mode of attack whereis the intermetallic is classified as a cathode, the surrounding matrix tends to corrode leaving a ring shape around the particle or also called trenching. There is still some uncertainty on whether or not the trench itself is a result of microgalvanic coupling alone, or if the major contributor is local pH elevation, however a good treatise of this topic was given in a multi-part series of papers by the group at Virginia (Ilevbare, Schneider et al. 2004; Schneider, Ilevbare et al. 2004; Schneider, Ilevbare et al. 2007). In some cases damage may propagate to the base of the particle and eventually lead to particle fall out (Buchheit, Grant et al. 1997). The other mode of attack is when the intermetallic acts as anode and matrix as cathode; whereby the intermetallic will corrode leaving a cavity on the surface. Finally, in some microstructurally complex systems with ternary and above alloying additions, another type of attack found to not follow the traditional way of determining anode and cathode is incongruent dissolution commonly found in 2xxx and 7xxx due to the presence of *S* phase (Al_2CuMg) (Buchheit, Grant et al. 1997; Guillaumin and Mankowski 1998). This intermetallic contains Cu and Mg, whereby (Blanc, Lavelle et al. 1997; Buchheit, Martinez et al. 2000)., *S* phase experiences selective dissolution of the highly active Mg hence leaving Cu remnants at the bottom of the pit (Buchheit, Grant et al. 1997; Büchler, Watari et al. 2000) or redistrubuted near the particle site. Although the more detailed study of Boag et al. indicated that Al might be preferentially removed in the initial stages of attack on S-phase (first 2.5 minutes) as both Cu and Mg were observed to be present after Al dissolution at 5 minutes. This may have been due to a combination of the types of aluminium hydroxyl-chloro complexes that form and the partial switching of the areas of S-phase to Cu remnants where cathodic reactions lead to the formation of insoluble $Mg(OH)_2$ (Boag, Hughes et al. 2011). After 15 minutes the Mg is removed as well and no chloride was detected on these particles. The attack then continues with the dissolution of Al matrix (Buchheit, Grant et al. 1997; Guillaumin and Mankowski 1998). Localized attack of the intermetallic also influenced by the chloride content and pH of the electrolyte. Higher chloride content is widely reported to have higher pitting occurrence due to passive layer disruption of chloride ions (Seri 1994; Blanc, Lavelle et al. 1997; Ilevbare, Schneider et al. 2004) but this does not necessarily mean the pit will propagate deeper (Cavanaugh, Buchheit et al. 2010). There exists a dedicated monograph on this topic (Muster, Hughes et al. 2009). It has also been noted that the intermetallic Mg_2Si can undergo similar incongruent dissolution, whereby Si enrichment occurs at the expense of dissolving Mg (Birbilis and Buchheit 2005; Jain, 2006; Eckermann, Suter et al. 2008; Gupta, Sukiman et al. 2012).

The revelation of a large number of microstructure vs. corrosion micrographs will not be done herein, however if the readers are interested, a nice expose of such images exists in

Cavanaugh (Cavanaugh 2009). Instead, a demonstration is given here. Figure 6 shows the micron scale microstructure for 2024-T3 and 5083-H116 before and after corrosion exposure in 0.1M NaCl for 14 days. These relatively low magnification images do not reveal the precipitate structure in AA2024-T3, instead showing the coarse intermetallics that exist in the alloys. What is observed is that a higher intermetallic density gives more possible sites for localised attack. In the relatively widely studied 2024-T3 (Guillaumin and Mankowski 1998; Schmutz and Frankel 1998; Campestrini, van Westing et al. 2000; Leblanc and Frankel 2002; Boag, Hughes et al. 2009; Hughes, MacRae et al. 2010; Ralston, Birbilis et al. 2010; Boag, Hughes et al. 2011; Hughes, Boag et al. 2011; Zhou, Luo et al. 2012). pitting attack is associated with the Cu containing intermetallic such as *S* phase (coarse and precipitated) and the intermetallic density in AA2024-T3 is high, owing to the very large number of alloying elements which can leave the possibility of forming constituent particles based on Fe, or Si, or Cu, or Mn, etc. In contrast to AA2024, the alloy 5083-H116 with a smaller number of alloying elements, and Cu free, has a lower constituent number density. in addition to the Mg remaining in solid solution and reduces the susceptibility to localised attack (Vetrano, Williford et al. 1997; Aballe, Bethencourt et al. 2001; Yasakau, Zheludkevich et al. 2007).

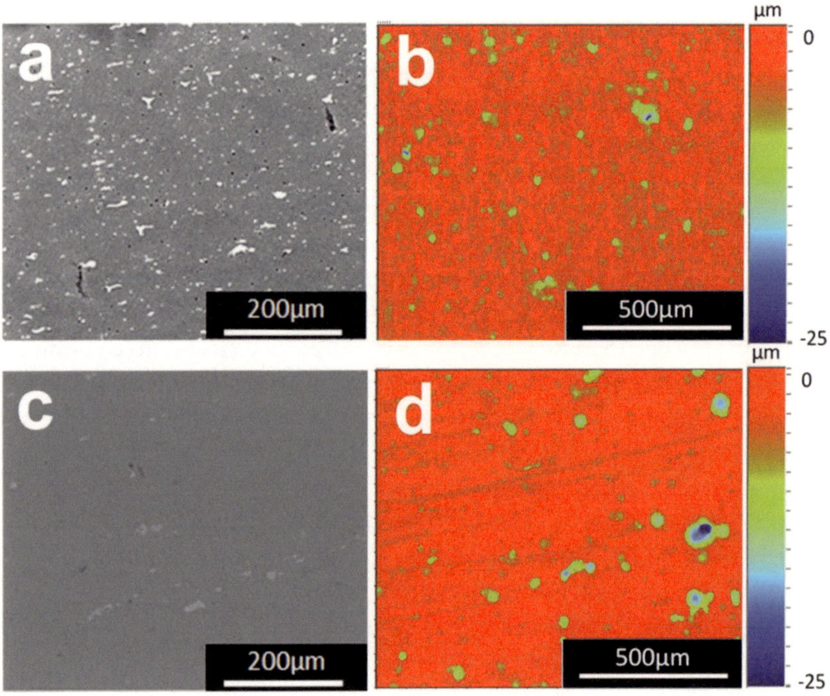

Figure 6. SEM images for AA2024-T3 and AA5083-H116 before exposure in (a) and (c) and after exposure in 0.1M NaCl for 14 days in (b) and (d). These are comparative images to show the extent of damage following immersion.

4. Corrosion protection

This section covers general approaches to protection of aluminium alloys in view of recent advances in the understanding of alloy microstructure. It includes an overview of pretreatment processes such as anodising, conversion coating and organic coatings (barrier and inhibitor combinations). It will examine recent advances in inhibitor design such as building in multifunctionality and touch upon self-healing coating systems. Approaches using multifunctionality can target anodic and cathodic reactions more effectively than using individual monofunctional inhibitors.

Standard metal finishing processes, which have been used for many years, are likely to continue to be used into the future unless they contain chemicals that are targeted for replacement such as chromium. The function of these coatings is primarily to provide better adhesion properties for paint coatings and a secondary role is to provide corrosion protection. The general approach for applying these coatings relies on metal finishing treatments (treatment prior to painting involving immersion in acidic and alkaline baths)) with the objective of reducing the heterogeneous nature of the metal surface such as removing the NSDL and second phase particles (Muster 2009). This is achieved in multistep treatment processes for metal protection (Twite and Bierwagen 1998; Buchheit 2003; Muster 2009) as for instance:

- selective deoxidation (IM particle removal and surface etching);
- deposition or growth of a manufactured oxide via electrochemical (anodising) or chemical (conversion coating) means;
- use of an organic coating for specific applications, normally including a primer and a topcoat.

On aluminium, most anodised coating processes produce an outer oxide with a cellular structure on top of a thin barrier layer that provides some protection against corrosion. Inhibitors can be incorporated into the outer porous layer of the anodized layer during formation or as a seal after formation to offer some extra protection upon damage. Chromic acid anodizing is one of a number of processes that are available for electrochemical growth of surface protective oxides. More environmentally friendly alternatives to chromic acid anodizing such as sulfuric, sulfuric-boric, sulphuric-tataric and phosphoric based processes have been available for a long time. There have been a number of recent advances in reducing the energy consumption of anodizing processes as well as improving coating properties. These advances are based on an improved understanding of the alloy microstructure described above and involve selective removal of second phase particles as part of the anodising process.

An alternative approach to anodizing is to precipitate a coating on a surface through chemical means called conversion coatings. For high strength Al-alloys such as 2xxx and 7xxx series chromate conversion coating (CrCC) is still the preferred process. Replacements for the toxic chromate-based conversion coatings include a range of treatments based on self-assembled monolayers, sol-gel chemistries, Ti/Z oxyfluorides, rare earth, cobalt, vanadates, molybdates and permanganate processes (Twite and Bierwagen 1998; Buchheit 2003; Kendig and Buchheit 2003). These processes are widely developed for chemically pretreated surfaces that have nearly all the IM particles removed (by chemical pretreatment) and are not specifically designed to address electrochemical and compositional variations found for a heterogeneous surface such as when the IM phases are present. Work like that in (Buchheit and Birbilis 2010) depicting the reaction rate variation across the surface, however, opens an avenue to start designing inhibitors where the initial reaction rate distribution across a surface can be significantly reduced to limit the overall activity of the surface. In this context reaction of inhibitive phases with manufactured IM compounds as well as IM particles within the alloy have been studied for a number of systems (Juffs, Hughes et al. 2001; Juffs 2002; Juffs, Hughes et al. 2002; Birbilis, Buchheit et al. 2005; Scholes, Hughes et al. 2009).

Once the anodised or conversion coating is applied, the surface is ready to receive the organic coating. There are many different types of organic coatings, however because of the focus on 2xxx and 7xxx alloy used in the aerospace industry this section will only deal with that application area. The organic coating system usually consists of a primer and a topcoat. The primer is the main protective layer including corrosion inhibitors that can be released when corrosive species or water reach the metal. From the perspective of providing protection for the underlying aluminium alloy, the inhibitor needs to be available during a corrosion event at a concentration higher than the minimum concentration at which the inhibitor stops corrosion (critical concentration). While this sounds obvious, the critical inhibitor concentration needs to be maintained over many years for structures such as airframes, where maintenance may not be possible in parts of the aircraft because of poor access. The chromate systems itself provide continuous protection and repair to the surface for as long as the dose of chromate remains above the critical concentration. This mechanism of inhibitor release and metal protection is recognized as a self-healing mechanism, since the release of the active species recovers the protective layer on top of the metal.

The search for green inhibitors as replacements for chromate has been driven by legislative imperatives for a number of years. Needless to say, replacement inhibitors do not have the same intrinsic inhibitive power at low solubility as chromate. Thus solubility, inhibitive power and transport within the primer system (which consists of a number of inorganic phases as well as the epoxy) ultimately mean that finding a replacement for chromate is difficult. This means that alternatives must be present at higher concentration leading to the use of more soluble compounds and consequently encapsulation as a method of regulating the response to external or internal triggers emerges as a prospective way to achieve this objective.

Many current inhibitors are water soluble salts and thus ionic. Consequently, they exist as either anions or cation in solution and perform the single function of anodic or cathodic inhibition. So the simplest improvement to inhibitor design is to increase the functionality by finding compounds which play both a cationic and anionic inhibitive role. A large range of cations including Zn, Ca, and rare earths (Bohm, McMurray et al. 2001; Du, Damron et al. 2001; Kendig and Buchheit 2003; Taylor and Chambers 2008; Muster, Hughes et al. 2009) have been combined with either organic (Osborne, Blohowiak et al. 2001; Sinko 2001; Voevodin, Balbyshev et al. 2003; Khramov, Voevodin et al. 2004; Blin, Koutsoukos et al. 2007; Taylor and Chambers 2008; Muster, Hughes et al. 2009) or inorganic (oxyanions, carbonates, phosphates, phosphites, nitrates, nitrites, silicate (Bohm, McMurray et al. 2001; Sinko 2001; Blin, Koutsoukos et al. 2007; Taylor and Chambers 2008) compounds.

Anions with dual functionality, such as some of the transition metal oxyanions which are both oxidants and anions, have been investigated extensively. The oxidizing oyxanions or some organophosphates have some degree of bio-inhibition required for some applications. Substitution of different organophosphates into rare earth-based inhibitors provide versatility in designing inhibitors for specific applications (Birbilis, Buchheit et al. 2005; Hinton, Dubrule et al. 2006; Ho, Brack et al. 2006; Blin, Koutsoukos et al. 2007; Markley, Forsyth et al. 2007; Markley, Hughes et al. 2007; Forsyth, Markley et al. 2008; Deacon, Forsyth et al. 2009; Scholes, Hughes et al. 2009). Thus Ce(di-butyl phosphate)$_3$ is a good inhibitor and relatively "green" whereas Ce(di-phenyl phosphate)$_3$ is also a good inhibitor, but the diphenyl phosphate also has strong bio-inhibition characteristics (García 2011). However, good bio-inhibition usually means that there are increased environmental and health risks. Obviously the number of cathodic and anodic inhibitors means that there are an enormous number of possible combinations, particularly if ternary and quaternary combinations are considered. Hence high-throughput techniques are being used to assess new inhibitor.

As pointed out above, the kinetics of inhibitor release are of the utmost importance since the inhibitor should be available at levels above the critical inhibitor concentration. Optimization of the release kinetics by novel delivery systems becomes integral to incorporation of new inhibitors.

There are a number of different mechanisms investigated for release of healing agents or corrosion inhibitors which can be incorporated into organic coatings. Both mechanical damage and water are triggers for inhibitor release. In the former case mechanical damage breaks capsules containing water soluble inhibitors. In the latter case water dissolves inhibitor directly incorporated in the primer. Droplet formation within defects such as scratches means that the inhibitor is only released when required i.e., when the defect is moist (Furman, Scholes et al. 2006). There is some evidence to suggest that initial high release of inhibitors may be facilitated through atmospheric exposure of the intact paint where penetration of water into the film "prepares" the inhibitor, probably via surface hydrolysis reactions, within the paint, for diffusion and release into the defect (Joshua Du, Damron et al. 2001; Furman, Scholes et al. 2006; Scholes, Furman et al. 2006; Souto, González-García et al. 2010). The presence of water in the film allows soluble inhibitor species to be released into the paint system and diffuse to the metal/coating interface to provide in-situ corrosion prevention or repair called pre-emptive healing (Zin, Howard et al. 1998; Osborne, Blohowiak et al. 2001; Mardel, Garcia et al. 2011). Thus it has been demonstrated that water can trigger cerium dibuthylphosphate (Ce(dbp)$_3$) release into an epoxy matrix resulting in improved adhesion and resistance to filiform corrosion attack through interfacial modification (Mardel, Garcia et al. 2011)

In terms of delivery systems, hard capsules, which have been used in polymer healing (Dry 1996; White, Sottos et al. 2001; Mookhoek, Mayo et al. 2010) need to be smaller for paint systems particularly in the aerospace industry where coatings are typically 20 μm or less (Yin, Rong et al. 2007; Fischer 2010; Hughes, Coles et al. 2010; Mookhoek, Mayo et al. 2010). In polymer applications, capsules up to a few hundred microns can be accommodated (Yin, Rong et al. 2007; Wu, Meure et al. 2008; Tedim, Poznyak et al. 2010). The concept of encapsulation has already been successfully applied to protective organic coatings under different concepts: i) liquids filling completely the void created by the damage by adopting a bi-component systems where one component is encapsulated and the other distributed in the matrix (Cho, White et al. 2009), or single based components with water reactive oils like linseed and tung oils (Suryanarayana, Rao et al. 2008; Samadzadeh, Boura et al. 2010)and ii) liquids (i.e. silyl esters) forming a hydrophobic and highly adhesive layer covering the metallic surface by reaction with the underlying metal and the humidity in air (García, Fischer et al. 2011). One adaption for capsules is to increase the volume of self-healing material by manufacturing rods instead of spheres. Rods with the same cross-sections as spheres can deliver larger volumes of material (Bon, Mookhoek et al. 2007; Mookhoek, Fischer et al. 2009). For inhibitors, their role is to prevent a surface reaction (corrosion) and therefore, the volume of material required is much smaller than that required to actually fill the defect. Consequently, there has been considerable effort looking at "nano-containers" (Voevodin, Balbyshev et al. 2003; Raps, Hack et al. 2009; Tedim, Poznyak et al. 2010).

Water is the most obvious trigger since it can permeate most polymers. pH variations are more specific and respond to the pH excursions that occur in corrosion reactions and by an understanding reactions that occur at different sites in the alloy microstructure. The presence of chloride ions (and other anions) within the coating can be used as specific triggers for the release of corrosion inhibitors and uptake of corrodents using anion exchange materials, such as layered double hydroxides (e.g. hydrotalcites) (Tedim, Poznyak et al. 2010) (Bohm, McMurray et al. 2001; Buchheit, Guan et al. 2003; Williams and McMurray 2003; Zheludkevich, Salvado et al. 2005; Mahajanarn and Buchheit 2008). In this context hydrotalcites have been loaded with vanadate, chromate, nitrate and carbonate which exchange for chloride ions and prevent interfacial damage (Bohm, McMurray et al. 2001; Williams and McMurray 2003; Mahajanarn and Buchheit 2008). The incorporation of Mg particles into paint act as sacrificial anodes to protect Al alloys and steels (Battocchi, Simoes et al. 2006).

5. Recent advances in aspects related to corrosion of aluminium alloys

The search for new multifunctional inhibitors has led to the development of high throughput and combinatorial assessment of new combination of inhibitors. These include multielectrode techniques, and high throughput versions of standard corrosion tests. A range of new electrochemical techniques including AC/DC/AC, SVET, LEIS, SECM and SIET will also be described.

Since corrosion of aluminium alloys tends to be dominated by electrochemical processes, most of the techniques employed for the evaluation of corrosion and protection are based on electrochemical approaches. Furthermore, combining electrochemical techniques with other microscopic, analytical and spectroscopic techniques enables the identification of corrosion products in solution (such as inductive coupled plasma (ICP) and UV-Vis). This combination provides an even broader mechanistic understanding of the level of corrosion and/or corrosion protection.

The increasing number of corrosion inhibitor alternatives to chromates has boosted interest in developing high-throughput techniques and combinatorial assessment of new corrosion inhibitors in aqueous solution. At the same time, the traditional techniques (accelerated or not) employed in the evaluation of the performance of organic coatings require long evaluation periods and are relatively expensive to run, and only offer qualitative or semi-quantitative information at best (e.g. salt fog spray tests). For these reasons, new accelerated techniques for the evaluation of coating performance that offer quantitative results are needed.

Figure 7 shows a simplified flowchart for the formulation of anticorrosive (organic) coatings. The chart includes some of the most common techniques employed in corrosion inhibitor and coating performance evaluation. In the figure, the parallel and complex line of the development of the polymeric matrix (i.e. organic coating) is not included, but awareness of its existence is important, since factors such as the corrosion inhibitor/coating matrix compatibility should be taken into account. For the development of anticorrosive organic coatings, several steps are proposed:

a. Formulation of inhibitors:

The number of corrosion inhibitor candidates is virtually unlimited, and is motivated by the urgent need to replace chromate based inhibitors by environmentally friendly and non-toxic ones, as well as the development of new concepts such as self-healing and synergies between anodic and cathodic inhibitors which open up the broad range of possibilities of organic chemistry. One example of the complexity of the introduction of organic compounds as corrosion inhibitors is the effect that the position of certain groups in a cyclic organic compound can have in the corrosion protection efficiency (Harvey, Hardin et al. 2011).

b. Evaluation in aqueous solution:

Once the inhibitors have been formulated, they can be tested by means of traditional aqueous solution tests such as electrochemical impedance spectroscopy (EIS), potentiodynamic polarisation (PP), immersion tests and weight loss/gain. Also local electrochemical techniques (see point 4-evalaution of organic coatings) give very important information of the mechanisms of corrosion protection offered by the different species in solution. Since traditional techniques require long periods of time and a large number of samples, the introduction of high- throughput techniques as a preceding step is important in order to reduce the number of inhibitors that enter further evaluation processes using traditional aqueous solution tests. It is necessary to highlight that high-throughput techniques are not aimed at replacing traditional tests but at complementing them in order to reduce cost and time in the corrosion inhibitor selection process. Some examples are: single metal, multielectrode array (Chambers, Taylor et al. 2005; Chambers and Taylor 2007), microchannels (White, Hughes et al. 2009) and multi metal multielectrode (Muster, Hughes et al. 2009; García, Muster et al. 2010; Kallip, Bastos et al. 2010).

c. Introduction into an organic matrix:

The introduction of pigments into organic coatings adds some extra difficulties to the whole process, leading to a lot of extra research to avoid undesirable reactions between the polymer matrix and the inhibitors. Some of the parameters to take into account are the ratio between the pigment volume concentration (PVC) and the critical pigment volume concentration (CPVC), the possible side reactions between the polymer and the pigment, with consequences in parameters like the barrier properties, gloss, active corrosion protection, and adhesion amongst others. At the same time, parameters such as contaminant reduction and parameters related to the polymeric matrix itself (such as adhesion and the glass transition temperature (Tg)) should be considered. In any case, once the pigments have been introduced into the organic coating, the coating's performance has to be tested for protection efficiency and if results are promising, then start the optimization process.

d. Evaluation of the performance of organic coatings:

As in the case of aqueous solution tests for evaluation of corrosion inhibitors, the introduction of accelerated tests to evaluate coatings performance is necessary to reduce the amount of time and number of samples that move into traditional assessment. Several techniques have been proposed in this direction, such as the technique AC/DC/AC (Hollaender 1997; Bethencourt, Botana et al. 2004; Rodriguez, Gracenea et al. 2004; Garcia and Suay 2006; Garcia and Suay 2006) (García, Rodríguez et al. 2007; Garcia and Suay 2007) (Garcia and Suay 2007; García and Suay 2007; García and Suay 2009), (Poelman, Olivier et al. 2005; Allahar, Bierwagen et al. 2010; Allahar, Wang et al. 2010), and thermal cycling (Bierwagen, He et al. 2000). Also new concepts like the use of flow induced degradation (Wang 2009) are interesting for developing accelerated testing techniques.

e. Optimisation:

Before entering the pre-commercialization phase, the final step of anticorrosive coatings formulation is the optimization, which is the improvement of the system by modifying pigments concentration, type, and delivery systems to improve and extend the service lifetime protection and compatibility with the matrix. This step is iterative as shown in Figure 7. The whole process from conception to commercialization of the system can take several years.

Due to the impossibility to cover the broad amount of existing (new) high-throughput techniques for selection of corrosion inhibitors for aluminium alloys and accelerated tests for evaluation of protective organic coatings on aluminium alloys, we focus on two electrochemical techniques that have attracted a broad interest most recently due to their high potential and relatively well understood evaluation procedure.

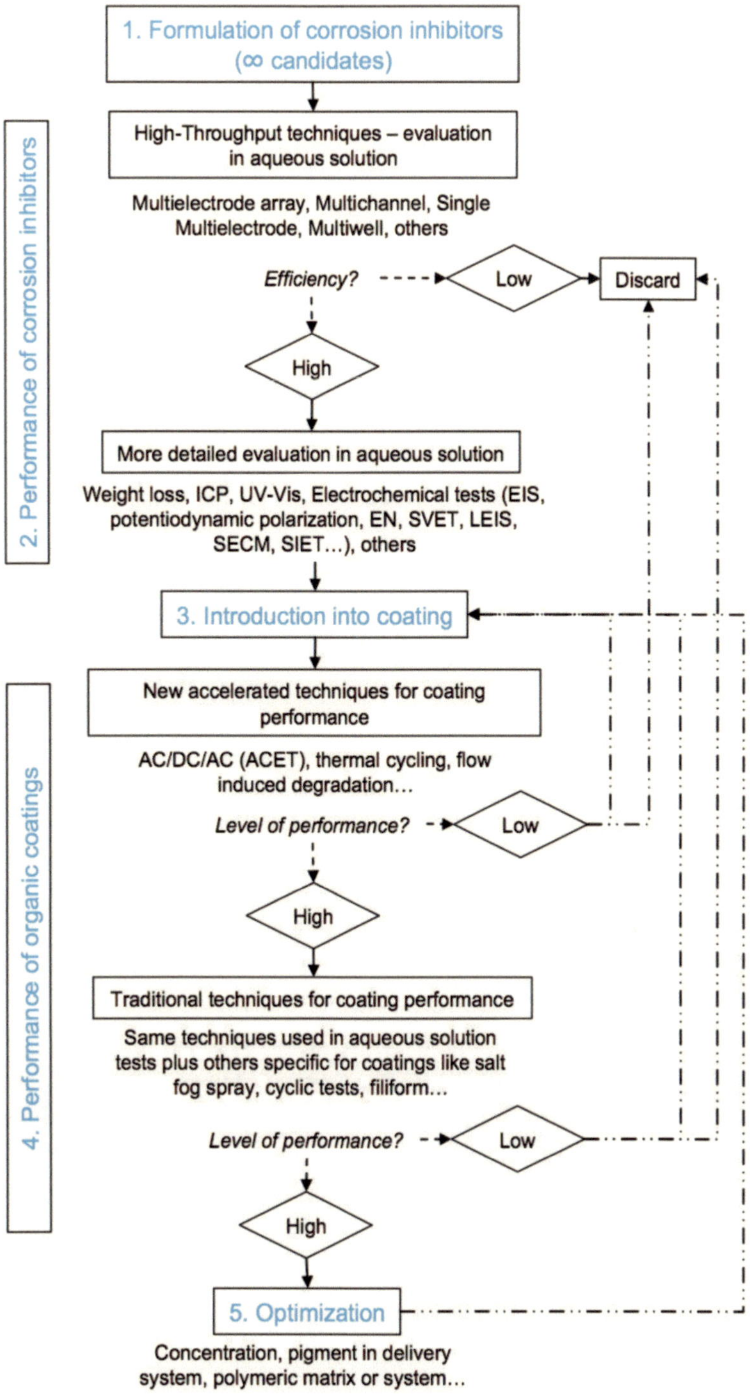

Figure 7. Simplified flowchart for anticorrosive coatings development departing from the inhibitor design or formulation.

5.1. High throughput assessment

In terms of high-throughput techniques, those based on electrochemical approaches are of most interest, since their measurement principles are closely related to familiar research techniques such as Electrochemical Impedance Spectroscopy (EIS) and potentiodynamic polarisation (PP), while at the same time offer quantitative information about the corrosion and corrosion protection mechanisms.

Chambers and Taylor (Chambers, Taylor et al. 2005; Chambers and Taylor 2007) first presented the use of multi-electrode arrays for rapid screening of corrosion inhibitors in different concentrations and pHs, putting identical pairs of AA2024-T3 wires into a large number of separate reaction cells containing different inhibitive solutions and applying a 100mV potential between the two wires in each cell (Figure 8(a)). The current between both electrodes was used to determine the polarisation resistance. With this set-up they were able to evaluate for one metal many different combinations of inhibitors and their synergistic behavior, while at the same time showing its possible application for evaluating the influence of the pH on inhibitors' efficiency.

Based on the work of Taylor and Chambers, Muster et al. (Muster, Hughes et al. 2009) proposed a variation of the method using a combination of different pairs of metals assembled together (Figure 8(b)) to form what was presented as a multielectrode (ME). The basic idea of this ME was to test a combination of nine pairs (or as many as are interesting) of different metals in the same electrode configuration connected by means of a multiplexer to a potentiostat/galvanostat. Measurements consist of applying 100mV between a selected pair of the same metals within the ME and measure the current flow between them, repeating afterwards the same procedure for the other metal pairs. This set up was employed to rapidly evaluate, without removing the ME from the solution, the concentration range at which a particular inhibitor or combination of inhibitors were offering corrosion protection. Hence, the setup allowed the determination of optimal metal-inhibitor combinations, while significantly reducing the evaluation time with respect to conventional PP tests, without the need of a reference electrode, which simplifies the experimental setup.

In a second paper (Garcia, Muster et al. 2010), the authors studied the effect of the pH and inhibitor type on the correlation between the ME and PP for AA2024-T3. The findings were promising due to the high level of correlation between the ME and traditional techniques, although some discrepancies were found for corrosion inhibitors that can speciate or precipitate at certain pHs. Nevertheless, the non- correlation was assumed to be dependent on the type of corrosion inhibitors and not due to conceptual or experimental mismatching between techniques.

A second concern with the ME was the possibility of cross-contamination due to the presence of several metals in one solution. Garcia et al. (Garcia, Muster et al. 2010) also addressed this problem studying the effect of cross-contamination for the AA2024-T3 wires within the ME. This study showed that if there was cross-contamination then it was not significant enough to influence the results. Despite these results, some more studies should be performed with the ME to check cross-contaminations for other metals such as AA7075-T6 which could be more susceptible to copper plating coming from other metals such as AA2024-T3.

The results obtained so far with multielectrodes (multielectrode array and ME) are very promising and relatively extended information can be found in literature. Nevertheless, more studies and data treatment simplification need to be performed to completely validate these techniques and lead them to an industrial application level.

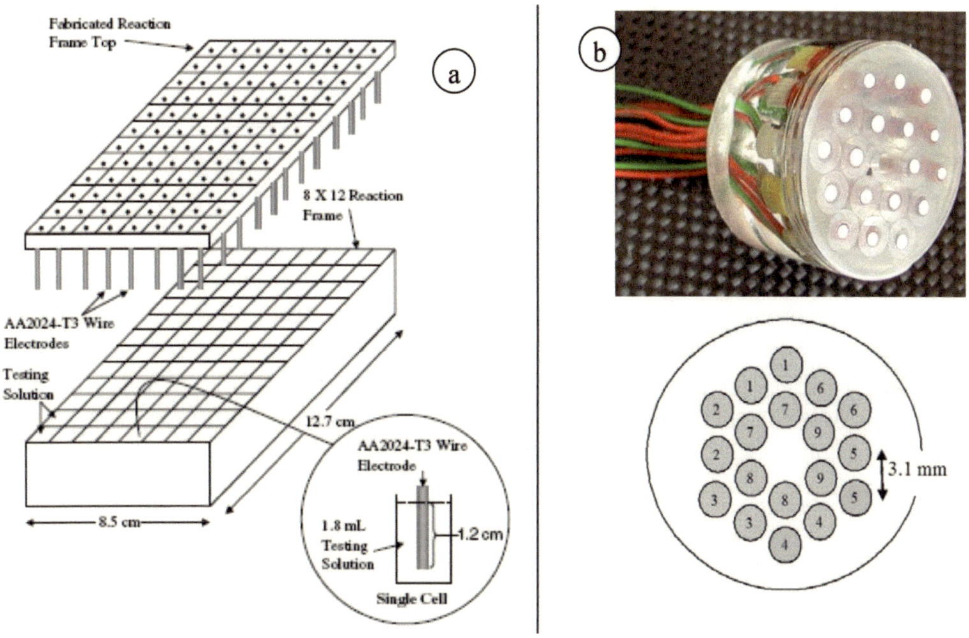

Figure 8. a) single metal multiarray (Chambers and Taylor 2007); b) multimetal multielectrode (Muster, Hughes et al. 2009)

5.2. AC/DC/AC accelerated technique for coating evaluation

This technique has already reached its maturity and is actually employed at an industrial level under the name of accelerated electrochemical technique (ACET) (Medco). Although the industrial application differs from the research technique (AC/DC/AC), the evaluation and testing are conceptually similar. The early version of the AC/DC/AC technique was performed for the packaging industry (Hollaender 1997). The accelerated technique developed by Hollaender (named AC/DC/AC) was based on the use of temporary stresses to accelerate degradation, and consisted of a first EIS test (AC) to evaluate the initial state of the coating, followed by a cathodic polarisation (DC) and a new EIS (AC) to detect the degradation of the lacquer.

The work initiated by Hollaender was further developed by Suay, Garcia and Rodriguez who successfully applied a modified version of the AC/DC/AC technique to evaluate the performance of organic coatings for carbon steel protection and compared the obtained results with EIS, salt-fog spray, and cyclic tests. The technique was then tested for liquid paints (Bethencourt, Botana et al. 2004; Rodríguez, Gracenea et al. 2004), powder coatings (García and Suay 2006; García and Suay 2006; García and Suay 2007; García and Suay 2007) and cataphoretic paints to optimize parameters such as cataphoretic potential and curing time (Poelman, Olivier et al. 2005; García, Rodríguez et al. 2007; García and Suay 2007; García and Suay 2009). The new version of the technique included a crucial step: the relaxation of potentials (open circuit potential relaxation) after the application of each cathodic polarisation. Furthermore, the potentials applied during the cathodic polarisation depended on the type of coating that was studied, although -4V for 20 minutes was preferred. Figure 9(a) shows a schematic of the AC/DC/AC technique procedure, including the relaxation step proposed by Garcia and Suay. The AC/DC/AC cycle is repeated 6 times (6 cycles) leading to a testing time per sample of around 24 hours, which is a significant improvement when compared to traditional EIS and salt-fog tests which require weeks or months. Nevertheless, depending on the quality of the coatings the number of cycles could be increased and the relaxation time reduced or extended.

The cathodic polarisation aims to degrade the coating and coating-metal interface (e.g. pore formation and delamination) due to hydrogen and OH− production. If the coating is good then it has a higher number of cycles to degrade and the effects in the impedance and potentials relaxation are less pronounced, while a lower quality coating will display a faster degradation. An example of these effects compared to traditional EIS and salt-fog spray is presented by Garcia et al. (García and Suay 2009)

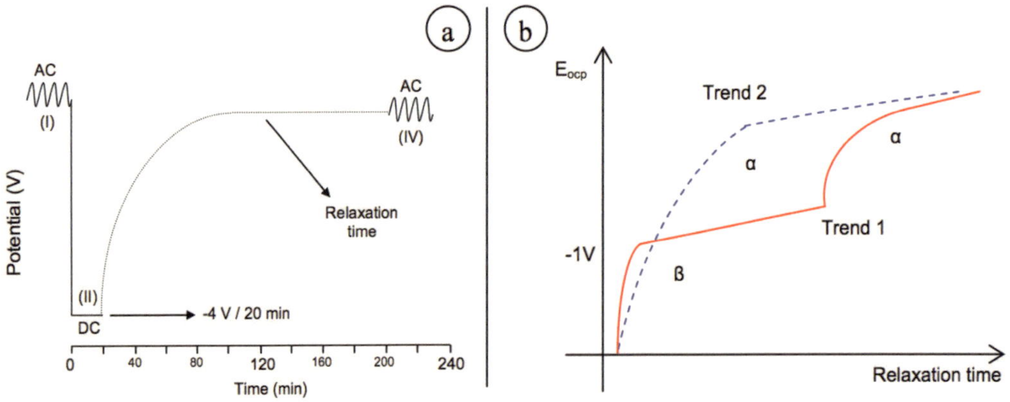

Figure 9. a) AC/DC/AC testing scheme (García and Suay 2009); b) Trends in relaxation of potentials.

Apart from the EIS spectral evolution with cycles, for which a higher drop of impedance is related to higher degradation, the evolution of the open circuit potential after polarisation (relaxation of potentials with time) provides extra crucial information about the degradation of the systems under study (García and Suay 2006). When the cathodic polarisation finishes, the potential of the system relaxes leading to two types of trends depending on the quality of the film (Figure 9(b)):

a. Strongly degraded systems (Trend 1, Figure 9(b)), show two time relaxations of the potential, namely α and β. The first relaxation in time (β) is related to the end of the cathodic reactions that took place at the metallic surface. This relaxation is observed as a quick relaxation around −1V (with small variations depending on the system). The second relaxation (α), which occurs later in time, corresponds to ions and electrolyte leaving the coating. The relaxation (β) could not be detected in all cases due to extremely long relaxation times, although if enough time was allowed this relaxation was then detected, supporting the idea of the two relaxation processes (García and Suay 2009)

b. Less degraded systems (Trend 2, Figure 9(b)), only show relaxation α which is detected at relatively short times of relaxation. This relaxation will take place at longer times as ions and electrolyte penetrate deeper into the film. More recently, Allahar et al. (Allahar, Wang et al. 2009; Allahar, Upadhyay et al. 2010; Wang, Battocchi et al. 2010) have performed an extensive study on the understanding of the AC/DC/AC technique and its principles, leading to a broader understanding of the relaxation of potentials and supporting previous theories, while at the same time further validating the technique.

In this section we have highlighted the complexity of the selection of corrosion inhibitors and anticorrosive coatings formulation leading to the design and validation of high-throughput and accelerated techniques, which at a certain stage can become commercial techniques. The interest in developing new techniques and understanding those already existing is indeed growing in recent years due to the need of developing environmentally friendly and non-toxic systems.

5.3. Staircase impedance

Staircase potentio-electrochemical impedance spectroscopy (SPEIS) is based on EIS, nowadays common for investigating electrochemical and corrosion system. The basis of EIS is by measuring impedance over a range of frequency and the data obtained is expressed graphically in Bode or Nyquist plot format. As for SPEIS, it is designed for impedance measurements over a range of frequency as well as over a range of voltages (similar to the signal sequence used for Mott-Schottky analysis (Barsoukov and Macdonald 2005; Orazem and Tribollet 2008). The potential range is divided into several potential steps and each step contains a DC relaxation period for a given rest time (T_R) to allow the current to stabilize; subsequently followed by an EIS test (Zhou, Birbilis et al. 2010). During the DC potential signal, the current at each potential is also recorded. Figure 10 demonstrates this process.

This method is used to assess the kinetic stability of electrochemical reactions over a range of potentials and pH, which gives a rather detailed insight into the corrosion behavior. In SPEIS however, by setting a range of voltage allows more details observation of how the system response at a particular voltage or 'step', even though it is not sufficient to explain the kinetic processes occurring. As reported by Zhou, SPEIS is able to illustrate the effect of pH and potential on the corrosion kinetics with clarity and able to obtain more information beyond the Pourbaix diagram (Zhou, Birbilis et al. 2010), including structural effects such as alloying, etc.

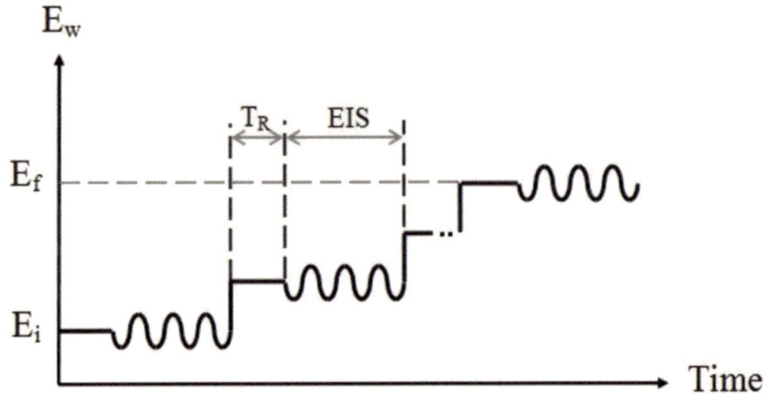

Figure 10. Illustration of polarizing signal during SPEIS

5.4. Potentiostatic transients for determination of metastable and stable pitting

Since pitting is the most common type of corrosion in Al alloys, researchers commonly default to the investigation of pitting potential, E_{pit} as a means for evaluating the corrosion performance. A more noble E_{pit} is often accepted as better resistance to pitting. However, there are some significant limitations in such simple assessments (Gupta, Sukiman et al. 2012). For example, the pitting potential yields no information regarding how many pits form or how large pits may be. Additionally, the environment plays a key role in the severity of pitting damage (i.e. pit depth, pit size) (Cavanaugh 2009).

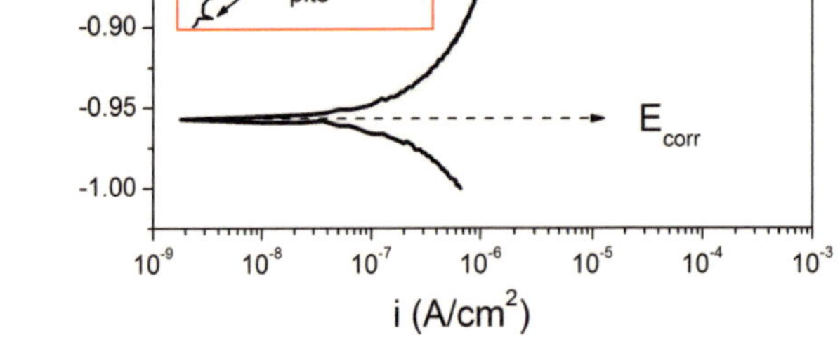

Figure 11. Polarisation curve indicating the corrosion potential E_{corr} and E_{pit} with metastable pits region prior to E_{pit} highlighted in the inset (adapted from (Cavanaugh 2009))

Pits that form at pitting potential are known as stable pits. However, at potentials slightly below pits stabilization (at E_{pit}), pitting like events occur that are known as metastable pit (small surges of current that repassivate). Wu suggested that individual metastable pitting events influence subsequent events (Wu, Scully et al. 1997), and such events are readily observed in Al-alloys. Figure 11 indicates the metastable pitting region in a typical potentiodynamic polarisation curve, where E_{pit} is also depicted.

In order to measure metastable pitting, a potentiostatic value is selected at a potential just below E_{pit} where the metastable pitting events are the most frequent (while still in the passive region) (Cavanaugh 2009). Metastable pitting events are measured by counting the current fluctuations when an alloy is held potentiostatically for a period of time. The transient currents as shown in Figure 12 depend on the nucleation, growth and repassivation of the metastable pits.

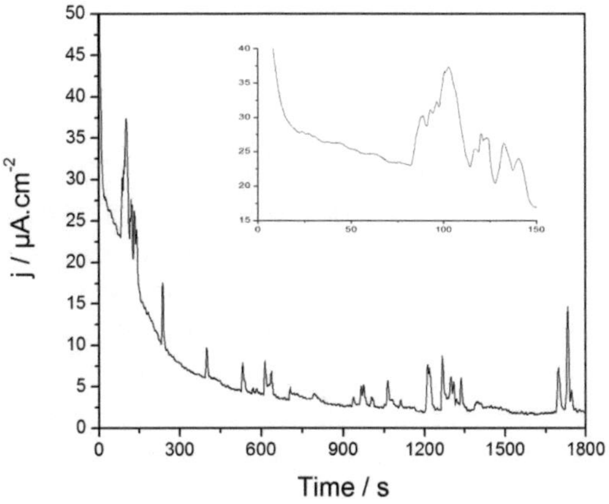

Figure 12. Current transient for AA5083-H116 obtained at 25mV below E_{pit}. Inset shows a zoom of a region of interest that represents the typical transient features (adapted from(Gupta, Sukiman et al. 2012))

The number of metastable pitting is expected to be proportional to the amount of pits formed (Williams, Stewart et al. 1994; Cavanaugh, Birbilis et al. 2012). This is confirmed by (Gupta, Sukiman et al. 2012), where the numbers of metastable pitting from potentiostatic test for various commercial Al alloys correlate with stable pitting tested in a long term immersion test. Ilevbare and Burstein however, stated that multiple metastable events may correspond to only one pit site (Burstein, Liu et al. 2004; Ilevbare, Schneider et al. 2004; Sasaki and Isaacs 2004; Trueman 2005; Speckert and Burstein 2011), however most of such analyses were not performed on Al-alloys. Since there are few studies on metastable pitting of Al alloy, the associated theory and principles will evolve in the coming decade. It is however obvious that metastable pitting analysis can be used to compare the pitting susceptibility between different environments [Cavanaugh 2009] and alloy systems [Gupta, Sukiman et al. 2012] more effectively than an examination of (E_{pit}) alone.

Aside from the methods described above, the use of modern tests does not exclude the use of other traditional tests that require longer testing times and samples such as salt fog spray tests, weight determination, outdoor exposure, or cyclic salt fog-climatic chamber tests, but they aim at providing unique insights that are relevant to specific problems at hand such as pitting germination and inhibitor selection.

It is also worth mentioning here, the great interest that local electrochemical techniques have recently attracted to evaluate the corrosion protection offered when the coatings are damaged (i.e. self-healing). These techniques have been successfully applied to several conceptually different self-healing systems like scanning vibrating electrode technique (SVET) for encapsulated agents (Hughes, Cole et al. 2010; García, Fischer et al. 2011), scanning electrochemical microscope (SECM) for shape memory polymers, selective ion electrode technique (SIET) to evaluate activity of metals (Lamaka, Karavai et al. 2008), and local electrochemical impedance spectroscopy (LEIS) for evaluation of corrosion protection by inhibitor release from coatings (Jorcin, Aragon et al. 2006)

6. Summary and challenges

This chapter has given an abridged and focused treatment of the corrosion behavior of Al alloys on the basis of chemistry, microstructure and environment. Understanding these parameters is crucial in deployment of Al-alloys, but also in the development of more durable Al-alloys. It is obvious that alloying (i.e. chemistry) and microstructure dictates not only mechanical strength – but also corrosion performance. In this chapter we stopped short of discussing corrosion propagation such as intergranular corrosion, exfoliation corrosion and stress corrosion cracking – given space constraints, however such topics (along with hydrogen embrittlement) remain critical for structural alloy deployment.

In regards to corrosion protection, we have attempted to cover some modern developments and present some new techniques to assess the corrosion behavior designed to meet the more complicated and challenging requirements of inhibitor selection / chromate replacement. The information gathered from these techniques is beneficial for future protection and alloy development.

Looking to the future, the current strategy to increase strength by precipitation hardening of a crystalline matrix may no longer be feasible if an alloy with strength higher than 1000MPa is required. Methods to meet this threshold include (the un-upscalable) severe plastic deformation method such as High Pressure Torsion (Liddicoat, Liao et al. 2010) – however the thermal stability of these structures is low. Alternatively the strength of Al based metals could be significantly enhanced (up to 1500MPa) when rapidly quenched to form amorphous alloys (Masumoto 1994; Inoue and Takeuchi 2004; Li, Li et al. 2009; Yang, Yao et al. 2009). Such alloys typically use transition metals such as Zr, Ti, Nb, La, etc (Inoue, Gook et al. 1995; Zawrah and Shaw 2003; Rizzi and Battezzati 2004; Samanta, Manna et al. 2007; Huang, Li et al. 2008; Li, Li et al. 2009), however, again – the stability of such structures is unknown, and the ability form large components is also a challenge. From a corrosion point of view, amorphous Al-alloys appear promising (Manna, Chattopadhyay et al. 2004; Lucente and Scully 2007; Lucente and Scully 2008; Tailleart, Huang et al. 2012) with the corrosion behavior of such alloys is not widely explored therefore leaves a lot of opportunities for new discoveries.

Author details

N. L. Sukiman[1,6], X. Zhou[1], N. Birbilis[1], A.E. Hughes[2], J. M. C. Mol[3], S. J. Garcia[4], X. Zhou[5] and G. E. Thompson[5]

1 Department of Materials Engineering, Monash University, Clayton, Australia

2 CSIRO Materials Science and Technology, Melbourne, Australia

3 TU Delft, Materials Department, Delft, Netherlands

4 TU Delft, Aerospace Engineering, Delft, Netherlands

5 School of Materials, The University of Manchester, Manchester, United Kingdom

6 Department of Mechanical Engineering, University of Malaya, Kuala Lumpur, Malaysia

References

[1] Aballe, A., M. Bethencourt, et al. (2001). "Localized alkaline corrosion of alloy AA5083 in neutral 3.5% NaCl solution." Corrosion Science 43(9): 1657-1674.

[2] Allahar, K. N., V. Upadhyay, et al. (2010). "Characterizing the relaxation of the open-circuit potential during an AC-DC-AC accelerated test." Corrosion 66(9): 0950011-09500111.

[3] Allahar, K. N., D. Wang, et al. (2009). Real time monitoring of an air force topcoat/mg-rich primer system in b117 exposure by -embedded electrodes.

[4] Ambat, R., A. J. Davenport, et al. (2006). "Effect of iron-containing intermetallic particles on the corrosion behaviour of aluminium." Corrosion Science 48(11): 3455-3471.

[5] Ambat, R. and E. S. Dwarakadasa (1992). "The Influence of PH on the Corrosion of Medium Strength Aerospace Alloy-8090, Alloy-2091 And Alloy-2014." Corrosion Science 33(5): 681-690.

[6] Anawati, S. Diplas, et al. (2011). "Effect of copper on anodic activity of aluminum-lead model alloy in chloride solution." Journal of The Electrochemical Society 158(5): C158-C163.

[7] Anawati, S. Diplas, et al. (2011). "Surface Characterization of Heat Treated AlPbCu Model Alloys." Journal of The Electrochemical Society 158(6): C178-C184.

[8] Anawati, B. Graver, et al. (2010). "Multilayer corrosion of aluminum activated by lead." Journal of The Electrochemical Society 157(10): C313-C320.

[9] Andreatta, F., H. Terryn, et al. (2004). "Corrosion behaviour of different tempers of AA7075 aluminium alloy." Electrochimica Acta 49(17–18): 2851-2862.

[10] Ashtari, P., H. Tezuka, et al. (2003). "Influence of Sr and Mn additions on intermetallic compound morphologies in Al-Si-Cu-Fe cast alloys." Materials Transactions 44(12): 2611-2616.

[11] Baer, D. R., C. F. Windisch, et al. (2000). "Influence of Mg on the corrosion of Al." Journal of Vacuum Science & Technology a-Vacuum Surfaces and Films 18(1): 131-136.

[12] Barsoukov, E. and J. R. Macdonald (2005). Solid State Devices, in Impedance spectroscopy: theory, experiment, and applications, Wiley-Interscience, Inc.

[13] Battocchi, D., A. M. Simoes, et al. (2006). "Comparison of testing solutions on the protection of Al-alloys using a Mg-rich primer." Corrosion Science 48(8): 2226-2240.

[14] Bethencourt, M., F. J. Botana, et al. (2004). "Lifetime prediction of waterborne acrylic paints with the AC-DC-AC method." Progress in Organic Coatings 49(3): 275-281.

[15] Birbilis, N. and R. G. Buchheit (2005). "Electrochemical Characteristics of Intermetallic Phases in Aluminum Alloys." Journal of The Electrochemical Society 152(4): B140.

[16] Birbilis, N., R. G. Buchheit, et al. (2005). "Inhibition of AA2024-T3 on a phase-by-phase basis using an environmentally benign inhibitor, cerium dibutyl phosphate." Electrochemical and Solid State Letters 8(11): C180-C183.

[17] Birbilis, N., M. K. Cavanaugh, et al. (2006). "Electrochemical behavior and localized corrosion associated with Al7Cu2Fe particles in aluminum alloy 7075-T651." Corrosion Science 48(12): 4202-4215.

[18] Birbilis, N., T. Muster, et al. (2011). Corrosion of Aluminum Alloys. Corrosion Mechanisms in Theory and Practice, Third Edition, CRC Press: 705-736.

[19] Blanc, C., B. Lavelle, et al. (1997). "The role of precipitates enriched with copper on the susceptibility to pitting corrosion of the 2024 aluminium alloy." Corrosion Science 39(3): 495-510.

[20] Blin, F., P. Koutsoukos, et al. (2007). "The corrosion inhibition mechanism of new rare earth cinnamate compounds - Electrochemical studies." Electrochimica Acta 52(21): 6212-6220.

[21] Boag, A., A. E. Hughes, et al. (2011). "Corrosion of AA2024-T3 Part I: Localised corrosion of isolated IM particles." Corrosion Science 53(1): 17-26.

[22] Boag, A., A. E. Hughes, et al. (2009). "How complex is the microstructure of AA2024-T3?" Corrosion Science 51(8): 1565-1568.

[23] Bohm, S., H. N. McMurray, et al. (2001). "Novel environment friendly corrosion inhibitor pigments based on naturally occurring clay minerals." Materials and Corrosion-Werkstoffe Und Korrosion 52(12): 896-903.

[24] Bon, S. A. F., S. D. Mookhoek, et al. (2007). "Route to stable non-spherical emulsion droplets." European Polymer Journal 43(11): 4839-4842.

[25] Brunner, J. G., J. May, et al. (2010). "Localized corrosion of ultrafine-grained Al–Mg model alloys." Electrochimica Acta 55(6): 1966-1970.

[26] Buchheit, R. G. (1995). "A Compilation of Corrosion Potentials Reported for Intermetallic Phases in Aluminum-Alloys." Journal of The Electrochemical Society 142(11): 3994-3996.

[27] Buchheit, R. G. (2000). "Electrochemistry of θ (Al2Cu), S (Al2CuMg) and T1 (Al2CuLi) and localized corrosion and environment assisted cracking in high strength Al alloys." Materials Science Forum 331: II/.

[28] Buchheit, R. G. and N. Birbilis (2010). "Electrochemical microscopy: An approach for understanding localized corrosion in microstructurally complex metallic alloys." Electrochimica Acta 55(27): 7853-7859.

[29] Buchheit, R. G., R. P. Grant, et al. (1997). "Local Dissolution Phenomena Associated with S Phase (Al_2CuMg) Particles in Aluminum Alloy 2024-T3." Journal of The Electrochemical Society 144(8): 2621-2628.

[30] Buchheit, R. G., H. Guan, et al. (2003). "Active corrosion protection and corrosion sensing in chromate-free organic coatings." Progress in Organic Coatings 47(3-4): 174-182.

[31] Buchheit, R. G., Hughes, A.E. (2003). Chromate and Chromate-Free Coatings. Corrosion: Fundamentals, Testing and Protection. C. Moosbrugger. Mterials Park, Oh, USA, ASM International. 13A: 720 -735.

[32] Buchheit, R. G., M. A. Martinez, et al. (2000). "Evidence for Cu Ion Formation by Dissolution and Dealloying the Al_2CuMg Intermetallic Compound in Rotating Ring-Disk Collection Experiments." Journal of The Electrochemical Society 147(1): 119-124.

[33] Buchheit, R. G., L. P. Montes, et al. (1999). "The Electrochemical Characteristics of Bulk-Synthesized Al[sub 2]CuMg." Journal of The Electrochemical Society 146(12): 4424-4428.

[34] Buchheit, R. G., F. D. Wall, et al. (1995). "Anodic Dissolution-Based Mechanism for the Rapid Cracking, Preexposure Phenomenon Demonstrated by Aluminum-Lithium-Copper Alloys." Corrosion 51(6): 417-428.

[35] Büchler, M., T. Watari, et al. (2000). "Investigation of the initiation of localized corrosion on aluminum alloys by using fluorescence microscopy." Corrosion Science 42(9): 1661-1668.

[36] Buis, A. and J. Schijve (1992). "Stress-Corrosion Cracking Behavior of Al Li-2090-T83 In Artificial Seawater." Corrosion 48(11): 898-909.

[37] Burstein, G. T., C. Liu, et al. (2004). "Origins of pitting corrosion." Corrosion Engineering Science and Technology 39(1): 25-30.

[38] Campestrini, P., E. P. M. van Westing, et al. (2000). "Relation between microstructural aspects of AA2024 and its corrosion behaviour investigated using AFM scanning potential technique." Corrosion Science 42(11): 1853-1861.

[39] Carroll, M. C., P. I. Gouma, et al. (2001). "Effects of minor Cu additions on a Zn-modified Al-5083 alloy." Materials Science and Engineering: A 319–321(0): 425-428.

[40] Carroll, M. C., P. I. Gouma, et al. (2000). "Effects of Zn additions on the grain boundary precipitation and corrosion of Al-5083." Scripta Materialia 42(4): 335-340.

[41] Cavanaugh, M., N. Birbilis, et al. (2007). "Investigating localized corrosion susceptibility arising from Sc containing intermetallic Al3Sc in high strength Al-alloys." Scripta Materialia 56(11): 995-998.

[42] Cavanaugh, M. K. (2009). Modelling the environmental dependence of localized corrosion evolution in AA7075-T651, Ohio State University.

[43] Cavanaugh, M., N. Birbilis, et al. (2007). "Investigating localized corrosion susceptibility arising from Sc containing intermetallic Al3Sc in high strength Al-alloys." Scripta Materialia 56(11): 995-998.

[44] Cavanaugh, M. K., N. Birbilis, et al. (2009). "A Quantitative Study on the Effects of Environment and Microstructure on Pit Initiation in Al-alloys." ECS Transactions 16(52): 1-11.

[45] Cavanaugh, M. K., N. Birbilis, et al. (2012). "Modeling pit initiation rate as a function of environment for Aluminum alloy 7075-T651." Electrochimica Acta 59: 336-345.

[46] Cavanaugh, M. K., R. G. Buchheit, et al. (2009). "Evaluation of a simple microstructural-electrochemical model for corrosion damage accumulation in microstructurally complex aluminum alloys." Engineering Fracture Mechanics 76(5): 641-650.

[47] Cavanaugh, M. K., R. G. Buchheit, et al. (2010). "Modeling the environmental dependence of pit growth using neural network approaches." Corrosion Science 52(9): 3070-3077.

[48] Chambers, B. D. and S. R. Taylor (2007). "High-Throughput Assessment of Inhibitor Synergies on Aluminum Alloy 2024-T3 Through Measurement of Surface Copper Enrichment." Corrosion 63(3): 268-276.

[49] Chambers, B. D., S. R. Taylor, et al. (2005). "Rapid Discovery of Corrosion Inhibitors and Synergistic Combinations Using High-Throughput Screening Methods." Corrosion 61(5): 480-489.

[50] Chen, G. S., M. Gao, et al. (1996). "Microconstituent-Induced Pitting Corrosion in Aluminum Alloy 2024-T3." Corrosion (Houston) 52(1): 8-15.

[51] Cho, S. H., S. R. White, et al. (2009). "Self-Healing Polymer Coatings." Advanced Materials 21(6): 645-+.

[52] Davenport, A. J., Y. Yuan, et al. (2006). "Intergranular Corrosion and Stress Corrosion Cracking of Sensitised AA5182." Materials Science Forum 519-521: 641-646.

[53] Davis, J. R. (1999). Corrosion of aluminum and aluminum alloys, Materials Park, OH : ASM International.

[54] Deacon, G. B., M. Forsyth, et al. (2009). "Synthesis and Characterisation of Rare Earth Complexes Supported by para-Substituted Cinnamate Ligands." Zeitschrift Fur Anorganische Und Allgemeine Chemie 635(6-7): 833-839.

[55] Dorward, R. C. and T. R. Pritchett (1988). "Advanced aluminium alloys for aircraft and aerospace applications." Materials & Design 9(2): 63-69.

[56] Dry, C. (1996). "Procedures developed for self-repair of polymer matrix composite materials." Composite Structures 35(3): 263-269.

[57] Du, Y. J., M. Damron, et al. (2001). "Inorganic/organic hybrid coatings for aircraft aluminum alloy substrates." Progress in Organic Coatings 41(4): 226-232.

[58] Eckermann, F., T. Suter, et al. (2008). "The influence of MgSi particle reactivity and dissolution processes on corrosion in Al–Mg–Si alloys." Electrochimica Acta 54(2): 844-855.

[59] Eidhed, W. (2008). "Effects of solution treatment time and Sr-modification on microstructure and mechanical property of Al-Si piston alloy." Journal of Materials Science and Technology 24(1): 29-32.

[60] Fang, H. C., K. H. Chen, et al. (2009). "Effect of Cr, Yb and Zr additions on localized corrosion of Al–Zn–Mg–Cu alloy." Corrosion Science 51(12): 2872-2877.

[61] Ferrer, C. P., M. G. Koul, et al. (2003). "Improvements in strength and stress corrosion cracking properties in aluminum alloy 7075 via low-temperature retrogression and re-aging heat treatments." Corrosion 59(6): 520-528.

[62] Fischer, H. R. (2010). natural Science 2: 873-901.

[63] Fleck, P., D. Calleros, et al. (2000). "Retrogression and reaging of 7075 T6 aluminum alloy." Materials Science Forum 331: I/.

[64] Forsyth, M., T. Markley, et al. (2008). "Inhibition of corrosion on AA2024-T3 by new environmentally friendly rare earth organophosphate compounds." Corrosion 64(3): 191-197.

[65] Frankel, G. S. (1998). "Pitting Corrosion of Metals." Journal of The Electrochemical Society 145(6): 2186-2198.

[66] Frankel, G. S., A. J. Davenport, et al. (1992). "X-ray absorption study of electrochemically grown oxide films on Al-Cr sputtered alloys." Journal of The Electrochemical Society 139(7): 1812-1820.

[67] Frankel, G. S., M. A. Russak, et al. (1989). "Pitting of Sputtered Aluminum Alloy Thin Films." Journal of The Electrochemical Society 136(4): 1243-1244.

[68] Fuller, C. B., A. R. Krause, et al. (2002). "Microstructure and mechanical properties of a 5754 aluminum alloy modified by Sc and Zr additions." Materials Science and Engineering A 338(1-2): 8-16.

[69] Furman, S. A., F. H. Scholes, et al. (2006). "Corrosion in artificial defects. II. Chromate reactions." Corrosion Science 48(7): 1827-1847.

[70] Galvele, J. R. (1976). "Transport Processes and the Mechanism of Pitting of Metals." Journal of The Electrochemical Society 123(4): 464-474.

[71] García, S. J., H. R. Fischer, et al. (2011). "Self-healing anticorrosive organic coating based on an encapsulated water reactive silyl ester: Synthesis and proof of concept." Progress in Organic Coatings 70(2-3): 142-149.

[72] García, S. J., Mol, J.M.C., Muster, T.H., Hughes, A.E., Mardel, J., Miller, T., Markely, T., Terryn, H., de Wit, J.H.W. (2011). Advances in the Selection and use of Rare-Earth-Based Inhibitors for Self Healing Organic Coatings, Accepted for publication in Self-Healing Properties of New Surface Treatments. Green Inhibitors. L. Fedrizzi, EFC-Maney Publishing. 58.

[73] García, S. J., T. H. Muster, et al. (2010). "The influence of pH on corrosion inhibitor selection for 2024-T3 aluminium alloy assessed by high-throughput multielectrode and potentiodynamic testing." Electrochimica Acta 55(7): 2457-2465.

[74] García, S. J., M. T. Rodríguez, et al. (2007). "Evaluation of cure temperature effects in cataphoretic automotive primers by electrochemical techniques." Progress in Organic Coatings 60(4): 303-311.

[75] García, S. J. and J. Suay (2006). "Anticorrosive properties of an epoxy-Meldrum acid cured system catalyzed by erbium III trifluromethanesulfonate." Progress in Organic Coatings 57(4): 319-331.

[76] García, S. J. and J. Suay (2006). "Application of electrochemical techniques to study the effect on the anticorrosive properties of the addition of ytterbium and erbium triflates as catalysts on a powder epoxy network." Progress in Organic Coatings 57(3): 273-281.

[77] García, S. J. and J. Suay (2007). "A comparative study between the results of different electrochemical techniques (EIS and AC/DC/AC). Application to the optimisation of

the cataphoretic and curing parameters of a primer for the automotive industry." Progress in Organic Coatings 59(3): 251-258.

[78] García, S. J. and J. Suay (2007). "Influence on the anticorrosive properties of the use of erbium (III) trifluoromethanesulfonate as initiator in an epoxy powder clearcoat." Corrosion Science 49(8): 3256-3275.

[79] García, S. J. and J. Suay (2009). "Optimization of deposition voltage of cataphoretic automotive primers assessed by EIS and AC/DC/AC." Progress in Organic Coatings 66(3): 306-313.

[80] Garrard, W. N. (1994). "Corrosion Behavior of Aluminum-Lithium Alloys." Corrosion 50(3): 215-225.

[81] Gimenez, P., J. J. Rameau, et al. (1981). "Experimental pH potential diagram of aluminium for seawater." Corrosion 37: 673-682.

[82] Giummarra, C., B. Thomas, et al. (2007). "New Al-Li Alloys for Aerospace Applications." Proceedings of the Light Metals Technology Conference.

[83] Goswami, R., G. Spanos, et al. (2010). "Precipitation behavior of the β phase in Al-5083." Materials Science and Engineering: A 527(4-5): 1089-1095.

[84] Graver, B., A. M. Pedersen, et al. (2009). "Anodic Activation of Aluminum by Trace Element Tin." ECS Transactions 16(52): 55-69.

[85] Graver, B., A. T. J. van Helvoort, et al. (2010). "Effect of heat treatment on anodic activation of aluminium by trace element indium." Corrosion Science 52(11): 3774-3781.

[86] Guillaumin, V. and G. Mankowski (1998). "Localized corrosion of 2024 T351 aluminium alloy in chloride media." Corrosion Science 41(3): 421-438.

[87] Guillaumin, V. and G. Mankowski (2000). "Localized corrosion of 6056 T6 aluminium alloy in chloride media." Corrosion Science 42(1): 105-125.

[88] Gundersen, J. T. B., A. Aytaç, et al. (2004). "Effect of heat treatment on electrochemical behaviour of binary aluminium model alloys." Corrosion Science 46(3): 697-714.

[89] Gupta, R. K., N. L. Sukiman, et al. (2012). "Metastable pitting characteristics of aluminium alloys measured using current transients during potentiostatic polarisation." Electrochimica Acta 66: 245-254.

[90] Gupta, R. K., N. L. Sukiman, et al. (2012). "Electrochemical Behavior and Localized Corrosion Associated with Mg2Si Particles in Al and Mg Alloys." ECS Electrochemistry Letters 1(1): B1-B3.

[91] Harvey, T. G., S. G. Hardin, et al. (2011). "The effect of inhibitor structure on the corrosion of AA2024 and AA7075." Corrosion Science 53(6): 2184-2190.

[92] Harvey, T. G., A. E. Hughes, et al. (2008). "Non-chromate deoxidation of AA2024-T3: Sodium bromate–nitric acid (20–60)." Applied Surface Science 254(11): 3562-3575.

[93] Hatch, J. E. (1984). Aluminum: properties and physical metallurgy, Metals Park, Ohio : American Society for Metals.

[94] He, Y., X. Zhang, et al. (2010). "Effect of minor Cr, Mn, Zr, Ti and B on grain refinement of as-cast Al-Zn-Mg-Cu alloys." Xiyou Jinshu Cailiao Yu Gongcheng/Rare Metal Materials and Engineering 39(7): 1135-1140.

[95] Hinton, B. R. W., N. Dubrule, et al. (2006). Raman, EDS and SEM studies of the interaction of corrosion inhibitor $Ce(dbp)_3$ with AA2024-T3. 4th International Symposium on Aluminium Surface Science and Technology. Beaune, France.

[96] Hirth, S. M., G. J. Marshall, et al. (2001). "Effects of Si on the aging behaviour and formability of aluminium alloys based on AA6016." Materials Science and Engineering A 319-321: 452-456.

[97] Ho, D., N. Brack, et al. (2006). "Cerium dibutylphosphate as a corrosion inhibitor for AA2024-T3 aluminum alloys." Journal of the Electrochemical Society 153(9): B392-B401.

[98] Hollaender, J. (1997). "Rapid assessment of food/package interactions by electrochemical impedance spectroscopy (EIS)." Food Additives and Contaminants 14(6-7): 617-626.

[99] Huang, Z. H., J. F. Li, et al. (2008). "Primary crystallization of Al–Ni–RE amorphous alloys with different type and content of RE." Materials Science and Engineering: A 489(1–2): 380-388.

[100] Hughes, A. E., A. Boag, et al. (2011). "Corrosion of AA2024-T3 Part II: Co-operative corrosion." Corrosion Science 53(1): 27-39.

[101] Hughes, A. E., A. M. Glenn, et al. (2012). A Consistent Description of Intermetallic Particle Composition: An analysis of 10 Batches of AA2024-T3. Aluminium Surface Science & Technology, Sorrento, Italy.

[102] Hughes, A. E., C. MacRae, et al. (2010). "Sheet AA2024-T3: A new investigation of microstructure and composition." Surface and Interface Analysis 42(4): 334-338.

[103] Hughes, A. E., I. S. Cole, et al. (2010). "Designing green, self-healing coatings for metal protection." NPG Asia Materials 2(4): 143-151.

[104] Hughes, A. E., I. S. Coles, et al. (2010). "Combining Green and Self Healing for a new Generation of Coatings for Metal Protection." Nature Asia Materials 2(4): 143-151.

[105] Ilevbare, G. O., O. Schneider, et al. (2004). "In Situ Confocal Laser Scanning Microscopy of AA 2024-T3 Corrosion Metrology." Journal of The Electrochemical Society 151(8): B453.

[106] Inoue, A., J. S. Gook, et al. (1995). "New amorphous alloys in Al-Mg-Ln (Ln = La, Ce or Nd) systems prepared by rapid solidification (rapid publication)." Materials Transactions, JIM 36(7): 794-796.

[107] Inoue, A. and A. Takeuchi (2004). "Recent progress in bulk glassy, nanoquasicrystalline and nanocrystalline alloys." Materials Science and Engineering: A 375–377(0): 16-30.

[108] Jain, S. (2006). Corrosion and protection of heterogeneous cast Al-Si (356) and Al-Si-Fe-Cu (380) alloys by chromate and cerium inhibitors, Ohio State University.

[109] Jain, S., M. L. C. Lim, et al. (2012). "Spreading of intergranular corrosion on the surface of sensitized Al-4.4Mg alloys: A general finding." Corrosion Science 59: 136-147.

[110] Jia, Z., B. Graver, et al. (2008). "Effect of magnesium on segregation of trace element lead and anodic activation in aluminum Alloys." Journal of The Electrochemical Society 155(1): C1-C7.

[111] Jones, R. H., D. R. Baer, et al. (2001). "Role of Mg in the stress corrosion cracking of an Al-Mg alloy." Metallurgical and Materials Transactions A: Physical Metallurgy and Materials Science 32(7): 1699-1711.

[112] Jorcin, J. B., E. Aragon, et al. (2006). "Delaminated areas beneath organic coating: A local electrochemical impedance approach." Corrosion Science 48(7): 1779-1790.

[113] Joshua Du, Y., M. Damron, et al. (2001). "Inorganic/organic hybrid coatings for aircraft aluminum alloy substrates." Progress in Organic Coatings 41(4): 226-232.

[114] Juffs, L. (2002). Investigation of Corrosion Coating Deposition on Microscopic and Macroscopic Intermetallic Phases of Aluminium Alloys. Master of Science, RMIT.

[115] Juffs, L., A. E. Hughes, et al. (2002). "The use of macroscopic modelling of intermetallic phases in aluminium alloys in the study of ferricyanide accelerated chromate conversion coatings." Corrosion Science 44(8): 1755-1781.

[116] Juffs, L., A. E. Hughes, et al. (2001). "The use of macroscopic modelling of intermetallic phases in aluminium alloys in the study of ferricyanide accelerated chromate conversion coatings." Micron 32(8): 777-787.

[117] Kallip, S., A. C. Bastos, et al. (2010). "A multi-electrode cell for high-throughput SVET screening of corrosion inhibitors." Corrosion Science 52(9): 3146-3149.

[118] Kannan, M. B. and V. S. Raja (2010). Enhancing the localized corrosion resistance of high strength 7010 Al-alloy. 138: 1-6.

[119] Kendig, M. W. and R. G. Buchheit (2003). "Corrosion Inhibition of Aluminum and Aluminum Alloys by Soluble Chromates, Chromate Coatings, and Chromate-Free Coatings." Corrosion 59(5): 379-400.

[120] Keuong, Y. W., J. H. Nordlien, et al. (2003). "Electrochemical activation of aluminum by trace element lead." Journal of The Electrochemical Society 150(11): B547-B551.

[121] Khramov, A. N., N. N. Voevodin, et al. (2004). "Hybrid organo-ceramic corrosion protection coatings with encapsulated organic corrosion inhibitors." Thin Solid Films 447–448(0): 549-557.

[122] Kim, K. T., J. M. Kim, et al. (2005). Effect of alloying elements on the strength and casting characteristics of high strength Al-Zn-Mg-Cu alloys. 475-479: 2539-2542.

[123] Kim, W. J., C. S. Chung, et al. (2003). "Optimization of strength and ductility of 2024 Al by equal channel angular pressing (ECAP) and post-ECAP aging." Scripta Materialia 49(4): 333-338.

[124] Koroleva, E. v., G. e. Thompson, et al. (1999). "Surface morphological changes of aluminium alloys in alkaline solution:: effect of second phase material." Corrosion Science 41(8): 1475-1495.

[125] Lamaka, S. V., O. V. Karavai, et al. (2008). "Monitoring local spatial distribution of Mg2+, pH and ionic currents." Electrochemistry Communications 10(2): 259-262.

[126] Larsen, M. H., J. C. Walmsley, et al. (2008). "Intergranular corrosion of copper-containing AA6xxx AlMgSi aluminum alloys." Journal of The Electrochemical Society 155(11): C550-C556.

[127] Lavernia, E. J. and N. J. Grant (1987). "Aluminum Lithium Alloys." Journal of Materials Science 22(5): 1521-1529.

[128] Leblanc, P. and G. S. Frankel (2002). "A study of corrosion and pitting initiation of AA2024-T3 using atomic force microscopy." Journal of The Electrochemical Society 149(6): B239-B247.

[129] Lee, Y. B., D. H. Shin, et al. (2004). "Effect of annealing temperature on microstructures and mechanical properties of a 5083 Al alloy deformed at cryogenic temperature." Scripta Materialia 51(4): 355-359.

[130] Li, C., D. Y. Li, et al. (2009). "Microstructure and mechanical properties of multicomponent aluminum alloy by rapid solidification." Journal of Materials Engineering and Performance 18(1): 79-82.

[131] Li, J. F., Z. Q. Zheng, et al. (2007). "Exfoliation corrosion and electrochemical impedance spectroscopy of an Al-Li alloy in EXCO solution." Materials and Corrosion 58(4): 273-279.

[132] Liddicoat, P. V., X.-z. Liao, et al. (2010). "Nanostructural hierarchy increases the strength of aluminium alloys." Nature Communications 1(6): 63-63.

[133] Lin, J. C., H. L. Liao, et al. (2006). "Effect of heat treatments on the tensile strength and SCC-resistance of AA7050 in an alkaline saline solution." Corrosion Science 48(10): 3139-3156.

[134] Liu, Y. and Y. F. Cheng (2010). "Role of second phase particles in pitting corrosion of 3003 Al alloy in NaCl solution." Materials and Corrosion 61(3): 211-217.

[135] Liu, Y. and Y. F. Cheng (2011). "Characterization of passivity and pitting corrosion of 3003 aluminum alloy in ethylene glycol-water solutions." Journal of Applied Electrochemistry 41(2): 151-159.

[136] Lucente, A. M. and J. R. Scully (2007). "Pitting of Al-based amorphous-nanocrystalline alloys with solute-lean nanocrystals." Electrochemical and Solid-State Letters 10(5): 39-43.

[137] Lucente, A. M. and J. R. Scully (2008). "Localized corrosion of Al-based amorphous-nanocrystalline alloys with solute-lean nanocrystals: Pit stabilization." Journal of The Electrochemical Society 155(5): C234-C243.

[138] Lynch, S. P., S. P. Knight, et al. (2009). Stress-corrosion cracking of Al-Zn-Mg-Cu alloys effects of composition and heat-treatment.

[139] Mahajanarn, S. P. V. and R. G. Buchheit (2008). "Characterization of inhibitor release from Zn-Al- V10O28 (6-) hydrotalcite pigments and corrosion protection from hydrotalcite-pigmented epoxy coatings." Corrosion 64(3): 230-240.

[140] Manna, I., P. P. Chattopadhyay, et al. (2004). "Development of amorphous and nanocrystalline Al65Cu35–xZrx alloys by mechanical alloying." Materials Science and Engineering: A 379(1–2): 360-365.

[141] Mardel, J., S. J. Garcia, et al. (2011). "The characterisation and performance of Ce(dbp)3-inhibited epoxy coatings." Progress in Organic Coatings 70(2–3): 91-101.

[142] Markley, T. A., M. Forsyth, et al. (2007). "Corrosion protection of AA2024-T3 using rare earth diphenyl phosphates." Electrochimica Acta 52(12): 4024-4031.

[143] Markley, T. A., A. E. Hughes, et al. (2007). "Influence of praseodymium - Synergistic corrosion inhibition in mixed rare-earth diphenyl phosphate systems." Electrochemical and Solid State Letters 10(12): C72-C75.

[144] Marlaud, T., A. Deschamps, et al. (2010). "Evolution of precipitate microstructures during the retrogression and re-ageing heat treatment of an Al-Zn-Mg-Cu alloy." Acta Materialia 58(14): 4814-4826.

[145] Martin, J. W. (1988). "Aluminum-Lithium Alloys." Annual Review of Materials Science 18: 101-119.

[146] Masumoto, T. (1994). "Recent progress in amorphous metallic materials in Japan." Materials Science and Engineering: A 179–180, Part 1(0): 8-16.

[147] Mazurkiewicz, B. and A. Piotrowski (1983). "The electrochemical behaviour of the Al2Cu intermetallic compound." Corrosion Science 23(7): 697-707.

[148] McCafferty, E. (2010). Passivity

[149] Introduction to Corrosion Science, Springer New York: 209-262.

[150] McKenzie, P. W. J. and R. Lapovok (2010). "ECAP with back pressure for optimum strength and ductility in aluminium alloy 6016. Part 2: Mechanical properties and texture." Acta Materialia 58(9): 3212-3222.

[151] Medco.

[152] Mondolfo, L. F. (1971). "Discussion of "grain-size refining of primary crystals in hypereutectic Al-Si and Al-Ge alloys"." Metallurgical Transactions 2(4): 1254.

[153] Mondolfo, L. F. and J. G. Barlock (1975). "Effect of superheating on structure of some aluminum alloys." Metallurgical Transactions B 6(4): 565-572.

[154] Mookhoek, S. D., H. R. Fischer, et al. (2009). "A numerical study into the effects of elongated capsules on the healing efficiency of liquid-based systems." Computational Materials Science 47(2): 506-511.

[155] Mookhoek, S. D., S. C. Mayo, et al. (2010). "Applying SEM-Based X-ray Microtomography to Observe Self-Healing in Solvent Encapsulated Thermoplastic Materials." Advanced Engineering Materials 12(3): 228-234.

[156] Muller, I. L. and J. R. Galvele (1977). "Pitting potential of high purity binary aluminium alloys—II. AlMg and AlZn alloys." Corrosion Science 17(12): 995-1007.

[157] Muster, T. H., A. E. Hughes, et al. (2009). "A rapid screening multi-electrode method for the evaluation of corrosion inhibitors." Electrochimica Acta 54(12): 3402-3411.

[158] Muster, T. H., A. E. Hughes, et al. (2009). Cu Distributions in Aluminium Alloys, New York, Nova Science Publishers.

[159] Muster, T. H., Hughes, A.E., Thompson. G.E. (2009). Cu Distributions in Aluminium Alloys. New York, Nova Science Publishers.

[160] Nisancioglu, K. (1990). "Electrochemical Behavior of Aluminum-Base Intermetallics Containing Iron." Journal of The Electrochemical Society 137(1): 69-77.

[161] Nisancioglu, K. and H. Holtan (1978). "Measurement of the critical pitting potential of aluminium." Corrosion Science 18(9): 835-849.

[162] Norova, M. T., I. N. Ganiev, et al. (2003). "Enhancement of the corrosion resistance of aluminum-lithium alloys by microalloying with calcium." Russian Journal of Applied Chemistry 76(4): 547-549.

[163] Oguocha, I., O. Adigun, et al. (2008). "Effect of sensitization heat treatment on properties of Al–Mg alloy AA5083-H116." Journal of Materials Science 43(12): 4208-4214.

[164] Oliveira Jr, A. F., M. C. de Barros, et al. (2004). "The effect of RRA on the strength and SCC resistance on AA7050 and AA7150 aluminium alloys." Materials Science and Engineering A 379(1-2): 321-326.

[165] Orazem, M. E. and B. Tribollet (2008). Semiconducting Systems, in Electrochemical Impedance Spectroscopy, John Wiley & Sons.

[166] Osborne, J. H., K. Y. Blohowiak, et al. (2001). "Testing and evaluation of nonchromated coating systems for aerospace applications." Progress in Organic Coatings 41(4): 217-225.

[167] Park, J. O., C. H. Paik, et al. (1999). "Influence of Fe-Rich Intermetallic Inclusions on Pit Initiation on Aluminum Alloys in Aerated NaCl." Journal of The Electrochemical Society 146(2): 517-523.

[168] Pedersen, L. and L. Arnberg (2001). "The effect of solution heat treatment and quenching rates on mechanical properties and microstructures in AlSiMg foundry alloys." Metallurgical and Materials Transactions A: Physical Metallurgy and Materials Science 32(3): 525-532.

[169] Perrault, G. G. (1979). "Role of Hydrides in the Equilibrium of Aluminum in Aqueous Solutions." J Electrochem Soc 126(2): 199-204.

[170] Poelman, M., M. G. Olivier, et al. (2005). "Electrochemical study of different ageing tests for the evaluation of a cataphoretic epoxy primer on aluminium." Progress in Organic Coatings 54(1): 55-62.

[171] Polmear, I. J. (2006). Light alloys : from traditional alloys to nanocrystals, Oxford ; Burlington, MA : Elsevier/Butterworth-Heinemann.

[172] Poole, W. J., J. A. Seter, et al. (2000). "A model for predicting the effect of deformation after solution treatment on the subsequent artificial aging behavior of AA7030 and AA7108 alloys." Metallurgical and Materials Transactions A: Physical Metallurgy and Materials Science 31(9): 2327-2338.

[173] Pourbaix, M. (1974). Atlas of Electrochemical Equilibria in Aqueous Solutions. National Association of Corrosion Engineers, Houston, TX, USA.

[174] Premendra, P., H. Terryn, et al. (2009). "A comparative electrochemical study of commercial and model aluminium alloy (AA5050)." Materials and Corrosion 60(6): 399-406.

[175] Puga, H., S. Costa, et al. (2011). "Influence of ultrasonic melt treatment on microstructure and mechanical properties of AlSi9Cu3 alloy." Journal of Materials Processing Technology 211(11): 1729-1735.

[176] Ralston, K. D., N. Birbilis, et al. (2010). "Role of nanostructure in pitting of Al-Cu-Mg alloys." Electrochimica Acta 55(27): 7834-7842.

[177] Ralston, K. D., N. Birbilis, et al. (2010). "Revealing the relationship between grain size and corrosion rate of metals." Scripta Materialia 63(12): 1201-1204.

[178] Raps, D., T. Hack, et al. (2009). "Electrochemical study of inhibitor-containing organic-inorganic hybrid coatings on AA2024." Corrosion Science 51(5): 1012-1021.

[179] Raviprasad, K., C. R. Hutchinson, et al. (2003). "Precipitation processes in an Al-2.5Cu-1.5Mg (wt. %) alloy microalloyed with Ag and Si." Acta Materialia 51(17): 5037-5050.

[180] Ringer, S. P., K. Hono, et al. (1996). "Nucleation of precipitates in aged AlCuMg(Ag) alloys with high Cu:Mg ratios." Acta Materialia 44(5): 1883-1898.

[181] Rizzi, P. and L. Battezzati (2004). "Mechanical properties of Al based amorphous and devitrified alloys containing different rare earth elements." Journal of Non-Crystalline Solids 344(1–2): 94-100.

[182] Rodríguez, M. T., J. J. Gracenea, et al. (2004). "Testing the influence of the plasticizers addition on the anticorrosive properties of an epoxy primer by means of electrochemical techniques." Progress in Organic Coatings 50(2): 123-131.

[183] Rosalbino, F., E. Angelini, et al. (2003). "Influence of the rare earth content on the electrochemical behaviour of Al-Mg-Er alloys." Intermetallics 11(5): 435-441.

[184] Sævik, Ø., Y. Yu, et al. (2005). "Characterization of lead enrichment on electrochemically active AlPb model alloy." Journal of The Electrochemical Society 152(9): B334-B341.

[185] Samadzadeh, M., S. H. Boura, et al. (2010). "A review on self-healing coatings based on micro/nanocapsules." Progress in Organic Coatings 68(3): 159-164.

[186] Samanta, A., I. Manna, et al. (2007). "Phase evolution in Al-Ni-(Ti, Nb, Zr) powder blends by mechanical alloying." Materials Science and Engineering A 464(1-2): 306-314.

[187] Sasaki, K. and H. S. Isaacs (2004). "Origins of Electrochemical Noise during Pitting Corrosion of Aluminum." Journal of The Electrochemical Society 151(3): B124-B133.

[188] Sato, N. (1990). "An overview on the passivity of metals." Corrosion Science 31(0): 1-19.

[189] Schmutz, P. and G. S. Frankel (1998). "Characterization of AA2024-T3 by scanning Kelvin probe force microscopy." Journal of The Electrochemical Society 145(7): 2285-2295.

[190] Schneider, O., G. O. Ilevbare, et al. (2007). "In situ confocal laser scanning microscopy of AA2024-T3 corrosion metrology: III. Underfilm corrosion of epoxy-coated AA2024-T3." Journal of The Electrochemical Society 154(8): C397-C410.

[191] Schneider, O., G. O. Ilevbare, et al. (2004). "In situ confocal laser scanning microscopy of AA 2024-T3 corrosion metrology II. Trench formation around particles." Journal of The Electrochemical Society 151(8): B465-B472.

[192] Scholes, F. H., S. A. Furman, et al. (2006). "Chromate leaching from inhibited primers: Part I. Characterisation of leaching." Progress in Organic Coatings 56(1): 23-32.

[193] Scholes, F. H., S. A. Furman, et al. (2006). "Corrosion in artificial defects. I: Development of corrosion." Corrosion Science 48(7): 1812-1826.

[194] Scholes, F. H., A. E. Hughes, et al. (2009). "Interaction of Ce(dbp)(3) with surface of aluminium alloy 2024-T3 using macroscopic models of intermetallic phases." Corrosion Engineering Science and Technology 44(6): 416-424.

[195] Scully, J. R., T. O. Knight, et al. (1993). "Electrochemical characteristics of the Al2Cu, Al3Ta and Al3Zr intermetallic phases and their relevancy to the localized corrosion of Al alloys." Corrosion Science 35(1–4): 185-195.

[196] Searles, J. L., P. I. Gouma, et al. (2002). "Stress Corrosion Cracking of Sensitized AA5083 (Al-4.5Mg-1.0Mn)." Materials Science Forum 396-402: 1437-1442.

[197] Semenov, A. M. (2001). "Effect of Mg additions and thermal treatment on corrosion properties of Al-Li-Cu-base alloys." Protection of Metals 37(2): 126-131.

[198] Senkov, O., R. Bhat, et al. (2005). "Microstructure and properties of cast ingots of Al-Zn-Mg-Cu alloys modified with Sc and Zr." Metallurgical and Materials Transactions A 36(8): 2115-2126.

[199] Seri, O. (1994). "The effect of NaCl concentration on the corrosion behavior of aluminum containing iron." Corrosion Science 36(10): 1789-1803.

[200] Seri, O. and K. Tagashira (1986). "Effect of Manganese Content on Corrosion Characteristics of Al-Mn Alloys." Keikinzoku/Journal of Japan Institute of Light Metals 36(12): 806-812.

[201] Sha, G. and A. Cerezo (2004). "Characterization of precipitates in an aged 7xxx series Al alloy." Surface and Interface Analysis 36(5-6): 564-568.

[202] Shaw, B. A., G. D. Davis, et al. (1991). "The Influence of Tungsten Alloying Additions on the Passivity of Aluminum." Journal of The Electrochemical Society 138(11): 3288-3295.

[203] Shaw, B. A., T. L. Fritz, et al. (1990). "The Influence of Tungsten on the Pitting of Aluminum Films." Journal of The Electrochemical Society 137(4): 1317-1318.

[204] Sinko, J. (2001). "Challenges of chromate inhibitor pigments replacement in organic coatings." Progress in Organic Coatings 42(3–4): 267-282.

[205] Song, R. G., W. Dietzel, et al. (2004). "Stress corrosion cracking and hydrogen embrittlement of an Al-Zn-Mg-Cu alloy." Acta Materialia 52(16): 4727-4743.

[206] Souto, R. M., Y. González-García, et al. (2010). "Examination of organic coatings on metallic substrates by scanning electrochemical microscopy in feedback mode: Revealing the early stages of coating breakdown in corrosive environments." Corrosion Science 52(3): 748-753.

[207] Speckert, L. and G. T. Burstein (2011). "Combined anodic/cathodic transient currents within nucleating pits on Al-Fe alloy surfaces." Corrosion Science 53(2): 534-539.

[208] Stelling, O., A. Irretier, et al. (2006). New light-weight aluminum alloys with high Mg2Si-content by spray forming. 519-521: 1245-1250.

[209] Sukiman, N. L., N. Birbilis, et al. (2010). Corrosion Maps For Aluminium Alloys. Proc. Conf. Corrosion And Prevention '10. Adelaide, Australia, Australasian Corrosion Association (ACA).

[210] Suryanarayana, C., K. C. Rao, et al. (2008). "Preparation and characterization of microcapsules containing linseed oil and its use in self-healing coatings." Progress in Organic Coatings 63(1): 72-78.

[211] Szklarska-Smialowska, Z. (1999). "Pitting corrosion of aluminum." Corrosion Science 41(9): 1743-1767.

[212] Szklarska-Smialowska, Z. (2002). "Mechanism of pit nucleation by electrical breakdown of the passive film." Corrosion Science 44(5): 1143-1149.

[213] Tahani, A., E. Chaieb, et al. (2003). "Electrochemical study of the influence of rolling on the resistance. The corrosion of alloy of aluminium 3003 in milieu 3% NaCl." Transactions of the SAEST (Society for Advancement of Electrochemical Science and Technology) 38(1): 43-46.

[214] Tailleart, N., B. Gauthier, et al. (2009). Metallurgical and physical factors in improving the corrosion resistance of thermally sprayed semi-amorphous Al-Co-Ce coatings.

[215] Tan, L. and T. R. Allen (2010). "Effect of thermomechanical treatment on the corrosion of AA5083." Corrosion Science 52(2): 548-554.

[216] Taylor, S. R. and B. D. Chambers (2008). "Identification and Characterization of Non-chromate Corrosion Inhibitor Synergies Using High-Throughput Methods." Corrosion 64(3): 255-270.

[217] Tedim, J., S. K. Poznyak, et al. (2010). "Enhancement of Active Corrosion Protection via Combination of Inhibitor-Loaded Nanocontainers." Acs Applied Materials & Interfaces 2(5): 1528-1535.

[218] Teixeira, J. D. C., L. Bourgeois, et al. (2007). A study of the microstructural evolution and strengthening effects of non-spherical precipitates in an al-cu-based alloy. 561-565: 2317-2320.

[219] Trueman, A. R. (2005). "Determining the probability of stable pit initiation on aluminium alloys using potentiostatic electrochemical measurements." Corrosion Science 47(9): 2240-2256.

[220] Twite, R. L. and G. P. Bierwagen (1998). "Review of Alternatives to Chromate for Corrosion Protection of Aluminum Aerospace Alloys." Progress in Organic Coatings 33(2): 91-100.

[221] Unocic, K. A., P. Kobe, et al. (2006). "Grain Boundary Precipitate Modification for Improved Intergranular Corrosion Resistance." Materials Science Forum 519-521: 327-332.

[222] Usta, M., M. E. Glicksman, et al. (2004). "The effect of heat treatment on Mg2Si coarsening in aluminum 6105 alloy." Metallurgical and Materials Transactions A: Physical Metallurgy and Materials Science 35 A(2): 435-438.

[223] Vetrano, J. S., C. H. Henager Jr, et al. (1998). Use of Sc, Zr and Mn for grain size control in Al-Mg alloys.

[224] Vetrano, J. S., R. E. Williford, et al. (1997). Influence of microstructure and thermal history on the corrosion susceptibility of AA5083. TMS Annual Meeting, Orlando, FL, USA.

[225] Voevodin, N. N., V. N. Balbyshev, et al. (2003). "Nanostructured coatings approach for corrosion protection." Progress in Organic Coatings 47(3–4): 416-423.

[226] Walmsley, J. C., Ø. Sævik, et al. (2007). "Nature of segregated lead on electrochemically active AlPb model alloy." Journal of The Electrochemical Society 154(1): C28-C35.

[227] Wang, D., D. Battocchi, et al. (2010). "In situ monitoring of a Mg-rich primer beneath a topcoat exposed to Prohesion conditions." Corrosion Science 52(2): 441-448.

[228] Wang, W. t., X. m. Zhang, et al. (2010). "Influences of Ce addition on the microstructures and mechanical properties of 2519A aluminum alloy plate." Journal of Alloys and Compounds 491(1-2): 366-371.

[229] Wei, R. P. (2001). "A model for particle-induced pit growth in aluminum alloys." Scripta Materialia 44(11): 2647-2652.

[230] Westermann, I., O. S. Hopperstad, et al. (2012). "Effect of alloying elements on stage-III work-hardening behaviour of Al -Zn-Mg(-Cu) alloys." International Journal of Materials Research 103(5): 603-608.

[231] White, P. A., A. E. Hughes, et al. (2009). "High-throughput channel arrays for inhibitor testing: Proof of concept for AA2024-T3." Corrosion Science 51(10): 2279-2290.

[232] White, S. R., N. R. Sottos, et al. (2001). "Autonomic healing of polymer composites." Nature 409(6822): 794-797.

[233] Williams, D. E., J. Stewart, et al. (1994). "The nucleation, growth and stability of micropits in stainless steel." Corrosion Science 36(7): 1213-1235.

[234] Williams, G. and H. N. McMurray (2003). "Anion-exchange inhibition of filiform corrosion on organic coated AA2024-T3 aluminum alloy by hydrotalcite-like pigments." Electrochemical and Solid State Letters 6(3): B9-B11.

[235] Wu, B., J. R. Scully, et al. (1997). "Cooperative stochastic behavior in localized corrosion: I. Model." Journal of The Electrochemical Society 144(5): 1614-1620.

[236] Wu, D. Y., S. Meure, et al. (2008). "Self-healing polymeric materials: A review of recent developments." Progress in Polymer Science 33(5): 479-522.

[237] Xu, D. K., N. Birbilis, et al. (2011). "Effect of solution treatment on the corrosion behaviour of aluminium alloy AA7150: Optimisation for corrosion resistance." Corrosion Science 53(1): 217-225.

[238] Yang, B. J., J. H. Yao, et al. (2009). "Al-rich bulk metallic glasses with plasticity and ultrahigh specific strength." Scripta Materialia 61(4): 423-426.

[239] Yasakau, K. A., M. L. Zheludkevich, et al. (2007). "Role of intermetallic phases in localized corrosion of AA5083." Electrochimica Acta 52(27 SPEC. ISS.): 7651-7659.

[240] Yin, T., M. Z. Rong, et al. (2007). "Self-healing epoxy composites - Preparation and effect of the healant consisting of microencapsulated epoxy and latent curing agent." Composites Science and Technology 67(2): 201-212.

[241] Yu, Y., Ø. Sævik, et al. (2005). "Effect of annealing temperature on anodic activation of rolled AA8006 aluminum alloy by trace element lead." Journal of The Electrochemical Society 152(9): B327-B333.

[242] Yu, Y., Ø. Saevlk, et al. (2004). "Characterization of lead enrichment on electrochemically activated binary Al-Pb model alloy." Materials Forum 28: 270-276.

[243] Zamin, M. (1981). "ROLE OF Mn In the Corrosion Behavior of Al-Mn Alloys." Corrosion 37(11): 627-632.

[244] Zawrah, M. and L. Shaw (2003). "Microstructure and hardness of nanostructured Al–Fe–Cr–Ti alloys through mechanical alloying." Materials Science and Engineering: A 355(1–2): 37-49.

[245] Zeng, F.-l., Z.-l. Wei, et al. (2011). "Corrosion mechanism associated with Mg2Si and Si particles in Al–Mg–Si alloys." Transactions of Nonferrous Metals Society of China 21(12): 2559-2567.

[246] Zhang, W. and G. S. Frankel (2003). "Transitions between pitting and intergranular corrosion in AA2024." Electrochimica Acta 48(9): 1193-1210.

[247] Zhao, Y. H., X. Z. Liao, et al. (2004). "Microstructures and mechanical properties of ultrafine grained 7075 Al alloy processed by ECAP and their evolutions during annealing." Acta Materialia 52(15): 4589-4599.

[248] Zheludkevich, M. L., I. M. Salvado, et al. (2005). "Sol-gel coatings for corrosion protection of metals." Journal of Materials Chemistry 15(48): 5099-5111.

[249] Zhong, J., K. Feng, et al. (2011). Effect of RE on the mechanical properties of 7075 Al alloy. 150-151: 1286-1289.

[250] Zhou, X., N. Birbilis, et al. (2010). "Kinetic Stability of Aluminium." Corrosion & Prevention Corrosion, Adelaide, Australia.

[251] Zhou, X., Y. Liu, et al. (2011). "Near-Surface Deformed Layers on Rolled Aluminum Alloys." Metallurgical and Materials Transactions A 42(5): 1373-1385.

[252] Zhou, X., C. Luo, et al. (2012). "Study of localized corrosion in AA2024 aluminium alloy using electron tomography." Corrosion Science 58: 299-306.

http://dx.doi.org/10.5772/53752

[253] Zieliński, A., J. Chrzanowski, et al. (2004). "Influence of retrogression and reaging on microstructure, mechanical properties and susceptibility to stress corrosion cracking of an Al-Zn-Mg alloy." Materials and Corrosion 55(2): 77-87.

[254] Zin, I. M., R. L. Howard, et al. (1998). "The mode of action of chromate inhibitor in epoxy primer on galvanized steel." Progress in Organic Coatings 33(3–4): 203-210.

[255] Zou, L., Q.-l. Pan, et al. (2007). "Effect of minor Sc and Zr addition on microstructures and mechanical properties of Al-Zn-Mg-Cu alloys." Transactions of Nonferrous Metals Society of China 17(2): 340-345.

Chapter 3

Influence of Structural Parameters on the Resistance on the Crack of Aluminium Alloy

Mohamed Mazari, Mohamed Benguediab,
Mokhtar Zemri and Benattou Bouchouicha

Additional information is available at the end of the chapter

http://dx.doi.org/10.5772/53292

1. Introduction

The propagation of the cracks occurs through the formation and the intensification of a plastic zone at the crack tip. This mechanism is accompanied by energy dissipation.

The energy dissipated per cycle is representative for the state of the material at the crack tip.

Many works based on energetic approach have been made by different authors in orders to characterise the behaviour of the material in propagation state.

Ikeda and al. [1] have used hysteresis loops near the crack in order to determine the energy for surface creation and compared them to those obtained on smooth specimens. They deducted the limits of the equivalent stresses corresponding to local loops; thus, the plastic work is given by:

$$U = \int_{-\infty}^{+\infty}\int_{-\infty}^{+\infty} U_{XY} dX dY \tag{1}$$

$$U_{XY} = \frac{1}{da/dN}\left\{\int_{\varphi 1}^{\varepsilon 2} \sigma_u d\varepsilon - \int_{\varphi 1}^{\varepsilon 2} \sigma_\theta d\varepsilon\right\} \tag{2}$$

U_{XY} is the local plastic work at the plastic zone of X and Y coordinates.

These authors have only taken into account the deformations which are perpendicular to the crack.

This is confirmed since Izumi and Fine [2] have demonstrated that the plastic work in the direction parallel to the crack represents only 10% of the plastic work in the perpendicular direction to the crack.

The micro gage measures are only possible for distances lesser than 100μm from the crack tip, Izumi and Fine have used an in interpolation function of the type :

$$\Delta\varepsilon_p \;\alpha\; \log \Delta r \tag{3}$$

Davidson and al. [3] elaborated another method for measuring U from sub grains size which are formed after the passing of the crack using the retro diffused electrons method in an electronic microscope.

The comparison of the two methods done by Liaw and al [4] show a difference between the plastic work values of only 15%.

These authors show that 70% of the total is dissipated in the zone located at distance greater than 100μm from the crack tip.

Weertman [5] developed a theoretical model with a crack growth rate, energy and stress intensity factor relationship such that:

$$\frac{da}{dN} = \frac{\pi \Delta K^4}{8\mu U \sigma_c} \tag{4}$$

Izumi and Fine [2] had expressed the crack growth rate with respect to the stress intensity factor and the energy by:

$$\frac{da}{dN} = \frac{A \Delta K^m}{\sigma^2 \mu U} \tag{5}$$

where m is Paris coefficient.

These authors note that in the case of alloys for which m=4, U is independent of ΔK and also when m<4, U is an increasing function of ΔK.

Recently Ranganathan [6] showed for, an aluminium alloy 2024 T351, that the specific energy Q is proportional to the stress intensity factor amplitude to the power 4 (ΔK^4) and that the crack growth rate is related according to a power to energy law.

2. Materials under study

The materials studied are aluminium- Zinc- Magnesium- Copper alloys under different forms and whose characteristic are given in table 1.

Alloy	Sense	Grain size (μm)	σE MPa	A%	K_{IC} MPa$\sqrt{m}$	σR MPa
X7075	LT	8000	392	16	76,9	464
	TL	350	395	13,4	59,9	474
7175F	LT	300	462	11,7		534
	TL	90				
7175M	LT	1700	472	12,9		550
	TL	550				

Table 1. Characteristics of 7075 alloys (high purity, with smaller and bigger grain size F and M)

3. Experimental method

The tests were conducted on CT40 specimen of thickness B=6 mm taken in the sense TL and LT. The specimen were paper polished then with diamond paste till 1μm to permit the optical follow up of the crack. All the test were conducted for a ration R=0,10 with a frequency from 10 to 40 Hz. The stress intensity factor is calculated according to the ASTM E647 [7]. Crack closure measurements had been made for different lengths of the crack using the global compliance variation method established by Kikukawa and al. [8] (fig. 1).

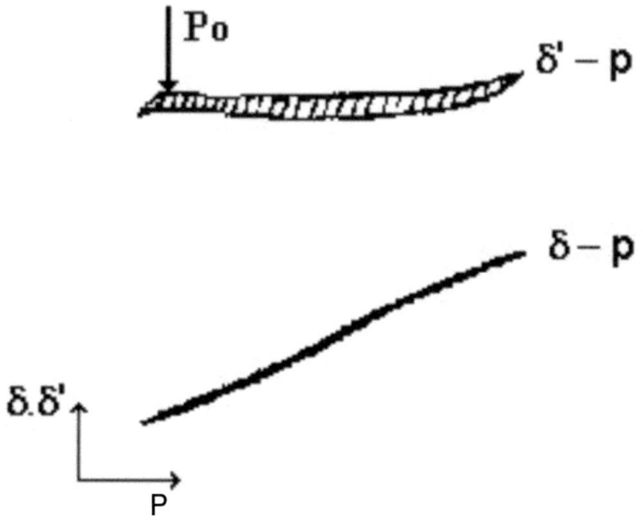

Figure 1. Curve Pδ (measurement of energy)

The hysteretic energy Q dissipated through one cycle is obtained by integration of the Pδ′ curves; and the specific energy U is given by the relation:

$$U = \frac{Q}{2B\frac{da}{dN}} \quad \left[\text{J/m}^2\text{/cycle}\right] \tag{6}$$

4. Experimental results and discussion

4.1. Influence of the material purity and the grain size

Fig. 2 shows the evolution of the specific energy U with respect to the stress intensity factor amplitude for the tree materials.

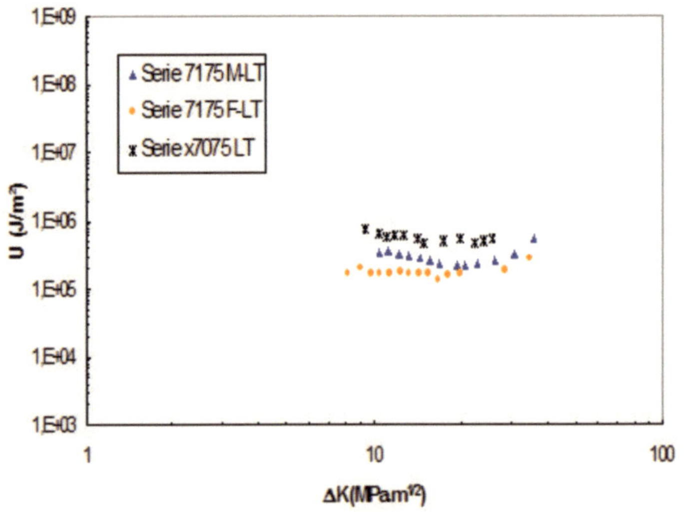

Figure 2. Evolution of energy U with respect to Δ K

We note that U is constant for crack material this result is comparable to that obtained by Ranganathan [6] for the aluminium 2024 T351.

We notice that for a pure material (X7075), the energy is four times greater than that obtained for material of ordinary purity (7175 M).

The evolution of the dissipated energy Q with respect to ΔK is given in fig. 3.

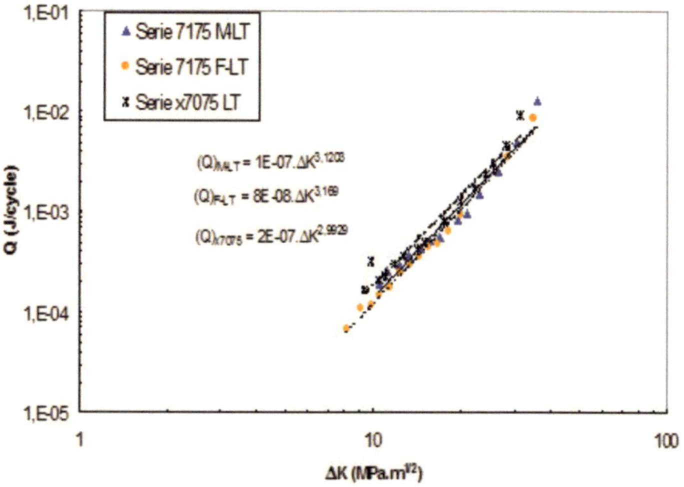

Figure 3. Evolution of energy Q with respect to Δ K

We note that Q obeys to a power law of the form $Q = A\ \Delta K^m$ for the three material. The variations of the crack growth rates with respect to Q show that growth rates obtained for the 7175 M and 7175 F materials (fig. 4).

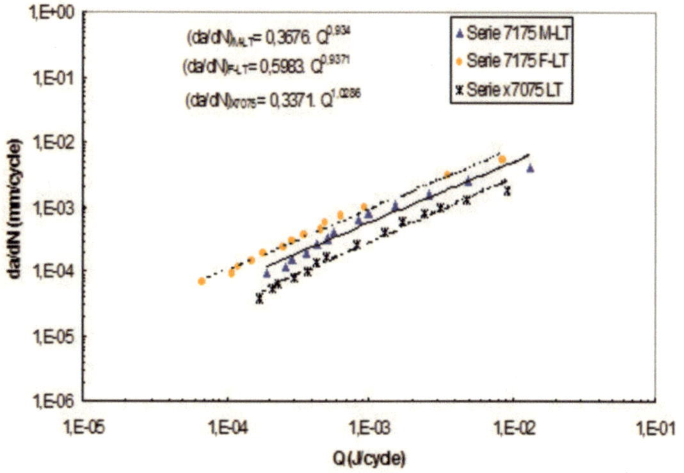

Figure 4. Evolution the speed of cracking with respect to Δ K

4.2. Influence of grains orientation

Figures 5 and 6 show the variation of the crack growth rates with respect to the energy Q and that of the specific energy U with respect to ΔK for 7175 F alloy in two different grain size orientations LT and TL.

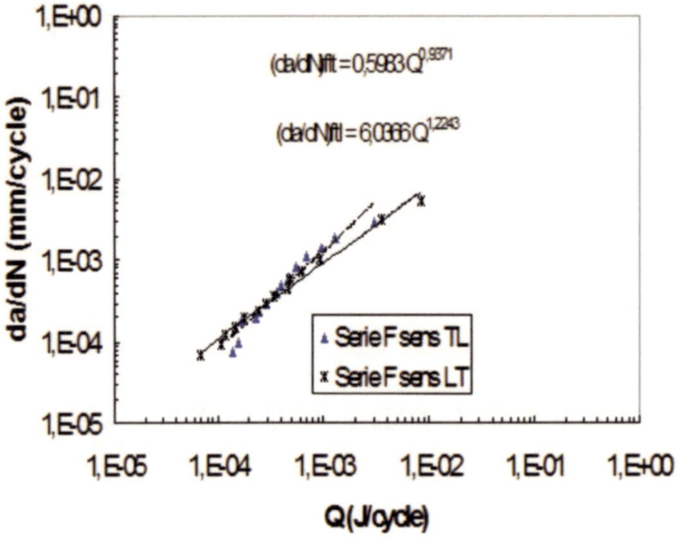

Figure 5. Evolution the crack growth rate with respect to Q

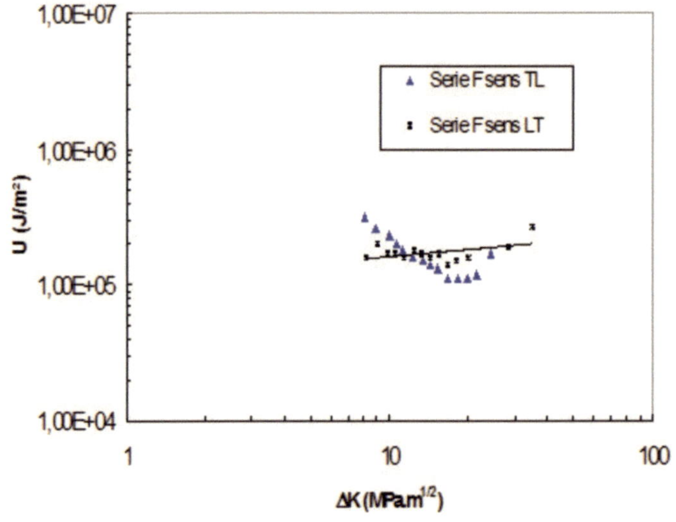

Figure 6. Variation of energy with respect to Δ K for two different orientations (TL and LT)

http://dx.doi.org/10.5772/53292

We note that the evolution of U is constant and equal to 2.10^5 for the LT orientation. The crack growth rates linearly vary with respect to Q.

5. Conclusion

The present study permits to investigate the influence of three characteristic parameters of materials on the evolution of the energy and on the crack growth rates.

From which the following conclusions can be drawn:

- pure alloys present a higher energy and lower crack growth rates;
- the grain size has a predominant influence on the specific energy;
- the grain orientation has not a consequent influence on these two parameters (energy and crack growth rates).

Author details

Mohamed Mazari*, Mohamed Benguediab, Mokhtar Zemri and Benattou Bouchouicha

*Address all correspondence to: lmsr_univsba@yahoo.fr

Materials and Reactive Systems Laboratory LMSR, Djillali Liabes University of Sidi Bel Abbes, Algeria

References

[1] Ikeda, S., and Al, Eng. Fract. Mech., vol. 9, 1977, pp. 123-136

[2] Izumi Y. and Fine M.E., Eng. Frac. Mech. 11, p. 791,1979

[3] Davidson, D.L., et al., Proc. Symp. Envir. Fract. of Eng. Matls, TMS ASME Warendale, P.A., the U.S.A., 1980, pp.59

[4] Liaw, P. K, and Al, Fract. of Eng. Matls and Str., vol. 3, 1980, pp.59

[5] Weertman, J, Int. Newspaper of Fracture, Flight. 9, 1973, pp. 125-131

[6] Ranganathan, NR, thesis Doctorate of the University of Poitiers No.419, 1985

[7] ASTM, Standard E 647, Philadelphia, U.s.a: American Society for Testing and Materials, 1983

Chapter 4

Effect of Micro Arc Oxidation Coatings on the Properties of Aluminium Alloys

Cenk Mısırlı, Mümin Şahin and Ufuk Sözer

Additional information is available at the end of the chapter

http://dx.doi.org/10.5772/53135

1. Introduction

Aluminium and its alloys are one of the most useful substances in industries such as aerospace industry where the demand for lighter, more durable, recyclable and replaceable materials is increasing because of economical and technical considerations. Their good corrosion resistance made them more important in use as a component of internal combustion engines for instance cylinder blocks cylinder heads and pistons [1].

By means of MAO process, which is also known as plasma electrolytic oxidation, spark anodisation or micro-arc discharge oxidation, with the combination of electrochemical oxidation and high voltage spark treatment, the engineering equipments which have excellent adhesive, strength, friction, corrosion, wear, electrical and thermal properties are obtained.

The micro-arc oxidation (MAO) process has been examined on dense metals in a wide range of structures [2], but its application to metallic foams has not been systematically investigated. PEO is a plasma-assisted electrochemical surface treatment that is used to convert surfaces of light metals, e.g. Al, Mg and Ti, into hard and well-adhered oxide layers [3]. The crucial problem in using these components are aluminium component's poor surface hardness and wear resistance which limits its use in applications and reducing the lifetime of the substance.

Previous treatments on aluminium such as anodising and thermal spraying have suffered from the low load support from the underlying material and insufficient adhesion. Also some coated aluminium alloys can achieve high surface hardness but when it comes to practice they often show poor performance under mechanical loading [4]. Furthermore, AA2014 alloy consists mainly of copper, magnesium, manganese, silicon and some other elements which make it usable for conditions where strength hardness and good workability is needed. If we define workability as the amount of deformation that a material can undergo without cracking or reaching undesirable conditions at a given temperature and strain rate, improving workability means to undo these undesirable conditions and increasing the processing ability of the of the materials and increasing the productivity. This situation can be achieved by optimum homogenization treatment [5].

PEO coatings on aluminium are characterised by complex mixtures of amorphous material and crystalline phases and consist mainly of two regions [6]: a dense and hard inner region which has increased a-Al_2O_3 content and a porous outer region containing more c-Al_2O_3, plus alumina-silicates and amorphous phases that incorporate some electrolyte species (e.g. Si).

This process combines electrochemical oxidation with a high voltage spark treatment in aqueous electrolytic bath which also contains modifying elements in the form of dissolved salts to be incorporated into resulting alumina coatings which are thick hard oxide ceramic coatings that are fabricated on aluminium alloys [7]. However, ceramic coatings were formed on a pure aluminium substrate in an aqueous solution of phosphate salt, using the micro-arc oxidation method. The result indicated that elements from the electrolyte like P could be incorporated into the film during MAO process [8].

In Aluminium and its alloys, depending on the production method and alloy, by means of some thermal processes and applications, important differences can be achieved on the tribologycal and surface properties of aluminium alloys by the heat treatment furnaces [9]. With particular attention, providing and controlling the necessity of heating, post phone, cooling process play a crucial role in the hardness and mechanical properties of aluminium and its alloys [10].

In this study, the authors investigated the best properties of AA2014 aluminium components and minimizing the undesirable conditions that occur in components by following the method of homogenization and micro-arc discharge oxidation process along the specimens.

2. Experimental details

Before coating and homogenization, the specimens are machined to fatigue test specimens and the dimensions of the specimens are given in Figure 1.

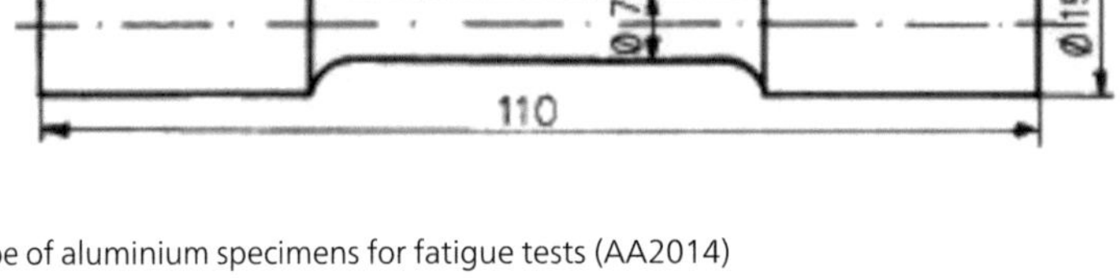

Figure 1. The shape of aluminium specimens for fatigue tests (AA2014)

The chemical composition of the AA2014 aluminium alloy used as specimens is given in the Table 1.

Cu (%)	Si (%)	Mg (%)	Mn (%)	Fe (%)	Cr (%)	Sn (%)
4.821	0.816	0.612	0.572	0.221	0.057	0.041

Table 1. The chemical composition of the AA2014 aluminum alloy

Before coating process, some of the specimens undergo homogenization heat treatment at 524 °C for 6 hours. The specimens are cooled on air after the treatment. The properties and the drawing of the furnace are shown in Figure 2.

Figure 2. The Photograph of the furnace that is used for homogenization treatment.

Coatings were formed by the use of micro arc oxidation technique where a voltage pulsed at mains frequency 50 Hz(hertz) was selected in the range 50–500 V(volts) and a predefined current density at the coating surface was maintained by control of the voltage pulses during the process. The electrolyte that is used in coating consists of sodium silicate in distilled water with other additives [11]. The electrolyte temperature is controlled to remain under 75°C during treatment. 3 treatment times were chosen to produce total coating thicknesses of 20, 60 and 120 μm(micronmeters) respectively as shown in Figures 3.a. and 3.b.

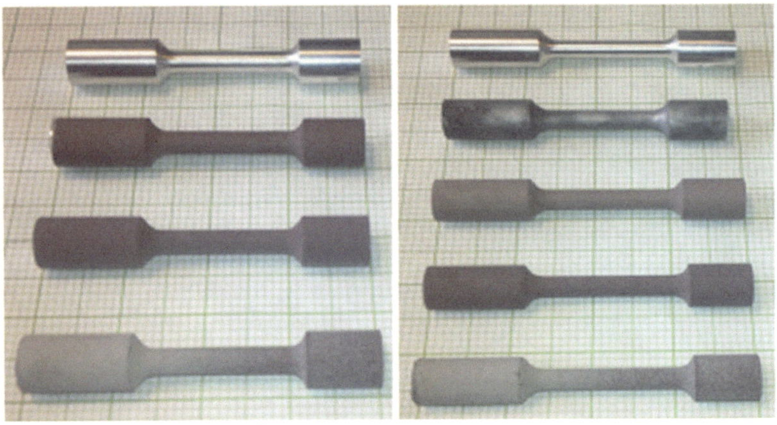

a. b.

Figure 3. Non-homogenized specimens (non-coated, 20, 60, 120 μm(micronmeters) coated respectively), homogenized specimens (non-coated, only homogenised, 20, 60, 120 μm(micronmeters) coated respectively)

The electrolytic solution consists of distilled water with potassium hydroxide and sodium silicate additives. The tank is filled with 200 litres of distilled water with 12 grams of sodium silicate and 2 grams of potassium hydroxide per litre. The required amount of sodium silicate and potassium hydroxide are dissolved in another tank with a stirring device before addition process. The used sodium silicate was in solid state and had 5 water molecules in their formula that is why in total, 4171 grams of sodium silicate was used in process. The Ph of the solution is measured as 12,4.

The MAO process was made with 200μf(micronfarads) interference because of the geometry of the samples. The voltage values in anode and cathode were measured in every 5 minutes during coating process. The current was also measured. The tables and graphs regarding to coating voltages are given in results section.

After the coating process, the substrates were polished with Si-C 800 mesh abrasive papers of decreasing grit size. A scanning electron microscope (Apollo 300) (SEM) was employed for observation of the microstructure and surface and fracture cross – section morphology of the coatings before and after the tests. The crystallographic characteristics of the coatings were investigated using an X-ray diffractiometer (Philips PANalytical) (CuKα radiation) with the film attachment, at grazing angles of 0.5 and 2.0°. The X-ray generator settings were 60 kV(kilovolts) and 55 mA(milliamperes), the step size 0.02°, step time 10 sec(seconds) and the scans were acquired from 20 to 90° (in 2Θ). The fatigue tests were performed on a 50 kN(kilonewtons) capacity of servo-hydraulic material testing machine (Instron 8051) having a frequency range of 0.5-100 Hz(hertz). [13].

3. Results and discussion

The specimens that are machined to fatigue test specimens are put into the furnace pre heated to 524°C and after the specimens are put into the furnace, the temperature is remained at 524°C [13]. The specimens are treated in the furnace for 6 hours and after that, they are cooled in air.

Regarding to the coating process, the current-voltage charactheristics and the coating thicknesses of the specimens are given in Table 2 and Figures 4, 5 and 6.

normal specimens 20 μm(micronmeters) coated				**homojenized specimens 20 μm(micronmeters) coated**			
Time (minute)	kathode voltage V(voltage)	anode voltage V(voltage)	Current A(ampere)	Time (minute)	kathode voltage V(voltage)	anode voltage V(voltage)	Current A(ampere)
0	80	400	8,10	0	300	150	11,75
5	75	490	9,18	5	60	440	9,81
10	80	495	8,92	10	70	460	9,15
15	80	495	8,93	15	70	460	9,04
normal specimens 60 μm(micronmeters) coated				**homojenized specimens 60 μm(micronmeters) coated**			
Time (minute)	kathode voltage V(voltage)	anode voltage V(voltage)	Current A(ampere)	Time (minute)	kathode voltage V(voltage)	anode voltage V(voltage)	Current A(ampere)
0	40	200	11,88	0	160	100	11,90
5	60	445	9,54	5	62	450	9,58
10	65	455	9,28	10	65	460	9,32
15	70	475	9,01	15	70	475	9,29
20	80	480	8,79	20	80	490	9,12
25	80	500	8,83	25	80	500	9,08
30	80	500	8,64	30	80	500	9,01
35	85	500	8,58	35	82	500	8,92
40	90	505	8,38	40	85	502	8,78
45	100	505	8,27	45	92	505	8,65
50	105	505	8,25	50	100	505	8,36

normal specimens 120 µm(micronmeters) coated				homojenized specimens 120 µm(micronmeters) coated			
Time (minute)	kathode voltage V(voltage)	anode voltage V(voltage)	Current A(ampere)	Time (minute)	kathode voltage V(voltage)	anode voltage V(voltage)	Current A(ampere)
0	180	120	12,14	0	40	150	12,05
5	62	448	9,72	5	60	445	9,61
10	65	455	9,49	10	62	452	9,43
15	75	490	9,36	15	70	475	9,19
20	80	498	9,07	20	78	495	8,96
25	80	500	8,96	25	80	500	8,98
30	80	500	8,89	30	80	500	8,92
35	80	500	8,80	35	82	502	8,75
40	90	505	8,83	40	90	505	8,71
45	100	505	8,55	45	100	505	8,59
50	102	505	8,42	50	112	505	8,38
55	120	505	8,10	55	130	505	8,18
60	140	515	8,06	60	140	515	8,22
65	145	525	8,03	65	140	525	8,19
70	155	525	8,13	70	145	530	8,22
75	160	540	7,97	75	160	525	8,28
80	160	540	8,15	80	170	530	8,22
85	160	540	7,89	85	180	550	8,20
90	180	550	7,85	90	160	525	8,34
95	182	550	7,79	95	220	530	8,13
100	220	550	7,59	100	180	530	8,18
103	210	555	7,81	103	190	540	8,04

Table 2. Voltage charactheristics of coated specimens

According to diagram it is observed that both homogenized and non- homogenized specimens have standart values of voltages in different coating thicknesses, from these results, it is noted that the homogenization treatment has nearly no effect on the voltage values during operation. The coating thicknesses are measured by Mitutoyo device used to measure coating thickness and it is seen that at the same coating time interval the coating thicknesses are seem to be the same. After the cooling operation and coating non- homogenized, homogenized, coated and both heat treated and coated specimens are put into fatigue tests.

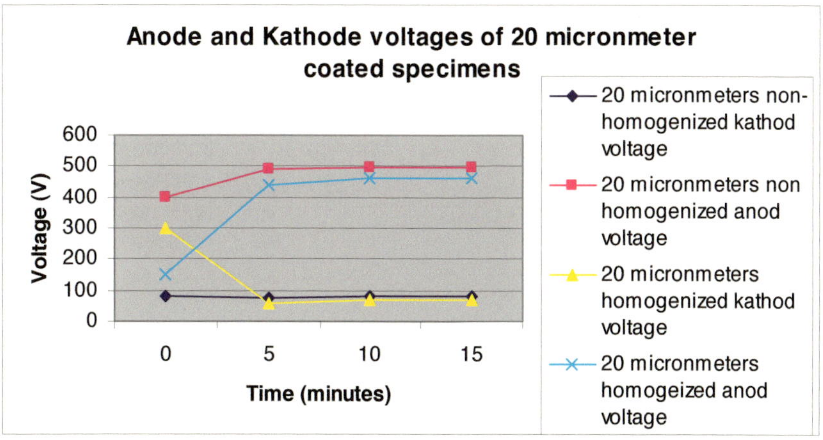

Figure 4. Anode and Kathode voltage graphs of 20 micronmeter coated specimens

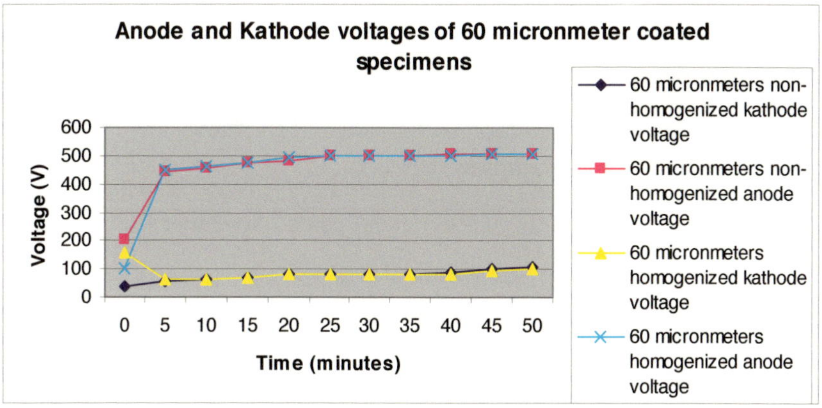

Figure 5. Anode and Kathode voltage graphs of 60 micronmeter coated specimens

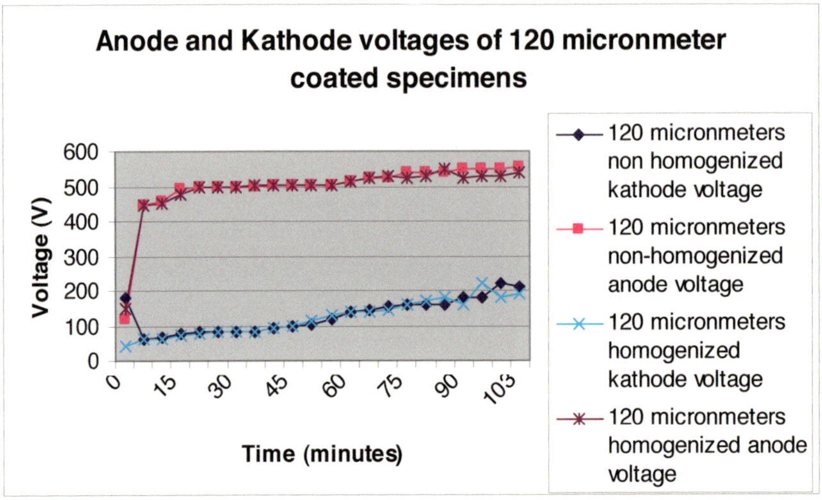

Figure 6. Anode and Kathode voltage graphs of 120 micronmeter coated specimens

The fatigue tests are taken into action in 8 groups. Non- homogenized and non- coated samples, only homogenized samples, non- homogenized but coated samples (20μm(micronmeters), 60μm(micronmeters) and 120μm(micronmeters) respectively) and homogenized and coated samples 20μm(micronmeters), 60μm(micronmeters) and 120μm(micronmeters) respectively). After several fatigue tests on these specimens we can find these results;

1. Homogenized and 20μm(micronmeters) coated specimens have the best results in fatigue tests.

2. Non homogenized and 20μm(micronmeters) coated sample comes after the first one. Considering this, it can be understood that thin coating in micro arc oxidation has the biggest positive effect on the surface properties of aluminum.

3. Homogenized and 120μm(micronmeters) coated specimens are coming after these 20μm(micronmeters) coated samples

4. Non homogenized and 120μm(micronmeters) coated specimens are coming after 3rd one.

5. Only homogenized specimens have nearly the same fatigue results as 3rd and 4th one,

6. Non homogenized and non coated specimens have the fatigue results that are known.

7. 60μm(micronmeters) coated specimens have the worse fatigue test results no matter how homogenized or non homogenized they are.

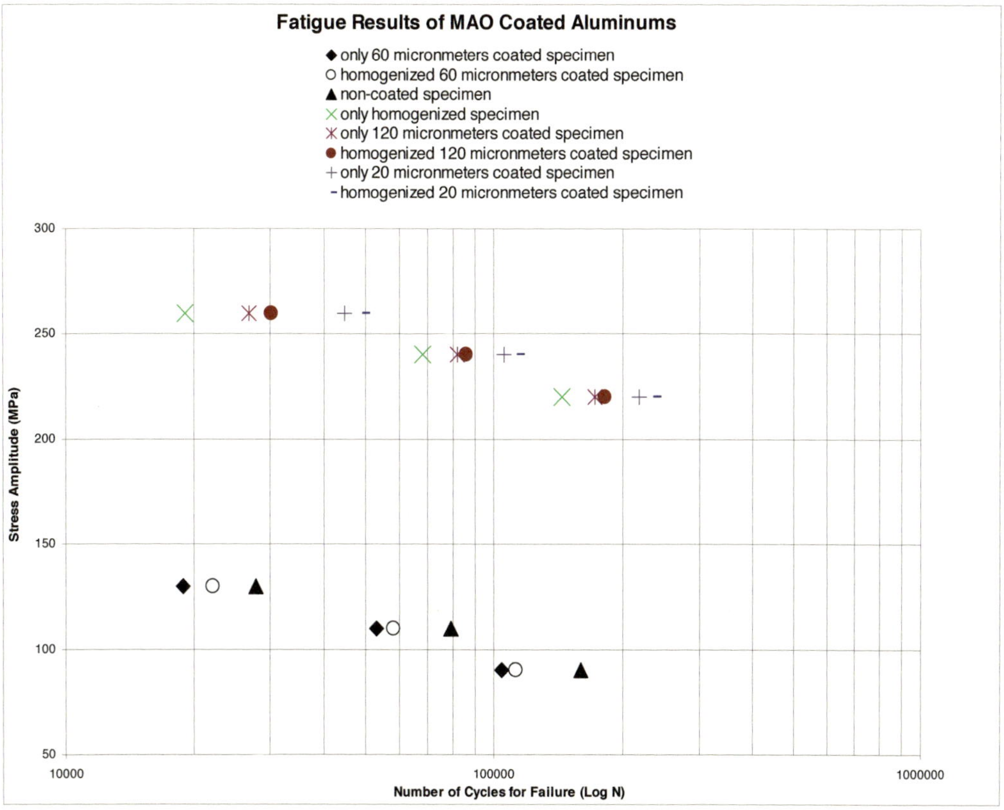

Figure 7. Fatigue test results of the specimens.

The results according to these tests are given in Figure 7. From Figure 7, we can understand that micro arc oxidation process has better effect on the surface properties of AA 2014 Aluminum alloys then homogenization, furthermore it can also be obtained that for surface properties, thin coatings are better than thick coatings in micro arc oxidation and the middle coatings are the worse coatings in micro-arc oxidation.

The SEM analyses give us an idea about the surface of the coating (Figure 8, 9 and 10.). According to the coating time the thickness of the coating on the material will increase but in MAO process, at first the quality of the coating increases with the increasing coating time but after the derigueur time the quality of the coating decreases with increasing coating time that is why finding the optimized coating time has a vital importance in MAO coatings, with the help of SEM analysis we can see the amount of pores on the coating and this will help us in understanding the quality of the coating. From the results shown above, it has been understood that the 120μm(micronmeters) coated specimens have more pores than the 60μm(micronmeters) coated specimens, but 120 μm(micronmeters) coated specimens showed better results on fatigue tests because of the high quality bonding between the base material and coating material. As expected before, 20 μm(micronmeters) coated specimens have the best results.

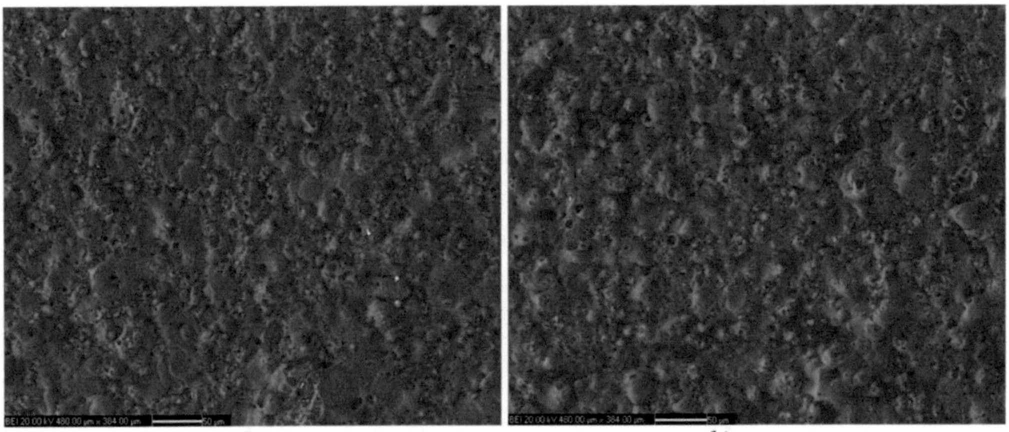

a) b)

Figure 8. SEM photographs of 20 μm(micronmeters) coated specimens. a) non-homogenized, b) homogenized

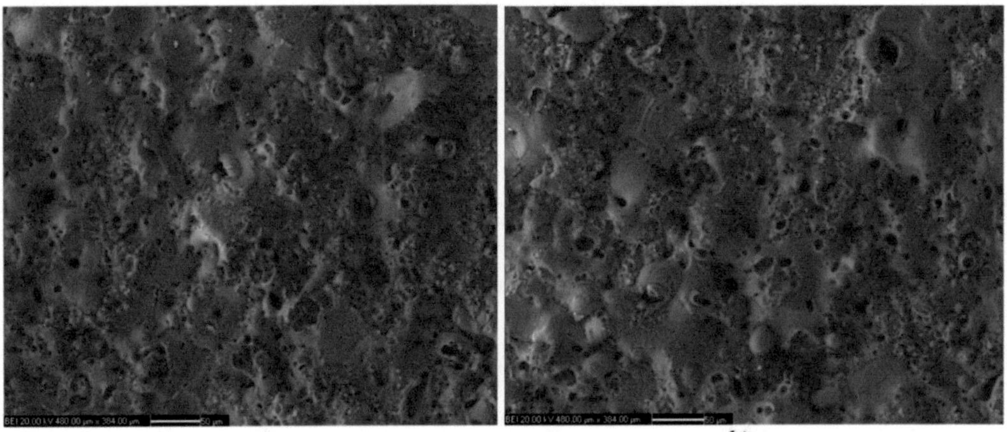

a) b)

Figure 9. SEM photographs of 60 μm(micronmeters) coated specimens. a) non-homogenized, b) homogenized

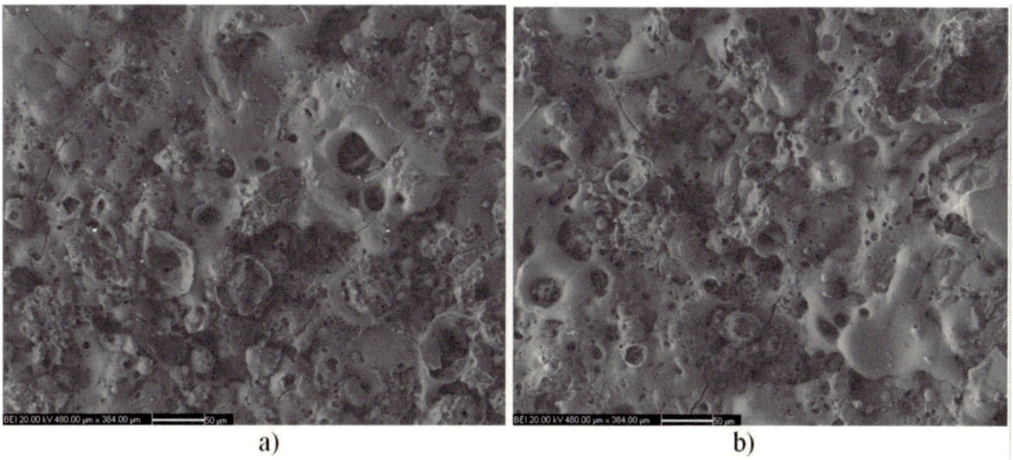

Figure 10. SEM photographs of 120 μm(micronmeters) coated specimens. a) non-homogenized, b) homogenized

X-ray diffraction test will help us understand the phase composition of the coating. Basically, this test will show the penetration depth of the coating in different angles and will give us some ideas about the composition of the coatings. Nie and Matthew used this test in angles 0.5° and 2.0° respectively and found out that in different depths of both porous and dense layer the coating composition remains same in mixture of both $\alpha\ Al_2O_3$ and γAl_2O_3. [3,12,14].

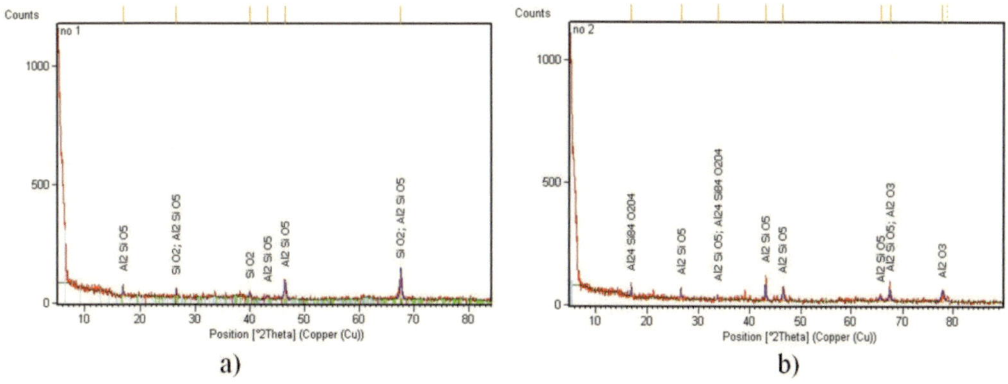

Figure 11. XRD results of 20 μm(micronmeters) coated specimens, a)non-homogenized, b)homogenized

XRD results in Figures 11, 12 and 13 show that homogenization have no effect in phase composition, but have a positive coating effect because it increases the combining ratio. According to these results, there is the same combination of coating that includes

3μm(micronmeters) (in pours layer) and 10μm(micronmeters) (in dense layer) of α and γAl_2O_3 with another alumina silicate compounds shown in figure. XRD spectra and results collected from the coated specimens indicate a considerable increase in oxygen and silicon content, explaining that the coating has a combination of aluminium oxides and complex Al-Si-O alumina silicate compounds. These results show that all MAO coatings had a similar structure and composition.

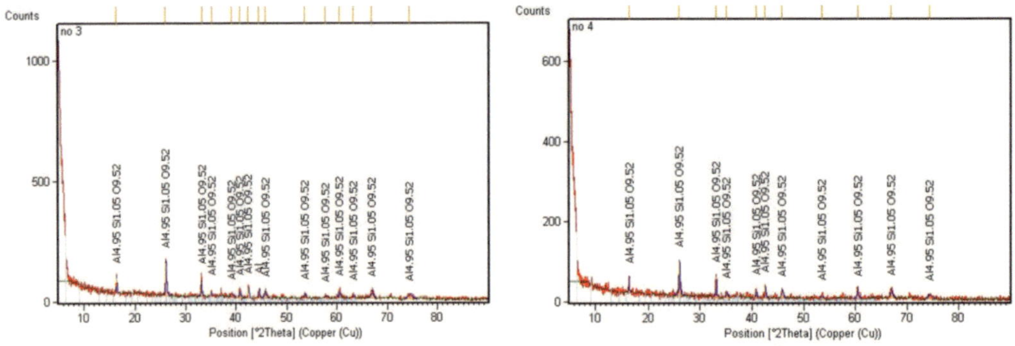

a) b)

Figure 12. XRD results of 60 μm(micronmeters) coated specimens, a)non-homogenized, b)homogenized

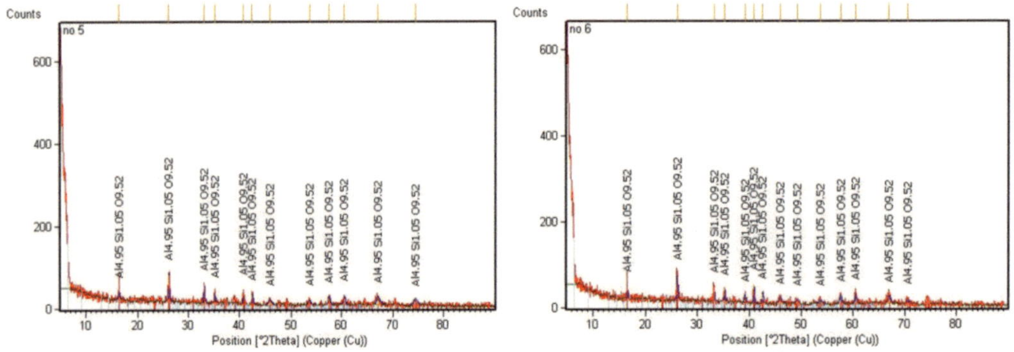

a) b)

Figure 13. XRD results of 120 μm(micronmeters) coated specimens, a)non-homogenized, b)homogenized

4. Conclusion

According to the SEM analysis, XRD results and spectrums, fatigue test results, the effect of the homogenization process on MAO coating process can be summarized as follows;

- The homogenization process has no effect on the phase composition of the specimens.

- According to the SEM analysis homogenization treatment has no effect on the number of pores on the coating material.
- According to the fatigue tests, it is found that homogenization has a positive effect on the quality and the durability of the material.

From these results it is explained that the homogenization treatment does not change the coating material porosity but definitely changes the final coatings durability so this treatment has a positive effect on bonding.

As a conclusion, the MAO process with its advantage is one of the most common coating methods for the light weight materials. Its corrosion resistance, thermal endurance, high hardness value makes this method more important for aerospace and automotive industries. If the investigations on this method come with solution for the expense of the process, it will be the best method for coating light materials.

Acknowledgements

In preparing this paper, micro arc oxidation coating was made in Gebze Institute of High Technology. SEM analysis and XRD analysis were made in Yildiz Technical University laboratories. The authors thank to these universities and scientists who helped us with these tests and analysis. Also, the authors would like to thank to Trakya University TUBAP 2010/28 project.

Author details

Cenk Mısırlı, Mümin Şahin and Ufuk Sözer

Mechanical Engineering Department, Trakya University, Edirne, Turkey

References

[1] Miller WS, Zhuang L, Bottema J, Wittebrood AJ, De Smet P, Haszler A, Vieregge A (2000). Recent development in aluminium alloys for the automotive industry. Mater. Sci. Eng.A. 280:37-49.

[2] Gu WC, Lu GH, Chen H, Chen GL, Feng WR, Yang SZ (2007). PEO protective coatings on inner surface of tubes. Surfaca and Coating Technology. 20:6619–22.

[3] Yerokhin AL, Nie X, Leyland A, Matthews A, Dowey SJ (1999). Plasma electrolysis for surface engineering. Surface and Coating Technology. 122:73–93.

[4] Curran JA, Clyne TW (2005). Thermo-physical properties of plasma electrolytic oxide coatings on aluminium. Surface & Coatings Technology. 199:168-76

[5] Gavgali M, Aksakal B (1998). Effect of various homogenisation treatments on the hot workability of ingot aluminium alloy AA 2014. Materials Science and Engineering A. 254:189-99.

[6] Guan YJ, Xia Y, Lia G (2008). Growth mechanism and corrosion behaviour of ceramic coatings on aluminum produced by auto control AC pulse PEO. Surface and Coating Technology. 202:4602–12.

[7] Gnedenkov SV, Khrisanfova OA, Zavidnaya AG, Sinebrukhov SL, Gordienko PS, Iwatsubo S, Matsui A (2001). Composition and adhesion of protective coatings on aluminium, Surface and Coatings technology. 145:146-51.

[8] Chen C, Dong Q, Wang D, Ji Q (2005). Microstructure and Growth Process of Al_2O_3 Film on Pure aluminium by Micro-Arc Oxidation. Surface Review and Letters. 12(5-6):781-785.

[9] Thompson DS, Singleton OG, McGowan RD, Spangler GE (1970). Practises and Equipment for Heat Treating Aluminium Alloys. Metal Progress. 78-83.

[10] ASM Metals Handbook (1985). Metals Park Ohio. 2:271-83.

[11] Nie X, Leyland A, Song HW, Yerokhin AL, Dowey SJ, Matthews A (1999). Thickness effects on the mechanical properties of micro-arc discharge oxide coatings on aluminium alloys. Surface and Coating Technology. 116-119:1055-60.

[12] Yerokhin AL, Dowey SJ (1999). Plasma electrolysis for Surface Engineering. Surface and Coating Technology. 122:73-93.

[13] Çan, Y., Mısırlı C (2008). Analysis of spur gear forms with tapered tooth profile. Materials&Design. 29:829-38.

[14] Sözer U. (2011). The coating of AA 2014 Aluminum Alloys by the means of Plasma Electrolytic Oxidation and investigation of their fatigue behaviour , Graduate School of Natural Applied Sciences of Trakya University Msc. Dissertation, Edirne- Turkey.

Section 2

Extrusion, Rolling and Machining

Chapter 5

Effects of Deep Rolling and Its Modification on Fatigue Performance of Aluminium Alloy AA6110

Patiphan Juijerm and Igor Altenberger

Additional information is available at the end of the chapter

http://dx.doi.org/10.5772/50651

1. Introduction

Recently, low-weight components are particularly required for environmental, ecological and economical aspects. Therefore, light-weight metals are frequently mentioned and selected for many applications where low density and high strength to weight ratios are an important consideration. Consequently, development and improvement in the field of light-weight metals can be seen continuously for advanced applications in automotive as well as aerospace industries. One of the most important light-weight metals is aluminium and its alloys which possess many attractive characteristics including excellent corrosion resistance in most environments, reflectivity, high strength and stiffness to weight ratio, good formability, weldability and recycling potential [1,2]. Certainly, these advantageous properties make them ideal candidates to replace heavier materials (steel or copper) for several industries. Therefore, mechanical behavior of aluminium alloys becomes more and more important, especially under cyclic loading due to failures occurring in machinery components are almost entirely fatigue failures. Accordingly, fatigue performance of aluminium alloys was investigated and also improved by mechanical surface treatments, e.g. shot peening, deep rolling and laser shock peening. Deep rolling is one of the most well-known mechanical surface treatment methods and exhibits a great depth of near-surface work hardening state and compressive residual stresses serving to inhibit or retard fatigue crack initiation as well as crack growth [3-5].

Ageing treatments combined with e.g. shot peening or deep rolling are methods to enhance the effects of conventional mechanical surface treatments for steels or materials having interstitial solute atoms. The fatigue performance of deep rolled steels SAE 1045, AISI 4140 and stainless steel AISI 304 can be considerably improved after annealing due to static/dynamic strain ageing. The formation of carbon atom clouds and very small carbide precipitates are the significant phenomena to impede dislocation movement and thus determine the mechanical properties. Normally, for precipitation-hardenable materials such as aluminium- and titanium alloys, mechanical surface treatment is performed after ageing treatments for fatigue lifetime enhancement. However, in some cases, mechanical surface treatment is performed on solution-heat-treated conditions with subsequent ageing treatments to produce increased hardness by precipitates especially in near-surface regions. The fine precipitates can be reasonably produced by prior plastic deformation due to preferential heterogeneous nucleation in the vicinity of dislocations in near-surface regions thus improving the cyclic deformation behavior [6,7]. On the other hand, the near-surface compressive residual stresses and work hardening induced by mechanical surface treatments decrease during annealing as well as ageing treatments. By the reduction of compressive residual stresses and near-surface work hardening, a detrimental effect for the fatigue lifetime is expected. The effects of decreased residual stress as well as work hardening together with the increased hardness by the ageing treatment on the fatigue performance are of particular interest, especially in smooth, soft and mechanically surface treated aluminium alloys because their fatigue lifetime depends significantly on the stability of near-surface work hardening as well as compressive residual stresses which can inhibit or retard surface fatigue crack initiation as well as fatigue crack growth. The purpose of this work is therefore to investigate systematically how optimized fatigue lifetimes by ageing treatments of deep rolled as-quenched AA6110 can be obtained. It should be noted that the optimize ageing treatment for the solution-heat-treated aluminium alloy AA6110 producing a maximum hardness value and also maximum yield stress is at a temperature of 160 °C for about 12 hr. Moreover, modern mechanical surface treatments have been developed from shot peening and deep rolling to give an optimized surface condition. At elevated temperatures, static/dynamic strain ageing is applied to shot peening and deep rolling. Consequently, warm shot peening and high-temperature deep rolling have become established techniques, particularly for materials that have interstitial solute atoms, e.g. steels containing carbon or nitrogen atoms [7]. Superior stability of macroscopic compressive residual stresses as well as work-hardening states during cyclic loading are observed due to dislocation pinning by interstitial solute atoms (so-called Cottrell clouds) or very fine carbides. Accordingly, superior fatigue performance was observed as compared to conventional mechanical surface treatments. It is very challenging when a modern mechanical surface treatment such as high temperature deep rolling is performed on precipitation-hardenable materials such as aluminium alloys that have mainly substitutional solute atoms where the full beneficial effects of static/dynamic strain ageing cannot be expected. Nevertheless, static/dynamic precipitation occurring during mechanical surface treatment at elevated temperatures may contribute to mechanical properties of the surface as well as the bulk, particularly for the solution-heat-treated condition. Thus, a newly developed mechanical surface treatment, high-temperature deep rolling, was performed on the solution-heat-treated aluminium alloy AA6110 for different deep-rolling temperatures up to 250 °C.

2. Methodology

The aluminium wrought alloy AA6110 was delivered from Alcoa Extrusions, Hannover, Germany as extruded bars with a diameter of 34 mm. The chemical composition of this alloy is 0.86 Si, 0.19 Fe, 0.45 Cu, 0.46 Mn, 0.78 Mg, 0.17 Cr, 0.02 Zn, 0.01 Ti and Al balance (all values in wt%). The alloy is a precipitation-hardenable Al–Mg–Si–Cu alloy, forming ′′ as well as Q ′ precipitates. Cylindrical specimens with a diameter of 7 mm and a gauge length of 15 mm were prepared. The loading direction during fatigue investigations corresponds to the extrusion direction of the bar. The specimens were solution heat treated in a furnace with argon atmosphere at a temperature of 525 °C for 30 min followed by water quenching to room temperature. The specimens were deep rolled at room temperature immediately after quenching. A hydraulic rolling device with a 6.6 mm spherical rolling element and a rolling pressure of 80 bar was used. The deep rolled as-quenched specimens were aged in the temperature range 50–300 °C with different ageing times up to approximately 28 and 148 h. For a high-temperature deep rolling, a pneumatic rolling device with a 40 mm diameter roller and a rolling force of 0.27 kN was applied at room temperature and at elevated temperatures up to 250 °C using inductive heating. On all specimens fatigue tests as well as thermal relaxation investigations were carried out. The non-mechanically surface treated specimens were electrolytically polished in the gauge length leading to a material removal of 100 μm before testing to avoid any influence of machining. Tension-compression fatigue tests were conducted with a servohydraulical testing device at stress control without mean stress (R = -1) and with a test frequency of 5 Hz. Strain was measured using capacitative extensometers. Residual stress- and FWHM-value-depth-profiles were determined by successive electrolytical material removal using the classical $\sin^2\Psi$-method with CuKα-radiation and the {333}-planes and 1/2 s_2 = 19.77 x 10^5 mm^2/N as elastic constant. All residual stresses and FWHM-values were measured in longitudinal direction of the specimens. No stress correction was carried out after electrolytical material removal of surface layers.

3. Results and discussion

3.1. Deep rolling followed by ageing treatment

Normally, for precipitation-hardenable materials, such as aluminium- and titanium alloys, mechanical surface treatment is performed after ageing treatments for fatigue lifetime enhancement as reported in [5,8-11]. However, in some cases, mechanical surface treatment is performed on solution heat treated (as-quenched) conditions with subsequent ageing treatments to produce increased hardness by precipitates especially in near-surface regions [12-15]. However, on the other hand, at the same time, near-surface macroscopic compressive residual stresses and work hardening states induced by mechanical surface treatments decrease during ageing treatments due to thermal residual stress relaxation. By the reduction of near-surface macroscopic compressive residual stresses and work hardening states, a detrimental effect for the fatigue lifetime can be expected. The effects of decreased residual stresses as well as work hardening states together with increased hardness values by the ageing treatment on the fatigue behavior of aluminium alloy AA6110 will therefore be thoroughly investigated and closely monitored in this section. Fig. 1 shows the hardness values at the surface of the deep rolled as-quenched condition after ageing treatments as a function of ageing times and temperatures. After ageing treatments in the temperature range 160-250 °C, the hardness values at the surface increased continuously with increasing ageing time until reaching a maximum value. Maximum hardness values of approximately 140, 133 and 123 HV were measured after ageing at temperatures of 160

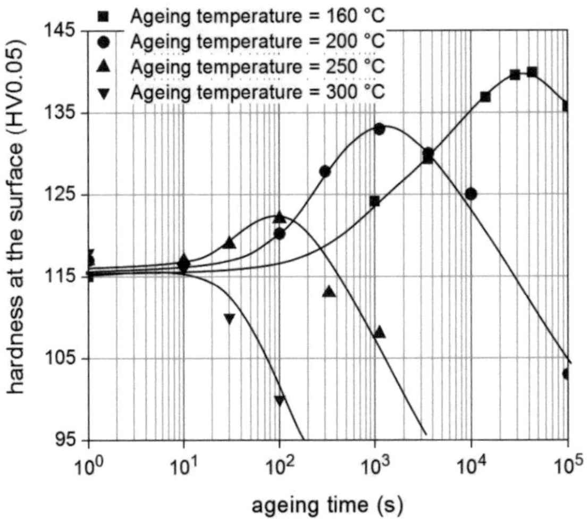

Figure 1. The hardness values at the surface of deep rolled as-quenched AA6110 during different ageing treatments.

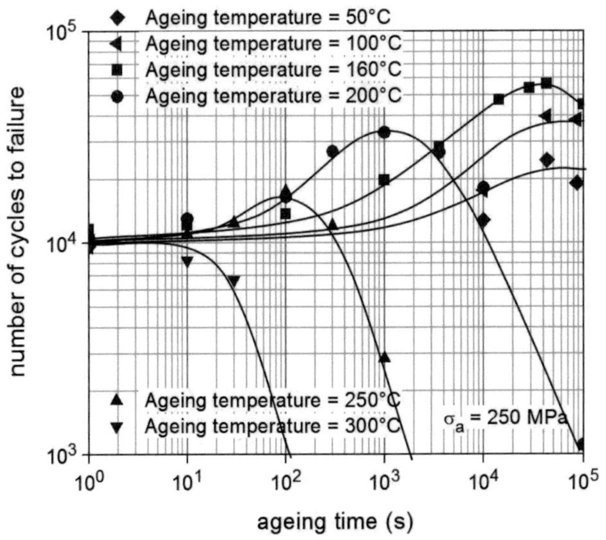

Figure 2. Fatigue lifetime as a function of ageing time and –temperature for an applied stress amplitude of 250 MPa

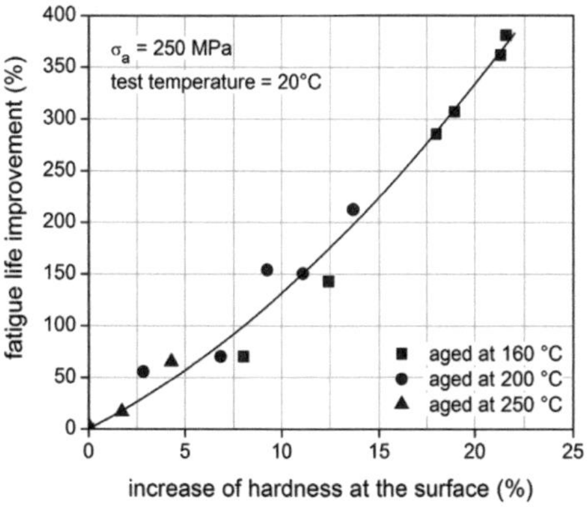

Figure 3. Fatigue life improvement as a function of hardness increase at the surface of deep rolled as-quenched AA6110 after ageing treatments at 160-250 °C.

200 and 250 °C for ageing times of approximately 12 hr, 1,000 and 100 seconds, respectively. For an ageing temperature of 300 °C, increased hardness values at the surface were not observed. The precipitated phases, '' as well as Q' lead to the increased hardness of copper-containing Al-Mg-Si aluminium alloys [1,2]. The maximum hardness of the deep rolled as-quenched AA6110 can be found after an ageing treatment at a temperature of 160 °C and an ageing time of 12 hr. For prolonged ageing treatments in the temperature range of 160-250 °C, the hardness values at the surface of deep rolled as-quenched specimens declined after having reached the peak hardness. The formation of coarse, semi-coherent, ' and/or Q' as well as incoherent precipitates, and/or Q at the surface as well as in near-surface regions is the reason for this observation. Conversely, for an ageing treatment at a temperature of 300 °C with a short ageing time, an increase of hardness was not observed. It might be due to the dominant recrystallization process taking place before occurring precipitation process [16].

Fatigue tests were also performed to evaluate the optimized fatigue lifetime. For this purpose, the relation between ageing time, temperature and fatigue lifetime was established at an applied stress amplitude of 250 MPa as shown in Fig. 2. After ageing treatments in the temperature range 160-250 °C, the fatigue lifetimes increased continuously with increasing ageing time until reaching a maximum lifetime. The optimized fatigue lifetime of the deep rolled as-quenched AA6110 was found after an ageing treatment at a temperature of 160 °C for about 12 hr. As expected, the increase of near-surface hardness after the ageing treatments resulted in an enhancement of the fatigue lifetime. Fig. 1 and Fig. 2 show that there is a clear correlation between hardness values and fatigue lifetimes of the deep rolled as-quenched AA6110 after ageing treatments. To clarify this correlation, a diagram of fatigue life improvement versus increase of hardness at the surface was constructed as shown in Fig. 3 using experimental data from Fig. 1 and Fig. 2. The residual stresses and FWHM-values decreased due to the relaxation as well as recovery processes. Residual stress and FWHM-values at the surface were reduced from -265 to -100 MPa and from 2.3° to 2.0°, respectively.

Because residual stress relaxation immediately at the surface is stronger than in subsurface layers, after ageing, a subsurface compressive residual stress maximum is formed as shown in Fig. 4. Non-statistically evaluated s/n-curves of the deep rolled as-quenched state after the optimized ageing treatment are shown as compared to the deep rolled as-quenched conditions in Fig. 5. At room temperature, the ageing treatment after mechanical surface treatment can enhance the fatigue lifetime especially in low cycle fatigue regime. However, in the high cycle fatigue regime, the difference in fatigue lifetimes was almost negligible. For elevated test temperature, the optimized-aged deep rolled condition exhibits also greater fatigue lifetimes as compared to the deep rolled as-quenched condition. Fatigue lifetimes of the deep rolled as-quenched condition were improved especially in the low cycle fatigue regime after the optimized ageing treatment (see Fig. 5) due to increased hardness values at the surface and in near-surface regions as well as in the bulk. However, the induced macroscopic compressive residual stresses as well as work hardening states seem to be essential in the high cycle fatigue regime, where relatively low stress amplitudes were applied and mechanical relaxation was not significant during cyclic loading at room temperature.

3.2. Comparison of conventional and modified deep rolling

The most important and interesting issue of the modified mechanical surface treatment (deep rolling followed by optimized ageing treatment) is the comparison with the conventional mechanical surface treatment (optimized ageing followed by deep rolling). First of all, important information of surface properties of the optimized-aged deep rolled as-quenched and deep rolled optimized condition is shown in Fig. 6 illustrating residual stress- and FWHM-depth-profiles of deep rolled as-quenched AA6110 after the optimized ageing treatment as compared to deep rolled optimized-aged AA6110. Hardness values of 137 and 161 HV are measured at the surface of deep rolled as-quenched AA6110 after the optimized ageing treatment and deep rolled optimized-aged AA6110, respectively. Noticeably, all important properties which are beneficial effects for fatigue lifetime enhancement, such as hardness, macroscopic compressive residual stress as well as the work hardening state of the optimized-aged deep rolled as-quenched condition are significantly less than of the deep rolled optimized/peak-aged condition. The greater near-surface macroscopic compressive residual stresses, FWHM-values as well as hardness values of the deep rolled optimized condition indicate that the deep rolling after an optimized ageing treatment results in an excellent combination of work- and precipitation hardening and thus excellent fatigue lifetime could be expected for the deep rolled optimized/peak-aged AA6110. To confirm this assumption, s/n-curves of the optimized-aged deep rolled as-quenched condition were plotted and compared to the deep rolled optimized/peak-aged condition in one diagram in Fig. 7. Obviously, fatigue lifetimes as well as -strength of the deep rolled optimized/peak-aged condition are superior to the optimized-aged deep rolled as-quenched condition. It can be concluded that both hardening effects, work- and precipitation hardening are required to yield the best fatigue lifetimes of AA6110. As known, the deep rolling treatment serves principally to induce near-surface work hardening and macroscopic compressive residual stress. For the ageing treatment after deep rolling, unfortunately, the work hardening and macroscopic compressive residual stresses were partially annealed out rapidly during the ageing treatment due to the relaxation process (mainly self diffusion). Therefore it can reasonably be assumed that fatigue lifetimes of the optimized-aged deep rolled as-quenched AA6110 were governed by the precipitation hardening and residually effective work hardening as well as compressive residual stresses at the surface and in near-surface regions. Therefore, the fatigue lifetime of the deep rolled as-quenched AA6110 followed by the optimized ageing treatment is better than of the deep rolled as-quenched AA6110 in the low cycle fatigue regime (see Fig. 5). However, the deep rolling after a suitable/optimized ageing treatment can completely combine the work- and precipitation hardening in near-surface regions of AA6110 into an optimized microstructure and thus result in the best surface properties and fatigue lifetime of the investigated AA6110 (see Fig. 7).

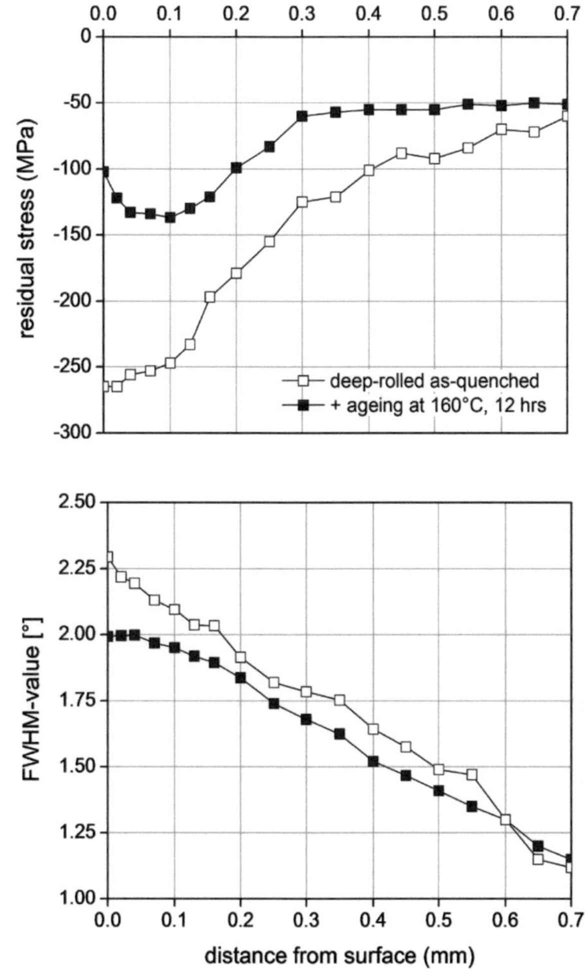

Figure 4. Residual stress- and FWHM-depth-profiles of deep rolled as-quenched AA6110 before and after the optimized ageing treatment.

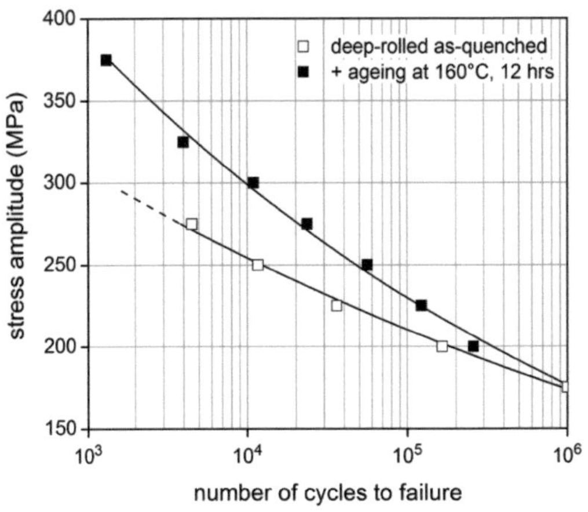

Figure 5. Non-statistically evaluated s/n curves of deep rolled as-quenched specimens before and after the optimized ageing treatment.

3.3. Deep rolling at elevated temperature

Thermomechanical surface treatment, the high-temperature deep rolling, has been successfully investigated for SAE 1045 as well as AISI 304 [20-24]. The fatigue behavior of these steels can be enhanced considerably due to static/dynamic strain ageing and together with very fine carbides at the surface and in near-surface regions [20-24]. Nevertheless, for aluminium alloys, it is still doubtful whether thermomechanical surface treatments can enhance the fatigue behavior more significantly than conventional mechanical surface treatments because aluminium alloys have mainly substitutional solute atoms. Thus, the fully beneficial effects of static/dynamic strain ageing can not be expected. However, static/dynamic precipitation during mechanical surface treatment at elevated temperature may contribute to mechanical properties of the surface as well as the bulk particularly for the as-quenched condition. Therefore, high-temperature deep rolling on the as-quenched aluminium alloy AA6110 was investigated. As-quenched specimens were deep rolled at different elevated temperatures of 160, 200 and 250 °C. Afterwards, near-surface properties and fatigue behavior were investigated and presented in this section. After deep rolling at elevated temperatures, near-surface residual stress, work hardening state and hardness-depth-profiles were measured as compared to the room-temperature deep rolled state as shown in Fig. 8 and Fig. 9. Obviously, macroscopic compressive residual stresses tend to decrease with increasing deep rolling temperature. Maximum macroscopic compressive residual stresses of -181, -152 and -59 MPa were measured at a depth of 20 μm after deep rolling at temperatures of 160, 200 and 250 °C, respectively. In contrast, after deep rolling at room temperature, a maximum macroscopic compressive residual stress value of -286 MPa was measured directly at the surface (see Fig. 6). After deep rolling at a temperature of 160 °C an approximately FWHM-value of 2.3° was measured which was to identical to the one observed after room-temperature deep rolling. However, FWHM-values tend to decrease at high temperature with increasing deep rolling temperature. The

FWHM-values about 2.1 and 1.6° were detected after deep rolling at temperatures of 200 and 250 °C, respectively. In addition, the case depth of work hardening after deep rolling at elevated temperatures seem to be greater than after room-temperature deep rolling. Near-surface hardness values, however, increased with increasing deep rolling temperature up to 200 °C as compared to deep rolling at room temperature. Hardness values in a depth of 25 μm of about 125 and 134.5 HV were measured after deep rolling at temperatures of 160 and 200 °C, respectively, whereas after deep rolling at room temperature, the hardness in a depth of 25 μm was approximately 113 HV. Conversely, after deep rolling at a temperature of 250 °C, a hardness in a depth of 25 μm of only about 104 HV was observed. As expected, due to occurring static/dynamic precipitation during deep rolling at elevated temperatures, near-surface hardness values after high-temperature deep rolling (160-200 °C) increased as compared to the room-temperature deep rolled as-quenched AA6110 (see Fig. 9). On the other hand, lower macroscopic compressive residual stresses and work hardening states

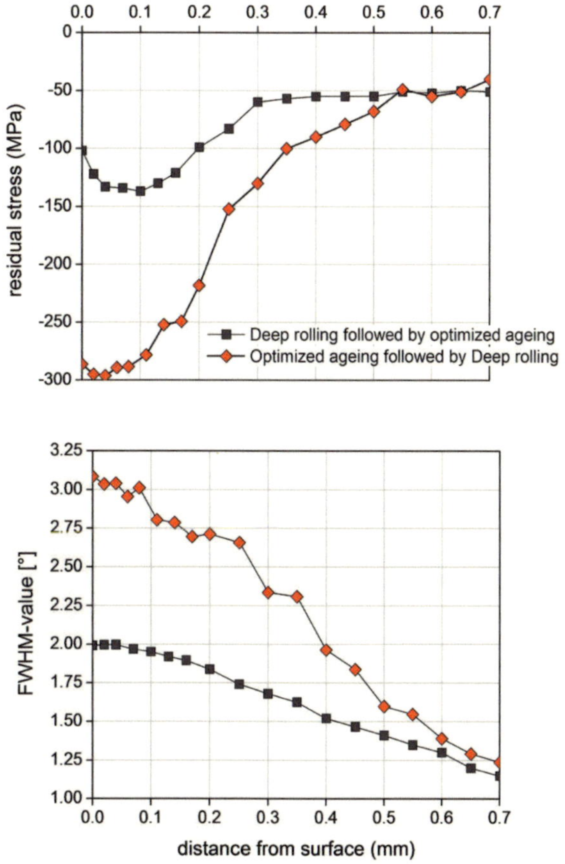

Figure 6. Residual stress- and FWHM-depth-profiles of deep rolled as-quenched AA6110 after the optimized ageing treatment as compared to deep rolled optimized-aged AA6110.

were measured as compared to the room-temperature deep rolled as-quenched condition (see Fig. 8) because static/dynamic recovery processes, which bring about relaxation phenomena, took place during deep rolling at elevated temperatures. The deep rolling treatment at a temperature of 250 °C produced detrimental effects on the near-surface properties, i.e. near-surface macroscopic compressive residual stresses, work hardening states and hardness values are considerably lower than of the room-temperature deep rolled as-quenched condition. That might be due to the fact that this temperature too high for the aluminium alloy AA6110 and leads to serve over-ageing effects and a high-rate static/dynamic recovery for this situation.

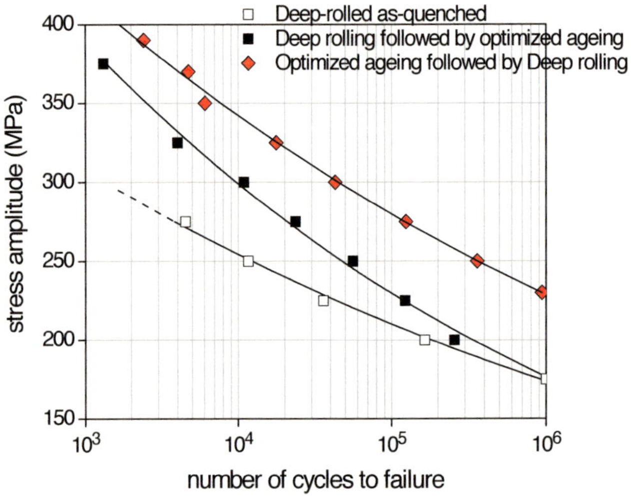

Figure 7. Non-statistically evaluated s/n curves of optimized-aged deep rolled as-quenched AA6110 as compared to deep rolled optimized-aged AA6110 as well as deep rolled as-quenched AA6110.

Non-statistically evaluated s/n-curves of the differently high-temperature deep rolled as-quenched condition at room temperature are presented as compared to the room-temperature deep rolled as-quenched condition in Fig. 10. The difference fatigue lifetimes for deep rolling treatments at temperatures between room temperature and 200 °C were insignificant. In the low cycle fatigue regime, fatigue lifetimes of as-quenched specimens deep rolled at a temperature of 200 °C seem to be slightly better than of the room-temperature deep rolled as-quenched condition, however in the high cycle fatigue regime, a contrary behavior was seen. The improvement of fatigue lifetimes at room temperature of the high-temperature deep rolled as-quenched AA6110 is not obvious as compared to the room-temperature deep rolled as-quenched AA6110. However, increased near-surface hardness values after deep rolling at a temperature of 200 °C slightly enhance fatigue lifetimes at room temperature in the low cycle fatigue regime. On the other hand, in the high cycle fatigue regime, the specimens deep rolled at a temperature of 200 °C show slightly lower fatigue lifetimes as compared to the room-temperature deep rolled as-quenched condition (see Fig. 10). This can be attributed to the lower near-surface macroscopic compressive residual stresses as well as work hardening states after deep rolling at a temperature of 200 °C (see Fig. 8).

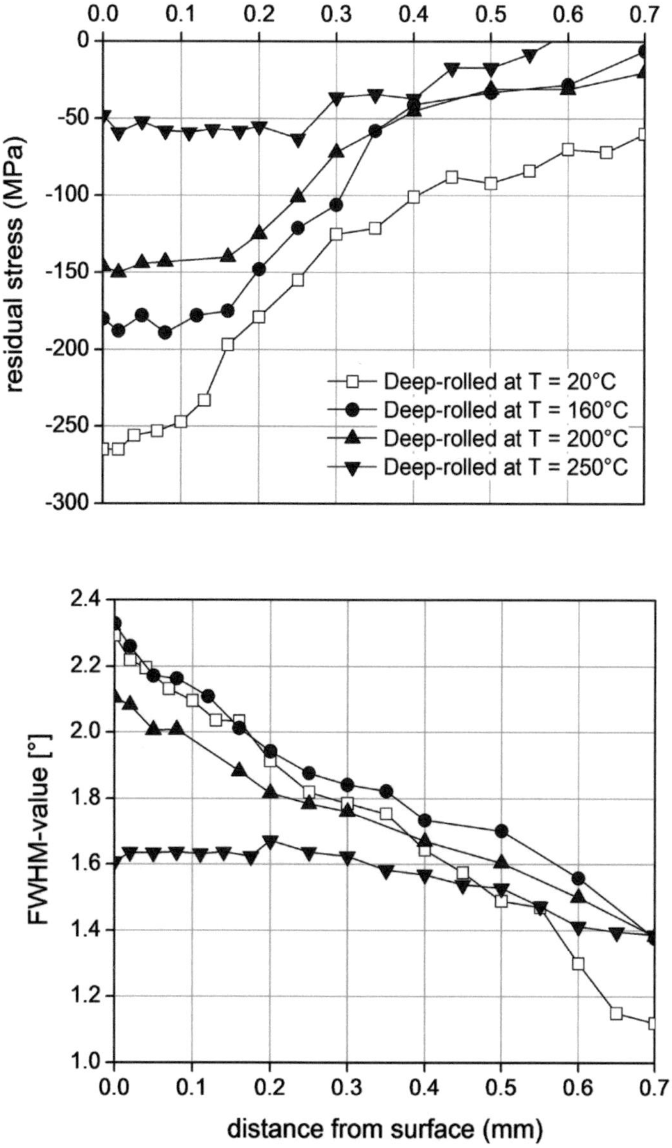

Figure 8. Depth-profiles of near-surface macroscopic compressive residual stresses and FWHM-values of high-temperature deep rolled as-quenched AA6110 for different deep rolling temperatures.

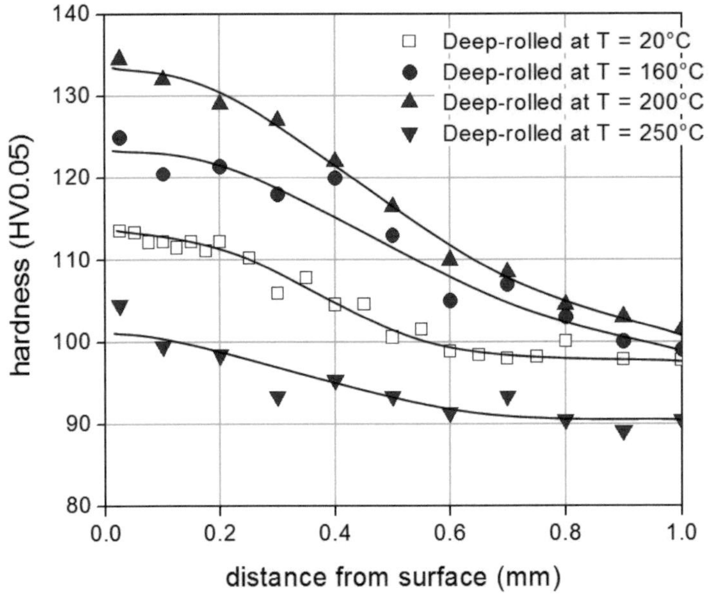

Figure 9. Depth-profiles of near-surface hardness values of high-temperature deep rolled as-quenched AA6110 for different deep rolling temperatures.

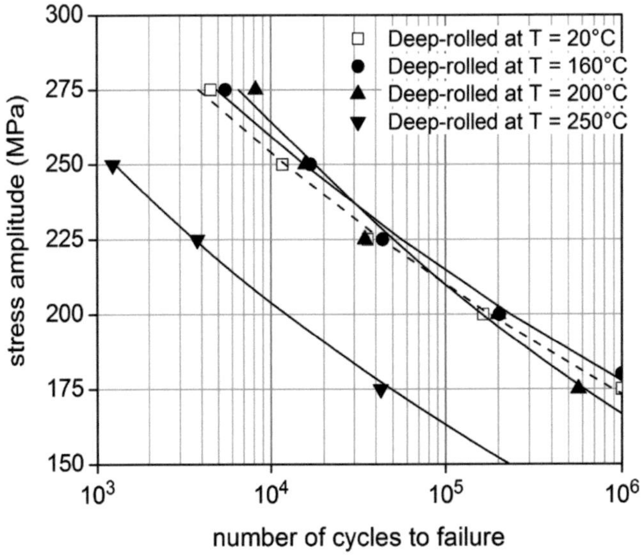

Figure 10. Non-statistically evaluated s/n-curves of high-temperature deep rolled as-quenched AA6110 for different deep rolling temperatures.

4. Conclusions

1. Deep rolling combined with an ageing treatment of as-quenched AA6110 resulted in an increase of near-surface hardness values. At the same time, a reduction of near-surface macroscopic compressive residual stresses and work hardening states takes place. The optimized ageing condition is an ageing temperature of 160 °C for about 12 hr.

2. Increased near-surface hardness values after the optimized ageing treatment seem to be very beneficial for the fatigue lifetimes in low cycle fatigue regime, whereas in the high cycle fatigue regime, macroscopic compressive residual stresses are additionally important.

3. Deep rolling at elevated temperatures up to approximately 200 °C resulted in an increase of near-surface hardness values in the as-quenched condition, whereas lower macroscopic compressive residual stresses and work hardening states were observed because static/dynamic recovery processes occurred. Deep rolling at a temperature of 250 °C exhibits detrimental effects on near-surface properties due to the over-ageing effects and high-rate static/dynamic recovery at this temperature.

4. An optimized ageing treatment followed by a conventional deep rolling treatment, can completely combine precipitation hardening and work hardening in near-surface regions of AA6110 into an optimized microstructure and residual stress state and thus result in the best surface properties and fatigue lifetime of the investigated AA6110.

Acknowledgements

The authors would like to express sincere thanks to the German Science Foundation (DFG) and to the Faculty of Engineering, Kasetsart University, Thailand, for financial support for Dr.-Ing. I. Altenberger and Dr.-Ing. P. Juijerm, respectively.

Author details

Patiphan Juijerm[1*] and Igor Altenberger[2]

*Address all correspondence to: juijerm@gmail.com

1 Materials Innovation Center, Department of Materials Engineering, Kasetsart University,, Thailand

2 WIELAND-WERKE AG, Central Laboratory, Research & Development,, Germany

References

[1] Polmear, I. J. (1996). Light Alloys: Metallurgy of the Light Metals. London Halsted Press.

[2] ASM. (1993). ASM Metal Handbook: 2 Properties and Selection: Nonferrous Alloys and Special-Purpose Materials. Ohio, ASM International.

[3] Niku-Lari, A. (1987). Advances in Surface Treatments. Oxford, Pergamon Press.

[4] Scholtes, B. (1997). Assessment of residual stresses. in: V. Hauk (Ed.). Structural and Residual Stress Analysis by Nondestructive Methods. Amsterdam Elsevier

[5] Altenberger, I., Nalla, R. K., Sano, Y., Wagner, L., & Ritchie, R. O. (2012). On the effect of deep-rolling and laser-peening on the stress-controlled low- and high-cycle fatigue behavior of Ti-6Al-4V at elevated temperatures up to 550 °C. Int. J. Fatigue; in press.

[6] Altenberger, I., & Scholtes, B. (1999). Improvement of fatigue behaviour of mechanically surface treated materials by annealing. *Scrip Mater*, 41(8), 873-881.

[7] Menig, R., Schulze, V., & Vöhringer, O. (2002). Effects of static aging on residual stress stability and alternating bending strength of shot peened AISI 4140. *Z Metallkd*, 93(7), 635-640.

[8] Juijerm, P., Noster, U., Altenberger, I., & Scholtes, B. (2004). *Mater. Sci. Eng.*, A379, 286.

[9] Juijerm, P., Altenberger, I., & Scholtes, B. (2006). *Mater. Sci. Eng*, A 426, 4.

[10] Juijerm, P., & Altenberger, I. (2006). *Scrip. Mater*, 55, 943.

[11] Juijerm, P., & Altenberger, I. (2006). *Scrip. Mater*, 55, 1111.

[12] Gregory, J. K., Müller, C., & Wagner, L. (1993). Bevorzugte Randschichtaushärtung: Neue Verfahren zur Verbesserung des Dauerschwingverhaltens mecha- nisch belasteter Bauteile,. *Metall*, 47, 915.

[13] Wagner, L., Berg, A., Dörr, T., & Hilpert, M. (2000). Kugelstrahlen und Festwalzen von Titan-, Auminium-, und Magnesiumlegierungen,. in: H. Wohlfahrt and P. Krull (Eds.), Mechanische Oberflächenbehandlungen Wiley-VCH Weinheim

[14] Berg, A., Kiese, J., & Wagner, L. (1998). Microstructural gradients in Ti-3Al-86Cr-4Zr -4Mo for excellent HCF strength and toughness. *Mater. Sci. Eng.*, A 243, 146 -149 .

[15] Gregory, J. K., & Wagner, L. (2003). Property improvement in light metals using shot peening,. in: L. Wagner (Ed.), Shot Peening, Wiley-VCH, Weinheim , 349.

[16] Hamphreys, F. J., & Haltherly, M. (1995). *Recrystallization and Related Annealing Phenomena,*, Pergamon, Oxford.

Chapter 6

Numerical Modelling in Designing Aluminium Extrusion

Wojciech Libura and Artur Rękas

Additional information is available at the end of the chapter

http://dx.doi.org/10.5772/51239

1. Introduction

Hot extrusion process is commonly used for producing many types of aluminium alloy profiles which are widely applied in various branches of industry, especially in vehicle construction and building. Extrusion process is characterized by very high deformation degree, although such deformations are usually non-uniform in product's both cross section and length. It results from specificity of metal flow in the container and lack of process conditions stability, especially including temperature of outflowing metal. Metal flow in extrusion process depends largely on die's geometry, especially if multi-hole or porthole die is used. Metal flow may be predicted through numerical modelling and using programs based on finite element method.

In this chapter, the procedure of designing dies for complex cases of aluminium alloys section extrusion is described. Firstly, general principles of designing dies for multi-hole and pocket extrusion is briefly described. Multi-hole dies enable increasing the process' efficiency and decreasing extrusion force, which has special significance when extruding hard-deformable alloys. Pocket dies are, in turn, commonly used for extruding solid sections with complex-shaped cross-section, as well as when applying economical method of extruding "billet – to – billet" method, which requires welding consecutive billets. Using numerical, and occasionally physical, modelling helps to significantly limit the range and costs of otherwise indispensable extrusion tests. The designing procedure is presented based on selected cases of extruding hollow sections from aluminium alloys. The first stage includes numerical modelling with the use of DEFORM program based on finite element method with the use of rigid-plastic model of the material. The software enables calculations of metal particle distribution in the container, but also distribution of strain, stress and temperature of extruded material. Analysis of different variants of the process helps determine optimum geometry of the die to guarantee obtaining good quality product. The second stage is testing the proposed dies in real-life conditions and, hence, it constitutes a form of validating correctness of the performed calculations.

2. Designing multi-hole dies

The presented work discusses cases of solid section extrusion, for which traditional flat dies are used. Multi-hole dies should be applied in hot extrusion process when deformation factor is to be limited. This, consequently, has profound impact on extrusion force. It is usually assumed that for soft aluminium alloys extreme extrusion ratio R, above which a multi-hole die has to be used, equals 40. For hard-deformable alloys, the ratio is significantly lower [1]. Excessive deformation factor results in increasing extrusion force, together with increasing velocity of outflowing metal. This may lead to forming fractures in hard-deformable alloys and make it necessary to decrease ram velocity. This, in turn, lowers extrusion effectiveness. Certainly, a multi-hole die should be used for extruding small-sized sections. Another case when it is required is extruding asymmetric sections, with considerably short diameter of circumscribed circle. Extruding such sections through single-hole die usually causes non-uniform metal flow and, thus, leads to geometrical instability of the product.

Figure 1. Entangled bars at press run-out, extruded through multi-hole die at non-uniform outflow velocity (Courtesy of Grupa Kęty).

Fig. 1 presents bars at press run-out, extruded through multi-hole die. Some of them are straight, but some are entangled as a result of non-uniform outflow velocity of individual bars from die holes. This, in turn, prevents a puller from being used at press run-out.

http://dx.doi.org/10.5772/51239

Designing a multi-hole die properly is a complex task, requiring focus on various factors. The criteria to consider include material aspects, shape of product, but also die durability issues. The most important quality factor in extrusion through multi-hole dies is to maintain uniform velocity of metal outflow at each hole. Only such conditions allow proper reception and running products on die run-out. Good quality is guaranteed only if the puller (running trolley) is able to keep extruded products parallel. Hence, when designing multi-hole dies, strongest attention should be paid to such distribution of holes which enables to maintain uniform conditions of metal flow in each hole.

Round small-sized sections may be extruded by dies with a very large number of holes. Depending on bar diameter, the holes may be distributed around one, two or three circles on the die surface (Fig. 2). Distances between circles should be such, that they guarantee uniform inflow of product in each hole. Holes are not placed in the die axis, because it may precipitate very unwanted non-uniformity of product outflow.

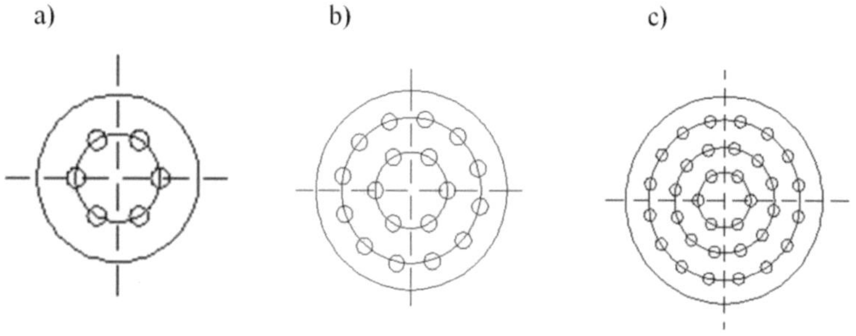

Figure 2. Distribution of round multi-hole dies, a) on one circle, b) on two circles, c) on three circles.

Sections which shape is other than round are usually extruded through holes placed on one circle on die surface [2].

Taking in to account the type of extruded material, one can state that practically every material could be subject to multi-hole die extrusion. The differences result from the number of holes used. Moreover, the type of extruded material influences the way it flows in container, that is, its geometry of its characteristic zones: main deformation zones (plastic zone), and dead zone. It is illustrated on Fig. 3 which schematically shows the area of main deformation zone and dead zones depending on number of holes in die. Fig 3a presents the size of main deformation zone and dead zone for hard-deformable material - both zones have significant volume, which results from high friction factor between the material and the tools. Significant changes were observed when a multi-hole die was used (Fig. 3b). There, main deformation zone seems to be divided into two smaller parts. An additional dead zone also appears between the holes. In this case, extrusion force will diminish as a result of both decreasing deformation degree as well as decreasing main deformation zone and dead zone.

By means of changing the holes' number and placement, one can influence the way material flows and, consequently, indirectly affect quality of product, characterized by its geometrical stability, structure and mechanical properties.

The third crucial factor for designing multi-hole dies is die durability. The distance between holes constitutes a very significant parameter. It should be as large as possible. Together with increasing the number of holes, die durability diminishes, although only seemingly. In fact, together with increasing the number of holes, extrusion forces significantly decrease, too. This means that the die load is high when the number of holes is small, whereas for larger number of holes, the load is lower. Another necessary condition is also symmetrical distribution of holes in die.

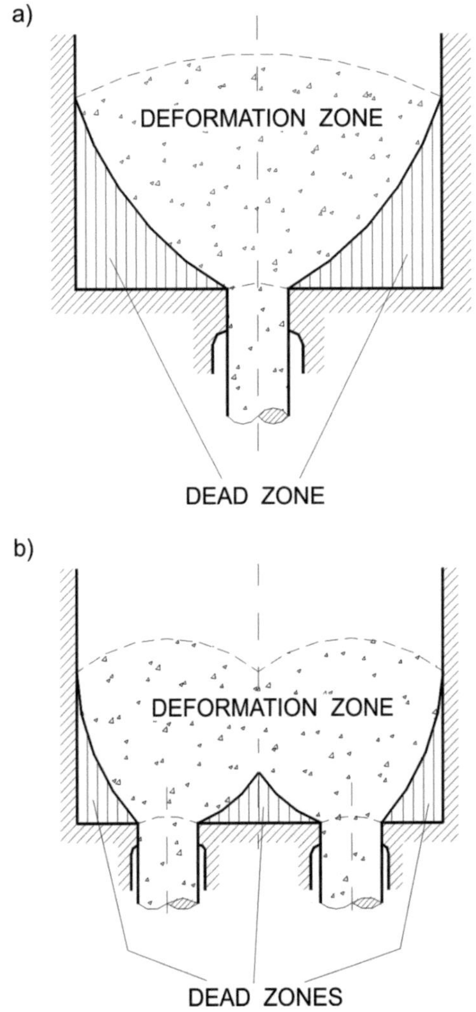

Figure 3. Flow character of hard alloy extruded through different dies, a) single-hole die, b) multi-hole die.

Supporting plates' structure also influences adjustments of distances between holes. They help to significantly limit die spring. Excessive die spring causes appearance of dangerous strain, which may end up destroying the die. Taking durability into consideration, when designing multi-hole dies one should also adapt the principle that holes laying on several circles are not distributed in one radial line. Appropriate hole distribution in such case is illustrated by Fig. 2 b and c.

3. Pocket dies

Pocket dies are used among others for obtaining products with highly differentiated section wall thickness, but also products with very small wall thickness [3]. They are also useful when applying "billet – to - billet" extrusion. In such case, there exists a need to weld two consecutive billets during process. Depending on pocket size, that is its size h (Fig. 4) and its width in relation to hole size, pocket may either facilitate or hinder metal flow into the hole, so as to obtain uniform metal particle velocity distribution in die hole. The works [3,4] present methodology of pocket shape selection for profiles with differentiated wall thickness based on local volume rates of metal flow criterion.

Volume rate of metal flow was defined as follows:

$$W = S \cdot v = const. \tag{1}$$

where, S - cross-section of given part of section, v - velocity of metal flowing into the given section part.

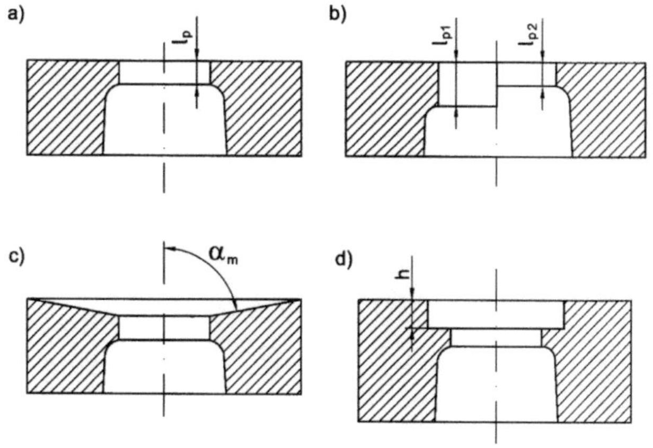

Figure 4. Schematic diagram of pocket die.

Pocket is also useful for multi-hole extrusion when one aims at equalizing velocities of metal flowing from holes distributed at different distance from die axis.

Currently, many problems connected with choosing extrusion die geometry are solved thanks to numerical modelling methods based on finite element method (FEM). There are many publications dealing with pocket die extrusion [3-9], but only a few works were devoted to application of numerical modelling for multi-hole die extrusion [10-12]. Calculation procedure using FEM, apart from die geometry, enables also strength parameters analysis, stress and strain factors, metal flow velocity, as well as temperature and time conditions.

The tests presented in this work concern various instances of extrusion in which die geometry was chosen considering numerical simulations in FEM-based program DEFORM™. The test procedure included:

- choosing multi-hole extrusion
- numerical modelling of different die variants
- experimental validation in industrial conditions.

For the purpose of the test, especially difficult instances were selected in which problems with quality of products were identified, as well as other technical difficulties. Materials used for the tests were medium and hard-deformable alloys. The data describing materials' rheology between 300 and 500°C and steel tool AISI H13 material characteristics was defined based on program DEFORM™'s material database.

4. Indirect extrusion through four-hole dies

Modelling indirect hot extrusion process of full AlMg1SiMn (6061) products with diameter 28 mm through four-hole dies aimed at analysing the influence of holes' distribution (Fig. 5) on metal flow and distribution of outflow velocity in die orifice. Inappropriate distribution of holes in the die results in unstable metal flow, and, as a result, extruded bars lean towards each other or away from each other. This hinders their movement at press run-out.

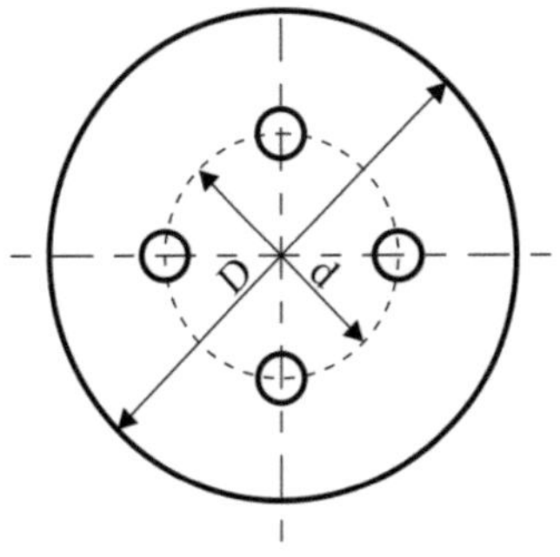

Figure 5. Schematic diagram of four-hole die.

For each analysed variant, a constant billet temperature was assumed at T_b = 340°C as well as constant tool temperature at T_t = 340°C. Calculations were performed at constant exit velocity v_e = 8 m/min, constant billet diameter of D_b = 246 mm and length L = 770 mm, assuming friction factor at m = 0,7. Extrusion ratio equalled 20,2.

At the beginning, calculations of indirect extrusion for flat dies were performed, analysing the influence of holes' distribution on quality of obtained products. Holes' distribution was determined taking into account a *k* indicator, defining relations between die surface area A_D and surface area of circle circumscribed on holes' axis, A_d, according to (2) (Fig. 5):

$$k = \frac{A_D - A_d}{A_d} = \frac{D^2 - d^2}{d^2} \quad (2)$$

where:

k – indicator defining location of holes in the die

D - diameter of die

d - diameter of circle circumscribed on holes' axis.

The change of diameter of circle circumscribed on holes' axis was determined according to (3):

$$d = \sqrt{\frac{D}{k+1}} \quad (3)$$

Indications in equation (3) as above.

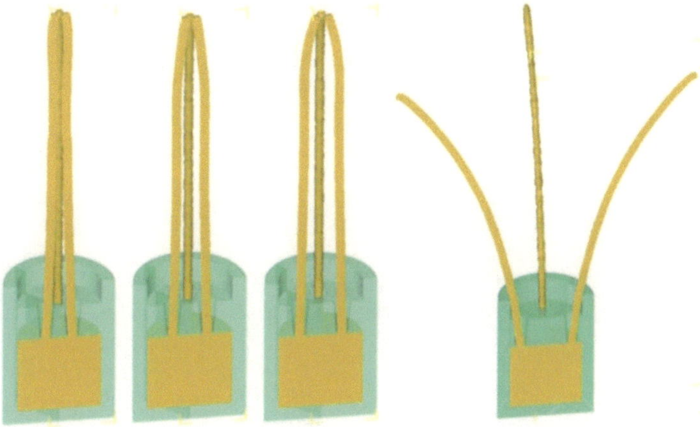

a) k = 9 b) k = 5 c) k = 3 d) k = 1

Figure 6. Metal flow at various holes' distribution in a flat die.

Numerical calculations for indirect extrusion through four-hole dies were carried out for four different hole distribution indicators, *k* = 1, 3, 5, 9. Results of modelling made it possible to observe product's reactions during extrusion process. For each *k* indicator value, diameter of the circle on which die holes are placed, changes. Low *k* value means high diameter, and conversely. Change in d diameter influences metal flow in deformation zone. Optimizing hole placement should result in uniform metal inflow from outer layer and from billet's inside so as to ensure that rod exits die holes parallel to extrusion axis. For the assumed hole distribution indicators, satisfactory geometrical stability of the product was not obtained (Fig. 6). In the case of extrusion with indicators $k = 3, 5, 9$ (Fig. 6a, b, c), bending the product towards die axis is observed, until they touch. Extrusion with indicator $k = 9$ causes bending at the very beginning of extrusion process and non-axial metal exit (Fig 6a). Distribution of holes in the die with indicators $k = 3$ and 5 initially causes a similar effect to indicator $k = 9$, and then, during extrusion process, product axis approximates die holes' axis. However, during extrusion with holes' distribution for indicator $k = 1$ (Fig. 6d), strong outwards bending of product appears.

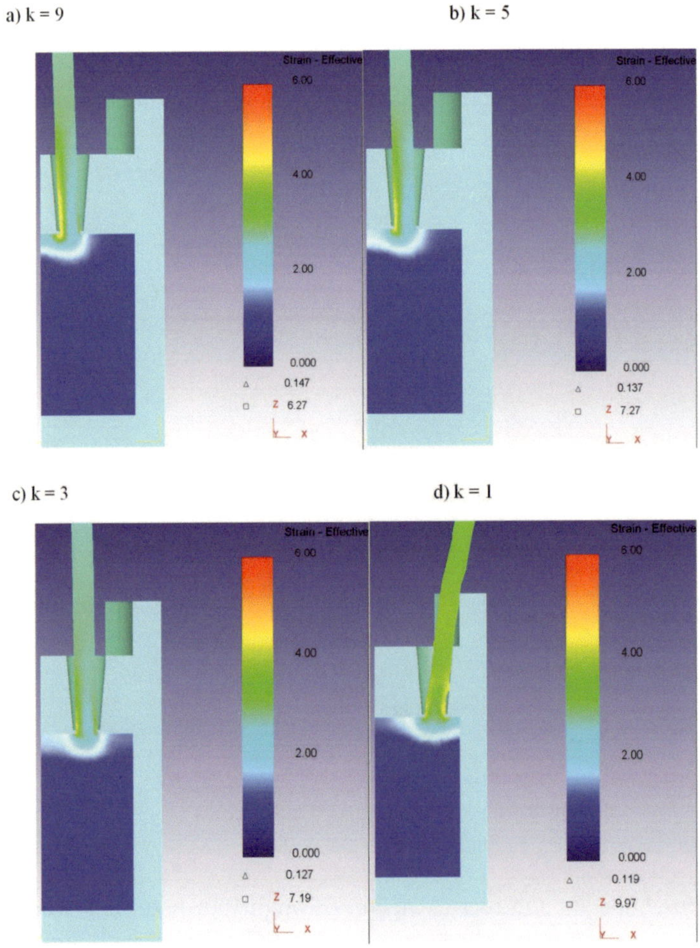

Figure 7. Distribution of effective strain for different hole's distribution indicators *k*.

Fig. 7 presents of effective strain for different indicators k. Diversification of effective strain is visible in cross-sections of extruded bars, which leads to their deformation after exiting a die hole.

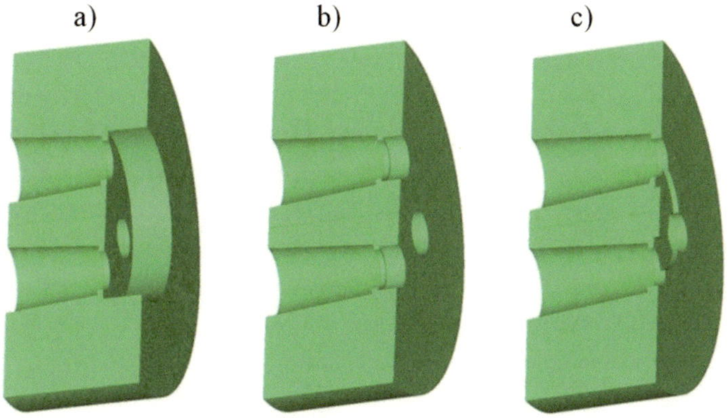

Figure 8. Four-hole pocket dies models for numerical calculations, a) central pocket die, b) local pocket die, c) ring-shaped pocket die.

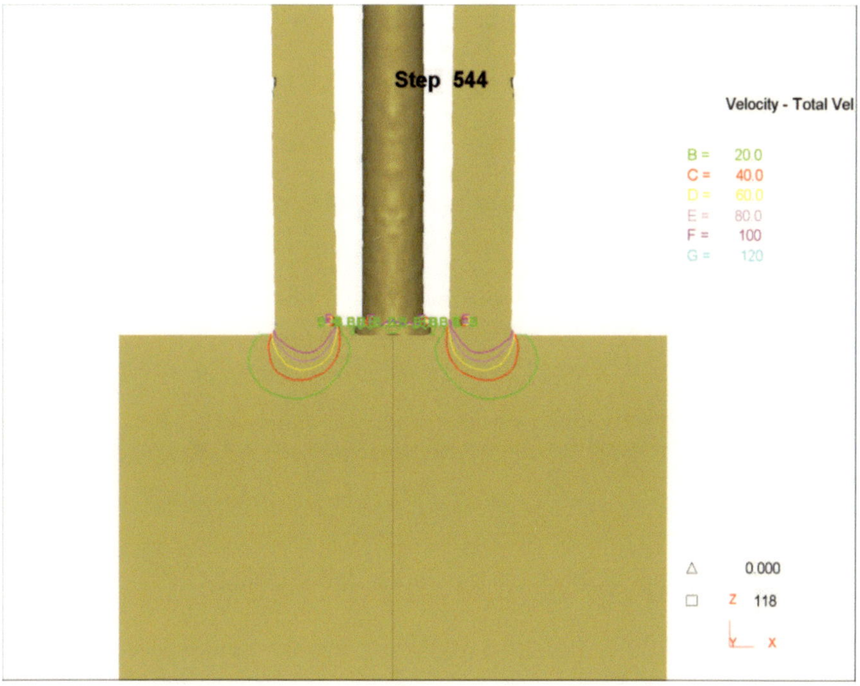

Figure 9. Distribution of metal particle exit velocity upon extrusion through ring-shaped pocket die.

Calculations have shown that, theoretically, in order to achieve the most beneficial extrusion, indicator k should equal $2 \div 3$. Unfortunately, die set structure of press does not provide for such wide setting of die holes. In this situation, calculations were carried out for several pocket dies with various geometry (Fig. 8a, b, c). Depending on its geometry, a pocket may be useful at controlling metal flow into individual die holes. In such case, pocket helps to regulate intensity of metal inflow into die holes. Central pocket (fig. 8a) intensifies metal flow in central part of ingot, whereas local pockets, i.e. individual for each hole, are supposed to reduce influence of non-uniformity of flowing in deformation zone. The ring-shaped pocket (fig. 8c) combines both abovementioned reactions. The final goal is to obtain uniform velocity distribution in die holes. For each of the analysed extrusion variants, a constant value of holes' distribution indicator was assumed at $k = 9$. It turned out that the optimum die featured a ring-shaped pocket with height $h = 2$ mm (Fig. 8c). Fig. 9 presents distribution of particle velocity upon extrusion through segmented pocket die, and Fig. 10 shows intensive strain distribution. Both distribution of particle velocity and distribution of intensive strain display satisfactory uniformity which helped design such die for industrial tests.

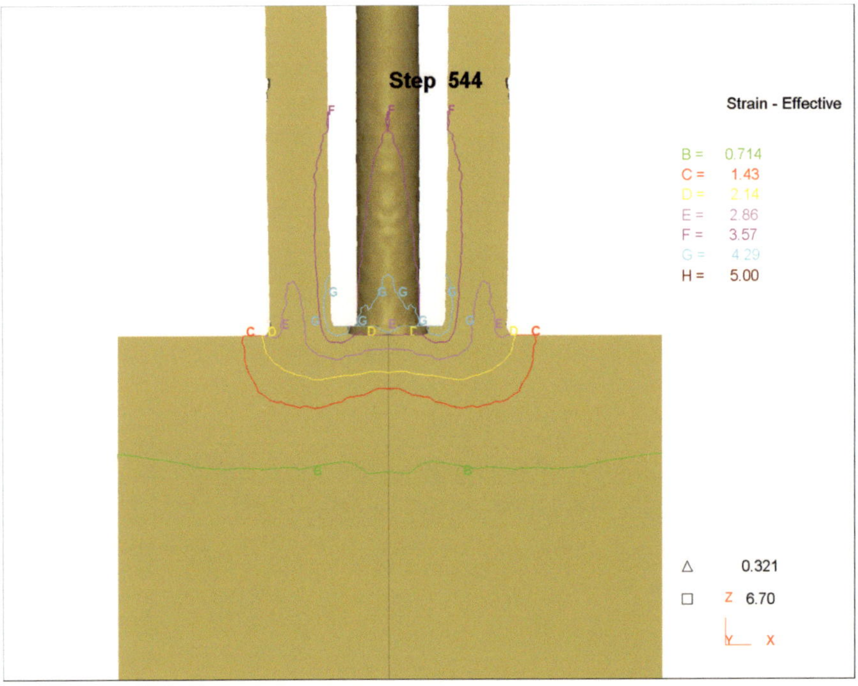

Figure 10. Distribution of intensive strain upon extrusion through ring-shaped pocket die.

Figure 11 presents a view of ring-shaped pocket die, designed for experimental tests. Extrusion tests of round 28 mm diameter 6082 aluminium bars through four-hole dies were performed on indirect press with nominal force of 28 MN. The press is a modern solution, featuring so called water wave designed for solutioning of extruded products directly at press run-out. Test parameters were identical as in numerical modelling, with the exception of temperature, as because of necessity to maintain isothermal extrusion conditions and product solutioning at press run-out, taper billet heating was applied 500/510/520/530°C, and container temperature was 500°C.

Upon extrusion through flat die, bars were bent towards extrusion axis, which, consequently, led to bars' joining at press run-out. Extrusion through ring-shaped pocket die caused expected effects, that is, good quality product together with maintaining shape stability. Fig. 11b features a photograph from the test performed, to show how product exits the die parallel to its axis from the start of the process.

a) b)

Figure 11. Extrusion tests in industrial conditions through four-hole die, view of ring-shaped pocket die, b) view of extruded bars at press run-out.

5. Indirect extrusion through 17-hole dies

In this case, the subject of analysis was indirect extrusion process of 9,5 mm diameter AlZn6MgCu (7010) alloy bars. This alloy is characterized by high strength, but also hard-deformability, that is why practically used extrusion velocity does not exceed 1,5 m/min. Based on observations of real process with the use of 17-hole flat die, it was stated that initially, middle bars exit die at higher velocity than bars located around circle edge. During the process, velocities are equalized, and later outer bars begin to exit much quicker than ones in the middle. As a result, high differentiation of inner and outer bars' exit velocity is observed at particular stages of extrusion.

In this case, numerical modelling procedure was applied for different die variants: a flat die and two pocket dies.A pocket, as it has been mentioned in chapter 2, may accelerate or decelerate metal flow into die holes in specified place. Pocket dimensions are also important. A narrow and tall pocket decelerates metal flow before die hole, whereas a wide and low one causes volume of metal entering die hole to increase. The first variant featured narrow pre-chambers used on outer holes in order to decelerate metal exit velocity from those holes. With such thin bars, pockets should not be too deep, as it hinders removing bars from die holes. In the second case, a wide ring-shaped pocket was used. It was 3 mm deep and was placed on inner holes (Fig. 12). Such a pocket facilitated metal flow at entering inner holes, which was supposed to result in equalizing velocities between inner and outer bars. In this case, experiences gained when analyzing four-hole die where taken into consideration. For each analysed variant, constant billet temperature was assumed at T_b = 350°C and constant tool temperature at T_t = 340°C. Calculations were performed for extrusion velocity at 2 mm/s (exit velocity v_e = 1,2 m/min). Billet diameter was 252 mm, length - 300 mm, extrusion ratio R = 41 and friction factor on the contact surface of billet and tools m = 0,7.

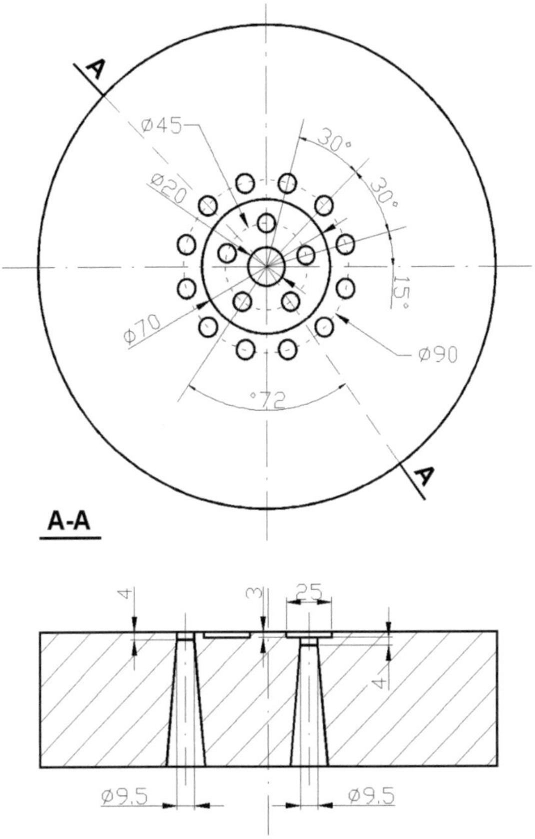

Figure 12. Schematic diagram of 17-hole die with ring-shaped pocket, dimensions in mm.

With the use of flat die, at the very beginning of the process, bars from inner holes exited slower than from the outer ones (Fig 13).

The aim of extrusion through 17-hole die with local pockets applied on outer holes was to hinder metal velocity within outer layer. However, the influence of local pockets was so strong, that in practice extrusion took place from the inner holes.A ring-shaped pocket die turned out to be a much better solution, though comparable with flat die, as upon its use, uniform metal outflow velocity was obtained from each die hole (Fig. 14).

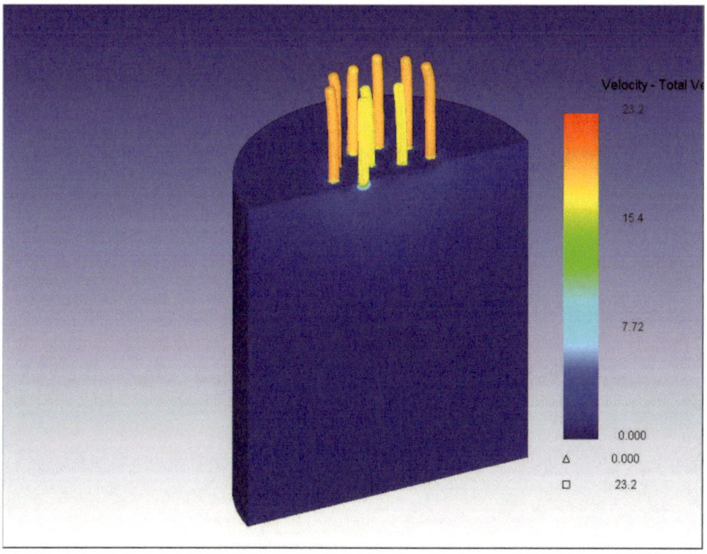

Figure 13. Distribution of metal particles exit velocity during extrusion through 17-hole flat die.

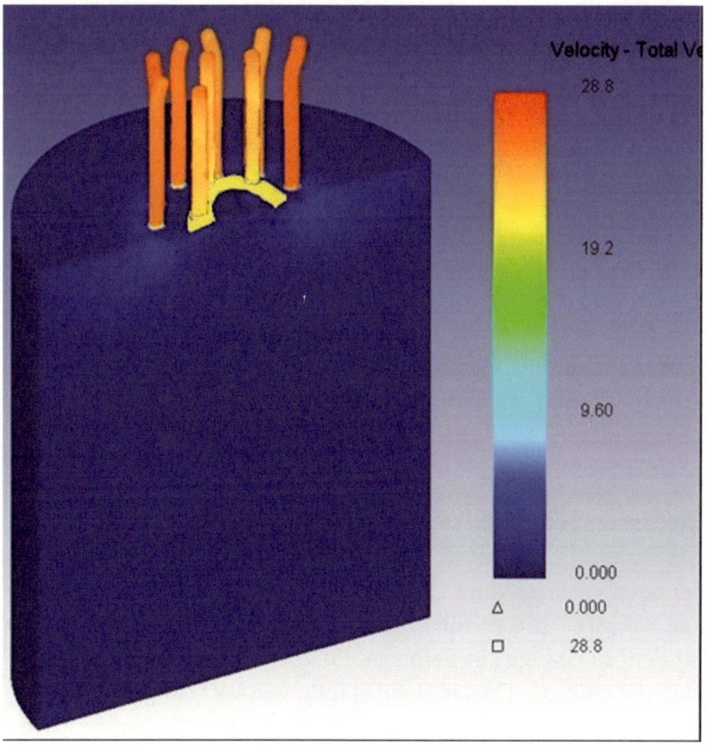

Figure 14. Distribution of metal particles' outflow velocity upon indirect extrusion through ring-shaped pocket die on outer holes.

Based on numerical calculations, two variants of 17-hole pocket die were designed. In the first variant, outer holes were equipped with pockets, whereas in the second one, a ring-shaped pocket was applied on inner holes. (Fig. 15).

Figure 15. Schematic diagram of 17-hole die with ring-shaped pocket.

Tests were performed with following basic parameters:

- billet diameter 246 mm,
- billet length 770 mm,
- container diameter 252 mm,
- billet temperature - taper heating 350/360/370°C
- container temperature 340°C
- ram velocity $v_0 = 0,5$ mm/s.

The extrusion tests were performed on a 28MN capacity indirect press. Positive results were achieved with the use of 17-hole die with ring-shaped pocket encompassing all inner die holes. 7010 alloy which was used for the test is a hard-deformable material which is extruded with very low velocity. As a result, one billet extrusion time is about 20 minutes. In such conditions, temperature changes in deformation zone which result in change of metal flow mechanics during the process. This, in turn, is followed by change in velocity of metal's exiting both inner and outer holes. According to flow mechanics during extrusion, inner bars move initially faster, but soon outer bars accelerate significantly. It happens so in case of traditional flat dies. Using ring-shaped pocket on inner holes proved to be a good solution, as it caused permanent acceleration of metal exiting inner holes, which lasted until the end of the process. Fig. 16 presents bars extruded through 17-hole die with ring-shaped pocket, initial phase (a) and advanced state when the use of puller was possible (b). The difference in total length of bars from inner and outer holes was about 2 m, which constituted about 7% of their length. Such results are acceptable from the point of view of effectiveness (small material loss). It is important, however, that in such conditions one can safely use a puller so that no rippling occurs on press run-out (Fig. 16b).

a) b)

Figure 16. Extrusion of bars through a 17-hole die with ring-shaped pocket,view of bars after exiting the die, b) guiding puller (courtesy of Grupa Kety).

In case of extrusion through a 17-hole die with local pockets placed on outer layer holes, the influence of pockets was so effective that metal exited the die practically through inner holesonly. It needs to be added that such result of the test turned out appropriate for indirect extrusion. In case of direct extrusion, when metal flow is significantly different and unfavourable, optimum die geometry solution might be different.

6. Direct extrusion of asymmetric S-type profile

Extruding asymmetric profiles is especially difficult because of complexity of metal flow which often leads to geometrical instability of the product in the form of twisting or bending, as well as its non-homogenous structure. In such case, either different lengths of die land are usually used or pocket dies. Such product is exemplified by AlMg4,5Mn (5083) alloy profile presented in Fig. 17.

This alloy belongs to hard-deformable materials extruded with low exit velocity. Extrusion of the profile presented in Fig. 17 in industrial conditions occurs on a flat die. Different lengths of die land were used there (5 and 12 mm) in order to regulate metal flow. Nevertheless, the hollow section was subjected to intense twisting upon extrusion. Torque was so strong that the product was torn from puller. Probably, the reason for such phenomenon was inappropriate metal flow velocity distribution in die hole. It was proposed to use a die with pocket of variable width. Reaction of metal in real conditions, as observed in this case, suggests that twisting is caused by significantly non-uniform metal inflow to die holes in the whole deformation zone. Even application of a two-hole die cannot considerably alter metal reaction in deformation zone, as it is impossible to precipitate symmetric flowing.

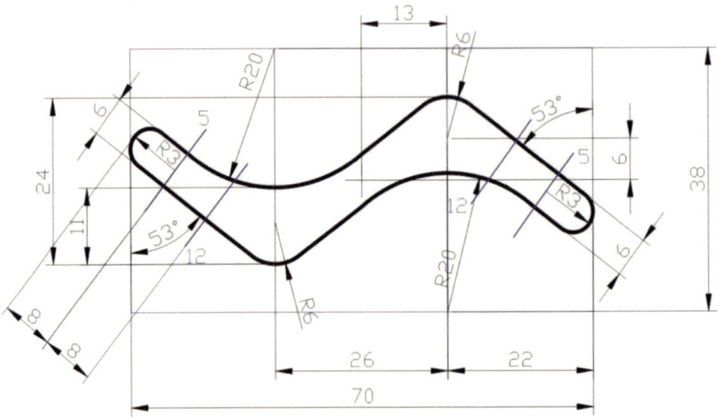

Figure 17. Asymmetric S-type AlMg4,5Mn extruded profile geometry.

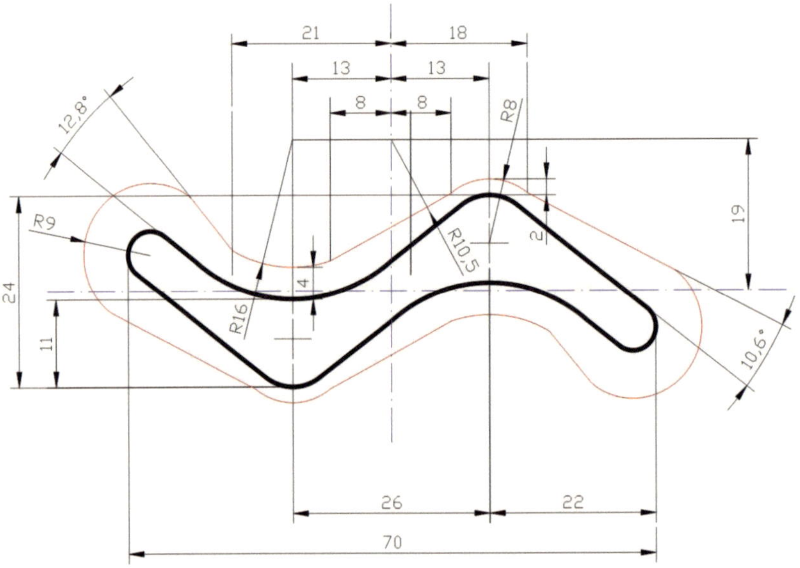

Figure 18. Design of pocket die for extruding asymmetric profiles.

Such reaction of metal may be corrected by means of applying a die with pocket of appropriate geometry. When designing the pocket die, the main parameter for choosing pocket dimensions was so called volume rates of metal flow criterion, as defined in equation (1). It

http://dx.doi.org/10.5772/51239

assumes that relation between velocities of metal flowing from container and entering individual parts of pocket corresponds to relationship between surface areas of cross-sections of extruded profile's fragments. In addition, pocket shape (Fig. 18) was corrected taking into account change in profile radius. Designing a die for such sections requires considerable experience as well as constructor's intuition. In the illustration, pocket shape was marked red. Depth of pocket was 8 mm.

In the first stage of tests, numerical modelling was performed for traditional flat die and pocket die. Fig. 19 presents distribution of metal velocity in die hole for both flat die and pocket die. One can notice strong diversification of velocity in flat die hole (Fig. 19a). Velocity at central part of the profile is almost three times higher than at its ends. As it may be seen, the use of die lands of much different length is not sufficient. For apocket die, diversification of metal flow velocity is relatively low (Fig. 19b). This confirms that the application of pocket is more efficient.

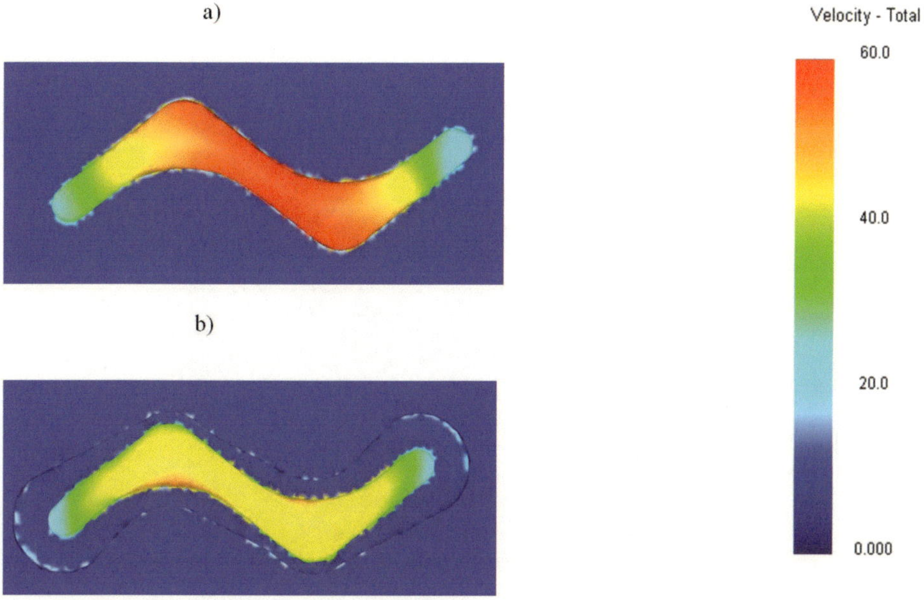

Figure 19. Distribution of metal exit velocity in the die orifice in case of, a) flat die, b) pocket die.

In order to establish how the obtained distribution of velocity influence section reaction after exiting die hole, visualization of this phenomenon was carried out in DEFORM program and was presented in Fig. 20. The images show twisting which practically occurs upon section extrusion (Fig. 20a). Pocket die makes it possible to minimize this unfavourable phenomenon, as, in this case, twisting is reduced to minimum.

In order to verify correctness of pocketdie design, profile extrusion tests were performed in industrial conditions. Tests of AlMg4,5 (5083) alloy profile extrusion were carried out on a direct press with nominal force of 25 MN with the use of designed pocket die. Extrusion conditions were as follows:

- billet diameter - 215 mm and length - 230 mm,
- extrusion ratio R = 58,
- billet temperature 470°C,
- exit velocity 3m/min.

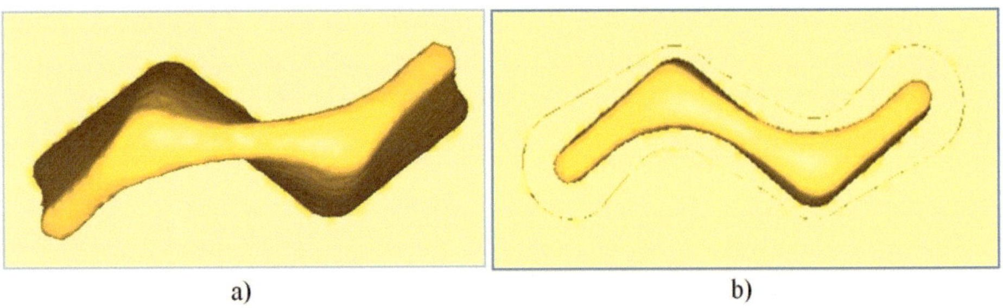

Figure 20. Visualization of twisting a profile extruded through, a) flat die and b) pocket die.

Figure 21. View of 5083 alloy profile at press run-out extruded through pocket die (Courtesy of Grupa Kety).

Fig. 21 presents a profile extruded through pocket die. Due to slight profile twisting, it was easily driven by puller at press run-out. Profile extruded through traditional flat die twisted several times during extrusion test. Thus, the results of industrial test confirmed the assumptions of numerical simulations.

Non-uniform metal flow velocity distribution in die hole usually results in non-uniform material structure on product's cross-section. In the case of hot extruded aluminum alloys, it is most often manifested by so called coarse grain layer. For this reason, extruded sections were subjected to metallographic tests in order to reveal their structure on cross-section. Macrostructure tests have shown differentiation of grain size on profile cross-section. More non-uniform macrostructure was present on cross-section of profiles extruded through flat die (Fig. 22a). In most part of profile, coarse grain layer covers the whole cross-section. In case of section extruded through pocket die, (Fig. 22b) coarse grain layer exists, but its thickness does not exceed 2mm. Still, this proves to be a more favorable structure than the one resulting from using flat dies, as inner cross-section is made of fine grain, which improves product strength. Such layer may be decreased by appropriate selection of extrusion conditions.

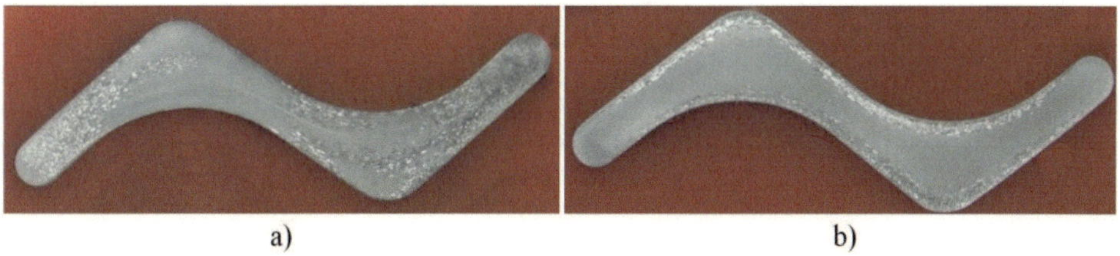

Figure 22. Macrostructure of cross-section of profile extruded through flat die (a) and pocket die (b), etched with Tucker's reagent.

7. Conclusion

The research method applied in the work was based on numerical modelling and facilitated designing multi-hole dies and pocket dies for extrusion of various aluminum alloys, especially hard-deformable alloys. Extrusion through multi-hole dies is an efficient way of improving extrusion process effectiveness. It has become key question to design multi-hole dies so that uniform metal flow velocity in each hole is guaranteed.

Analysis of three cases of extrusion helped identify variants displaying best expected conditions of real material flow. Particularly, in each analyzed case, application of pocket dies turned out to be the best solution. Upon extrusion through four-hole die, designed based on numerical simulations, the desired parallel flow of each 6082 alloy bar. It was original to use a ring-shaped pocket for 17-hole 7010 alloy extrusion. Similarly to calculations, almost identical bar flow velocity was observed from outer and inner holes, which enabled a puller to be used effectively on press run-out.

In an exceptionally difficult case, i.e. extruding asymmetric section from hard-deformable 5083 alloy, based on calculations, die variant with changeable width pocket was rightly selected. Its application helped reduce section twisting after exiting die holes. The results of the tests indicate that using pocket dies is a very good way of controlling metal flow, especially in case of asymmetric profiles. Another advantage of using such dies is that when applying them, one can avoid the difficulty of reconstruction the whole die set. Such necessity emerges when the optimum calculated hole distribution in the die would require increasing diameter of the die itself. However, this solution is possible, but quite costly and it requires recalculation of durability of individual die set elements. Die constructions designs have been validated in real conditions upon extrusion of selected profiles on an industrial press. Properly designed die also helps to avoid various defects of extruded products, related with their geometry and material structure.

Author details

Wojciech Libura* and Artur Rękas

*Address all correspondence to: libura@agh.edu.pl

AGH University of Science and Technology, Krakow, Poland

References

[1] Perlin, I. L. (1964). *Teoria pressovania metallov*, Metallurgia, Moskva.

[2] Laue, K., & Stenger, H. (1981). *Extrusion- Processes, Machinery*, Tooling. ASM Metals Park, Ohio.

[3] Leśniak, D., & Libura, W. (2007). Extrusion of sections with varying thickness through pocket dies. *Journal of Materials Processing Technology*, 194, 38-45.

[4] Leśniak, D., Libura, W., Podvysotskyy, V., & Milenin, A. (2002). Influence of pre-chamber die geometry on extrusion of solid section with different wall thickness. Kraków. *Proc. 5th Int. ESAFORM Conference.*

[5] Leśniak, D., Libura, W., Zasadziński, J., Buntoro Muller, K., & Fluhrer, J. (2004). Experimental and numerical investigations of aluminium extrusion through pocket dies. *Proc. 8th Int. Aluminum Extrusion Technology Seminar*, Orlando. USA, 191-198.

[6] Sinha, M. K., Deb, S., Das, R., & Dixit, U. S. (2009). Theoretical and experimental investigations on multi-hole extrusion process. *Materials and Design*, 30, 2386-2392.

[7] Duplancic, I., Mioc, M. Z., & Bracic, Z. (2000). Case studies on control of metal flow in pre-chamber dies. Chicago. *Proc. 7th Int. Aluminum Extrusion Technology Seminar*, II., 177-186.

[8] Prats, A. E., & Misiolek, W. Z. (1996). Analysis of Metal Flow in Weld Pocket Dies, Proc. 6th International Aluminum Extrusion Technology Seminar. vol. I. Chicago. ., 75-78.

[9] Nakanishi, K., Koba, H., & Kamitani, S. (1999). Metal flow control in hot extrusion of aluminum alloy using the pocket hole dies. Nürnberg. *Proc. 6th ICTP*, III., 1833-1838.

[10] Sheppard, T., & Peng, Z. (2004). Application of FEM to Modeling of Multi-hole Die Extrusion. *Proc. 8th Int. Aluminum Extrusion Technology Seminar*, I., 93-105.

[11] Peng, Z., & Sheppard, T. (2005). Effect of die pockets on multi-hole extrusion. *Mater. Sci. Eng. A*, 407(1-2), 89-97.

[12] Libura, W., Rękas, A., & Leśniak, D. (2010). Numerical Analysis of Four-Hole Extrusion of Aluminium Alloys. *Key Engineering Materials*, 424, 173-179.

Chapter 7

Linear Friction Based Processing Technologies for Aluminum Alloys: Surfacing, Stir Welding and Stir Channeling

Pedro Vilaça, João Gandra and Catarina Vidal

Additional information is available at the end of the chapter

http://dx.doi.org/10.5772/52026

1. Introduction

The friction based processing technologies encloses some of the most significant solid state manufacturing technologies for producing structural components in aluminium alloys. In the present chapter three processes will be depicted: i) Friction Surfacing (FS); ii) Friction Stir Welding (FSW) and iii) Friction Stir Channeling (FSC). These processes covers a large range of applications and share the characteristic of being a machine tool based processes having one rotating tool that travels over the surface of the components to be processed.

The three processes addressed in this chapter: FS, FSW and FSC are of major importance when considering the manufacturing of aluminium alloy. This statement can be justified considering the fact: i) aluminium alloys are one of the most relevant material group in engineer applications, and ii) aluminium alloys are particularly suitable for being solid state processed by most of the friction based manufacturing technologies. Actually, the zone undergoing direct mechanical processing goes over a severe thermo-physical cycle resulting in dynamic recrystallization and consequent grain refinement and homogeneous fine distribution of hardening particles [1]. Therefore the processed zone of aluminium alloys typically presents benefits when compared with original pre-processed condition [2,3].

The Figure 1 presents the technological interrelations of the three processes addressed in this chapter within the scope of the friction based manufacturing technologies.

On all the friction based manufacturing technologies a viscoplasticised solid state region is generated and processed into a new shape and properties. This region whilst remaining solid presents a three-dimensional material flow pattern almost as a liquid enabling easy mixing and blending between different materials. This phenomenon is generally referred to as the "third-body region" concept [4].

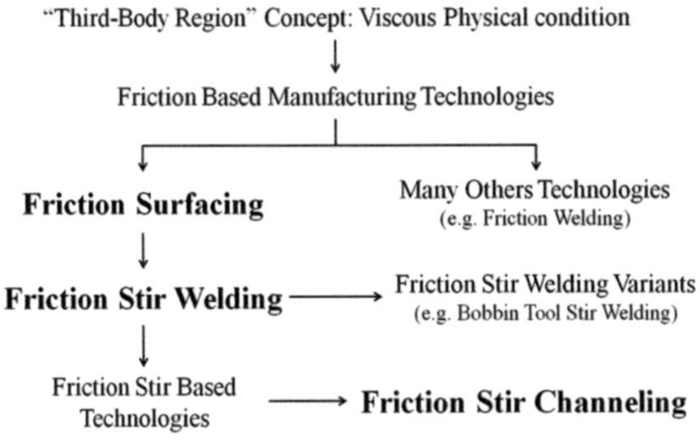

Figure 1. Friction Surfacing (FS), Friction Stir Welding (FSW) and Friction Stir Channeling (FSC) on the scope of the linear friction based manufacturing technologies.

In particularly for aluminium alloys this "third-body region" is characterized mechanically by a relatively low flow stress and by temperatures above recristalization temperature and below melting temperature of material. In the solid state processes governed exclusively by the introduction of mechanical energy (as it happens in friction based processing technologies), the heat is generated by friction dissipation during deformation at contacting interfaces and internally during material flow. Because the heat generated by friction dissipation tends to zero as the material gets near the fusion temperature the maximum temperature achieved within processed zone is physically limited by the fusion temperature and thus all the deformation is restricted to solid state condition.

One relevant property of the "third-body region" is to easily produce strong bonds at the interface with other similar solid state flowing material or even with material surfaces undergoing less severe deformation, e.g., elasto-plastic deformation. The solid state joining mechanisms involving the "third-body region" at temperatures well above recrystallization temperature are mainly diffusion, but approximation to interatomic equilibrium distances can also be found for lower temperature and higher pressure conditions at the joining interface. Thus, the "third-body region" is frequently used for promoting joining of similar and dissimilar materials in the various friction welding variants. This "easy to join" property is also used in other manufacturing technologies based on this "third-body region", e.g., extrusion of close hollow shapes where separate extruded components are joined together before exiting the dies.

The FS was firstly patented in 1941 [5] and is nowadays a well-established technology that applies one consumable tool to produce many possible combinations of coatings over a substrate. The FS is one of the many solutions developed based on the concepts of the friction manufacturing technology and the extension of its concepts gave rise to one very significant development in the history of welding technology: The FSW process that was firstly patented in 1991 [6]. The FSW is a process for joining components, using a non-consumable tool, with a suitably profiled shoulder and probe. FSW can be regarded as an autogenous keyhole welding technique in the solid-phase. The FSW technology has been subjected to the most demanding quality standard requirements and used in challenging industrial applications over a wide range of structural and non-structural components mainly in aluminium alloys. The disruptive character of the FSW process is emphasized by the numerous technic-scientific papers and patents published and the several friction stir based technologies that are being invented. The FSC was proposed for the first time in 2005 as a method of manufacturing heat exchanging devices [7] and is one of the most promising innovations based on the friction stir concepts with many industrial applications, some of them will be depicted in this chapter. The FSC uses a non-consumable tool to open channels along any path, shape and depth in monolithic aluminium alloys components.

There are friction based manufacturing technologies, such as, FSW where material flow within the "third-body region" is three-dimensionally enclosed and thus the process is flashless. In FSW the "third-body region" is constrained at top and bottom by the system: shoulder + anvil, and at the sides by the cold base material. But there are others friction based manufacturing technologies where the "third-body region" is open (or not fully constrained) and bulk flash is generated, e.g., FS. For the case of FSC one other intermediate condition is present: although the "third-body region" is three-dimensionally enclosed some of the inner material from the "third-bodyregion" is forced out by the tool features, producing an amount of flash volume correspondent to the volume of the inner channel produced. These conditions are depicted in Figure 2.

a) b) c)

Figure 2. Sample of finishing condition of friction based manufacturing technologies: a) Flash produced during FS due to unconstrained "third-body region"; b)Flashless condition of as-welded FSW due to complete enclosure of the material flow within the "third-body region"; c) Flash forced out of processing zone by the tool features during FSC.

The cost effectiveness of the three friction based manufacturing technologies addressed in this chapter is very promising: i) All the processes are environmentally friendly solutions, ii) Because of the low heat input demanded for the solid state processing all these solutions have high energy efficiency when compared with alternative/concurrent solutions; iii) The health and safety impact for human operators is insignificant due to no fumes emission and residual radiation and equipment with low operational hazards; iv) Since there are no consumables for FSW and FSC, and the only consumable in FS is the rod to be deposited over the substrate, any initial investment in the equipment typically have an early breakeven point, when compared with alternative/concurrent solutions; v) All these technologies are easy to automate with good repeatability of the results and high levels of quality assurance.

The Table 1 establish the composition and temper of the aluminium alloys involved in the several applications of the friction based manufacturing technologies addressed in this chapter. This Table 1 is intended to be used as support for the analysis of the results presented along the chapter.

	Composition (weight%)										
Material	**Al**	**Cu**	**Mg**	**Si**	**Zn**	**Mn**	**Fe**	**Cr**	**V/Ti**	**Other, each**	***Temper***
AA1050-O	≥99.5	≤0.05	≤0.05	≤0.25	≤0.05	≤0.05	≤0.40		≤0.05/	≤0.03	Annealed
AA2024-T3	remain	3.80 to 4.90	1.20 to 1.80	≤0.50	≤0.25	0.30 to 0.90	≤0.05	≤0.10	/≤0.15	0.05	Solution heat treated, cold worked, and naturally aged to a substantially stable condition
AA5083-H111	remain	≤0.10	4.0 to 4.90	≤0.40	≤0.25	≤0.40 to 1.0	≤0.40	0.05 to 0.25	/≤0.15	≤0.05	Small degree of strain-hardening only
AA6061-T4	remain	0.15 to 0.40	0.80 to 1.20	0.40 to 0.80	≤0.25	≤0.15	≤0.70	0.04 to 0.35	/≤0.15	≤0.05	Solution heat treated and naturally aged to a substantially stable condition
AA6082-T6	remain	≤0.10	0.60 to 1.20	0.70 to 1.30	≤0.20	0.40 to 1.0	≤0.50	≤0.25	/≤0.10	≤0.05	Solution heat treated and artificially aged
AA7178-T6	remain	1.60 to 2.40	2.40 to 3.10	≤0.40	6.30 to 7.30	≤0.30	≤0.50	0.18 to 0.28	/≤0.20	≤0.05	Solution heat treated and artificially aged

Table 1. Composition and temper designation of the aluminium alloys referenced within the present chapter.

2. Friction surfacing

2.1. Fundaments of the process

Friction surfacing (FS) is a solid state process used for the production of metallic coatings with metallurgical characteristics typical from hot forging operations. The process involves rubbing a rotating consumable rod (Figure 3a) against a substrate under an applied axial load. Heat generated by frictional dissipation promotes the viscoplastic deformation at the tip of the rod (Figure 3b).

As the consumable travels along the substrate (Figure3c), the viscoplastic material at the vicinity of the rubbing interface is transferred over onto the substrate surface, while pressure and heat conditions trigger a inter diffusion process that soundly bonds it. As the consuma-

http://dx.doi.org/10.5772/52026

ble rod material undergoes a thermo-mechanical process, a fine grained microstructure is produced by dynamic recrystallization. As shown in Figure 3d, FS enables the production of a continuous deposit from the progressive wear of the consumable rod. However, the coating cross section presents poorly bonded edges on both advancing and retreating sides, which are closely related to process parameters. The process is also characterized by the constant generation of a revolving flash of viscoplastic material at the rod tip, responsible for a smooth mushroom-shaped upset on the consumable rod [8,9].

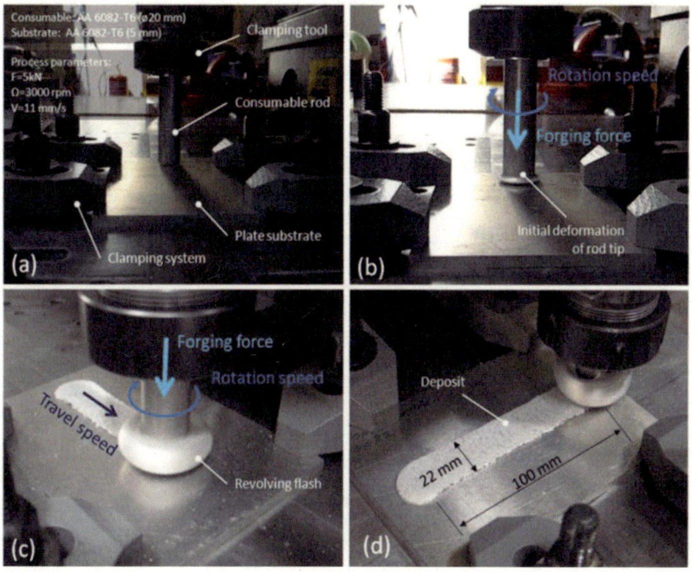

Figure 3. Deposition of AA6082-T651 aluminium alloy by friction surfacing. a) Experimental setup; b) Initial plunging phase; c) Deposition; d) Coating produced.

2.2. General features

The Figure 4 presents a cross section macrograph of both rod and coating, depicting the gradual transformations that the consumable material undergoes as it is deposited, as well as, the flash developed. Consumable rod microstructure is depicted in Figure 4a, presenting an anisotropic grain structure aligned along the rod extrusion direction. The hot working of the consumable rod tip and the deposit generates heat which is conducted along the consumable, pre-heating the material and enabling its plastic deformation by the colder material layers above in a compression/torsion process. Consumable rod heat affected microstructures presenting some precipitate coarsening and grain growth can be seen in Figure 4b, while evidences of plastic deformation are depicted in Figure 4c and d. The combination of plastic deformation and heat generation leads to a dynamic recrystallization which processes the material into a viscoplastic state with the nucleation and growth of a new set of undeformed grains (Figure 4e and f). FS enables the thermo-mechanical processing of the consumable material into a new metallurgical state.

Since this viscoplastic region is pressed against the substrate at temperatures approximately 50-90 % of the melting point, a diffusion bonding process takes place and a deposit of hot-worked consumable rod material is produced. Plastic deformation and friction can disrupt the relatively brittle oxide layers, establishing metal-to-metal contact and enabling the joining process [10]. The low heat input inherent to the FS process delivered locally over the substrate of an high thermal conductor material as the aluminium alloys are, will result in high cooling rates of the thermomechanically processed zone. Considering that fast cooling prevent grain growth, this may be the reason for the typical fine equiaxial recrystallized microstructure of the deposited material, depicted in Figure 4g. Heat is lost mainly by conduction to the substrate, originating the heat affected zone. A fully bonded interface can be seen in Figure 4h.

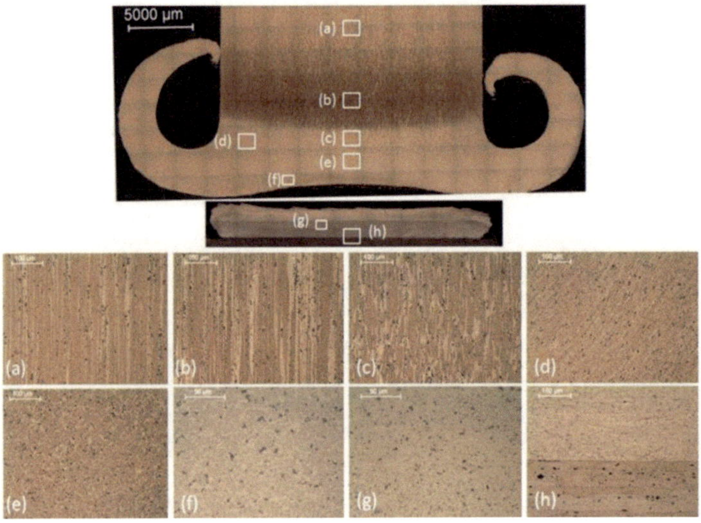

Figure 4. Microstructural transformations during the FS of AA6082-T6 over AA2024-T3.a) Consumable base material; b) Heat affected zone; c) Compression-driven TMAZ; d) Torsion-driven TMAZ; e-f) Fully recrystallized microstructure; g) Deposited material; h) Bonding interface.

The Figure 5 illustrates a proposed model for the global thermal and mechanical processes involved during friction surfacing, based on the metallurgical transformations described above. The speed difference between the viscoplastic material, which is rotating along with the rod at Vxy, and the material effectively joined to the substrate (Vxy = 0), causes the deposit to detach from the consumable. This viscous shearing friction between the deposit and the consumable is the most significant heat source in the process.

Since the highly plasticized material at the lower end is pressed without restraint, it flows outside the consumable diameter, resulting into a revolving flash attached to the tip of the consumable rod and poorly bonded coating edges. Unbonded regions are also related to the higher tangential speed at which the material flows at the outer radius of the consumable rod, because the relative speed between the deposit and the substrate shears the bonding interface and disrupts the ongoing diffusion bonding process. Nevertheless, flash and unbonded regions play an important role as temperature and pressure boundary conditions for the joining process.

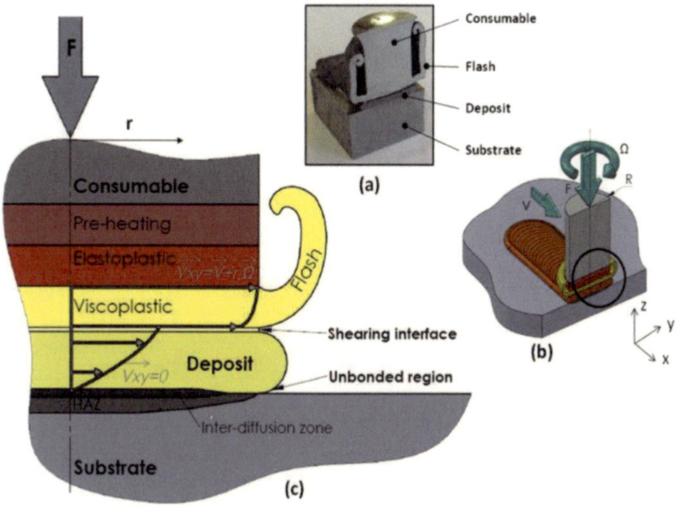

Figure 5. Thermo-mechanics of friction surfacing.a) Sectioned consumable; b) Process parameters and c) Thermo-mechanical transformations and speed profile. Nomenclature: F – forging force; Ω – rotation speed; V -travel speed; Vxy – rod tangential speed in-plan xy given by composition of rotation and travel movements.

2.3. Parameters of the process

Coatings are evaluated based on thickness, width and bond strength/extension which depend on controllable process parameters, such as, i) forging force; ii) rotation speed and iii) travel speed. Substrate thickness, rod diameter and material properties define the thermo-mechanical system thus determining process parameters:

- Forging force - improves bonding extension and results in wider and thinner deposits. However, excessive loads result in non-uniform deposition with a depression at the middle of the pass due to material expelling from the region beyond consumable rod diameter. Insufficient forging forces result in poor consolidated interfaces;

- Rotation speed - influences the bonding quality and coating width. While lower to intermediate rotation speeds enhance bonding quality, higher rotation speeds produce a more flat and regular deposit, with a more effective forging effect shaping the coating;

- Travel speed - strongly influences coating thickness and width, since it determines the rate at which material is deposited. As such, higher travel speeds result in thinner deposits. Faster travel speeds lead to shorter heat exposure periods, resulting in less grain growth and finer microstructures. Thinner deposits also cool more rapidly. The substrate heat affected zone decreases for higher travel speeds. Bonding at coating edges deteriorates for faster travel speeds;

- Tilt angle - A small tilting the consumable rod, in less than 3, has proven to reduce the unbounded extension of the deposit at the coating edges, by enabling a gradual increase of forging pressure applied by the consumable rod on the substrate, from the tip to the tail zone being thermomechanically processed, at each instant.

The Table 2 depicts the range of process parameters best suited for the friction surfacing of AA6082-T6, using 20 mm diameter consumable rods over various aluminium alloy substrate plates and resulting coating thickness, width and bonded width. In general, the fully bonded width of the deposit rarely exceeds the diameter of the consumable rod used, while the coating width extends beyond it, as evidenced by Figure 4.

Materials		**Process parameters**				**Coating characteristics**		
Consumable rod	**Plate substrate**	**Forging force [kN]**	**Rotation speed [rpm]**	**Travel speed [mm/s]**	**Tilt angle [°]**	**Thickness [mm]**	**Deposit width [mm]**	**Bonded width [mm]**
AA6082-T6 (20 mm)	AA2024-T3 AA7178-T6 AA5083-H111 AA1050-O	5 to 7	2500 to 3000	6 to 12	0 to 3	1 to 2	20 to 25	15 to 20

Table 2. Typical range of process parameter in the FS of AA6082-T6 over several aluminium substrate plates.

2.4. Advantages and limitations

Being a solid state process, FS allows depositing various dissimilar material combinations. Investigations report the deposition of stainless steel, tool steel and nickel-based alloys (Inconel) on mild steel substrates, as well as, stainless steel, mild steel and inconel consumables on aluminium substrates [11,12].

FS is best suited for applications with material compatibility issues. The process involves a hot forging action, which refines significantly the microstructure of the deposited material. The deposit is inherently homogenous and has good mechanical strength. The process can be automated, providing good reproducibility and doesn't depend on operator skill [13].

Since FS is mainly based on plastic deformation, this process presents some advantages over other coating technologies based on fusion welding or heat-spraying processes. Apart from avoiding defects commonly associated to fusion and solidification mechanisms (coarse microstructures, intermetallic formation, porosities, hot cracking or inclusions, e.g., slag), the heat input in FS is minimum and localized, preventing part distortion and minimizing the heat affected zone extension and dilution. This also makes FS suitable to process thermal sensitive materials, such as, aluminium alloys. Additionally, the absence of spatter, toxic fumes and emission of radiation makes this process cleaner and environmentally friendly. The absence of fusion and fast cooling rates enable to FS in a great variety of positions [14].

However, FS struggles with some technical and productivity issues which contribute to a limited range of engineering applications. One of the main process disadvantages is the poor bonding at the coating edges, as post-processing operations are often required to remove them. Moreover, the generation of a revolving flash at the consumable rod tip contributes to a decrease in mass transfer efficiency, as it represents material that does not bond to the substrate. Friction surfacing enables a limited control over the deposited thickness and width, as coating geometry is determined by a very narrow range of process parameters.

2.5. Properties and applications

The FS allows the dissimilar joining of materials that would be metallurgical incompatible otherwise. It allows assembling in a single composite component, tailored material property combinations which are difficult to gather in a single monolithic material. This enables an advanced and detailed design, adjusting the required material properties according to different loading areas of a part and precluding the use of more expensive and specific materials capable of assembling all functional requirements.

Although FS has limited large-overlay capabilities, this process is ideal for localized repair and cladding. The FS has been used in the production of long-life industrial blades, wear resistant components, anti-corrosion coatings and in the rehabilitation of worn or damaged parts such as, turbine blade tips and agricultural machinery. Other applications feature the hard facing of valve seats, brake disks and tools such as punches, guillotine blades and drills. The surfacing of pipe flange contact faces, the reclamation of worn railway points and the hermetic sealing of containers has also been reported as promising applications [15,16].

The FS can be performed over a great variety of substrate configurations and along complex trajectories. Some examples of FS path case studies can be seen in Figure 6. Figure 6a displays a single FS curvilinear path, while Figure 6b depicts a continuous cylindrical build-up, as the consumable rod moves along a 3D helicoidally trajectory.

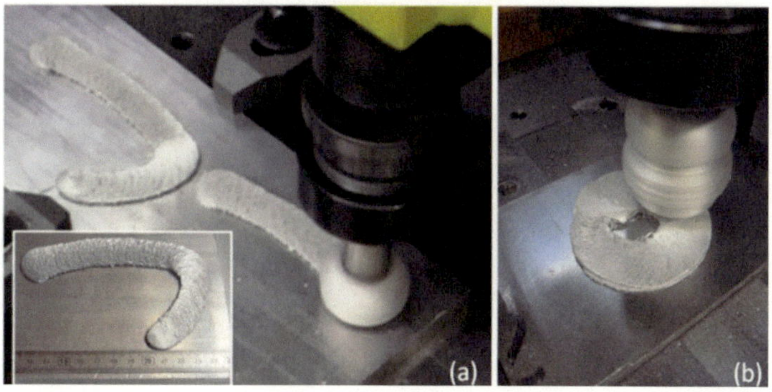

Figure 6. Examples of friction surfacing trajectories. a) Curve path; b) cylindrical build-up.

Given the rough coating surface, FS is often followed by post processing operations in order to achieve the desired geometry and surface finish. The Figure 7 depicts the milling surface finish of a friction surfaced AA6082-T6 deposit presenting a fully bonded defect-free layer and a smooth surface finish.

Figure 7. Surface finish by milling. FS of AA6082-T6 over AA7178-T6.

Another promising application consists on the vertical build-up of structures by performing successive fully overlapped depositions. Figure 8 depicts the manufacturing of a trapezoidal linear rail milled from the build-up of several AA6082-T6 friction surfaced passes. Figure 8d depicts the soundly bonded deposited layers, as FS allows the production of bulk layered composite materials from which parts or component features can be manufactured.

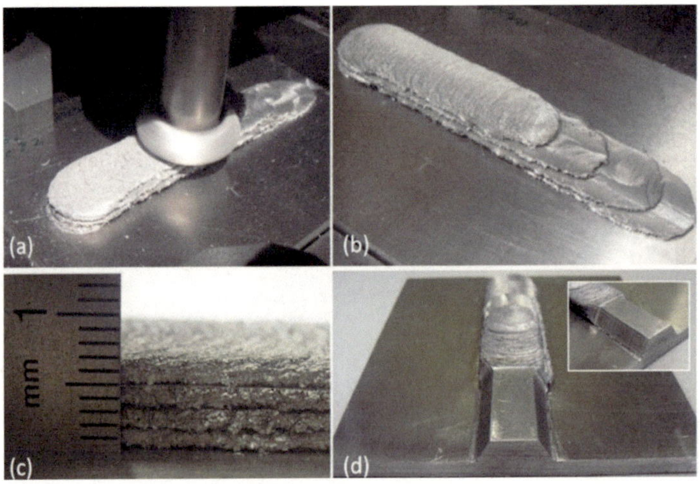

Figure 8. Build-up by friction surfacing. a) Successive deposition; b) Bulk produced from four overlapped passes; c) Detail of final thickness achieved; d) Milling of linear rail.

The mechanisms involved in friction surfacing enable an alternative process to produce surface composites, as the inherent severe plastic deformation can be used to promote the dispersion and mixture of reinforcement particles within the deposit matrix. Figure 9 depicts the joining interface between an aluminium silicon carbide reinforced composite layer over an AA2024-T3 plate substrate. AA6082-T6 aluminium rods were packed with 12.3 μm silicon carbine particles and used as consumables. A uniform distribution of reinforcements was achieved as the composite layer becomes soundly bonded to the substrate. Surface hardness severely increased and gradually decreased along the thickness. By increasing the volume of reinforcements packed inside the consumable rod, higher concentration distributions can be achieved, as shown by Figure 9b.

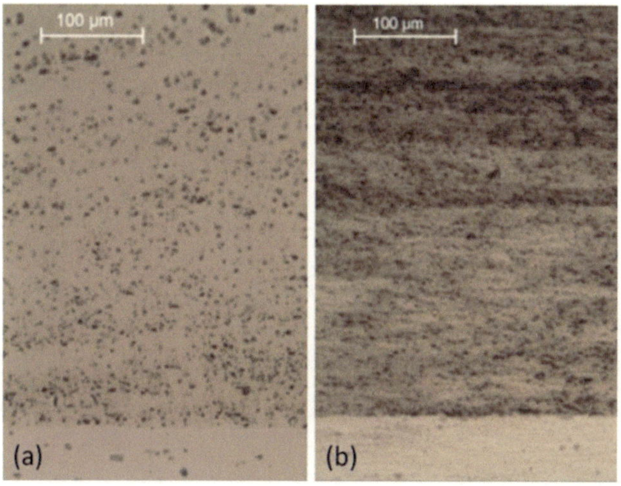

Figure 9. Joining interface of SiC reinforced AA6082-T6 coatings produced by FS over AA2024-T3 substrate.

2.6. Analytical modeling: establishment of performance analysis

Performance criteria regarding the material deposition rate and specific energy consumption were established inreference [17] for the characterization of friction surfacing, thereby contributing to establish a realistic comparison with other coating technologies.

2.6.1. Mass transfer

The Figure 4 depicts the material flow from the consumable rod to the deposit bonded to the substrate. Volumetric rod consumption rate (*CRvol*) is determined by multiplying the rod plunging speed (V_z) by its cross section area (A_r), where *r* is the rod radius (1).

$$CRvol[m^3/s] = A_r V_z = \pi r^2 V_z \tag{1}$$

Likewise, the product between the travel speed (V) and the deposited cross section area (Ad), expresses the volumetric deposition rate (DRvol) throughout the friction surfacing process (2).

$$DRvol[m^3/s] = A_d V \qquad (2)$$

Considering the consumable rod material density (ϱ), CR and DR can be rewritten in order to express the mass flow, as depicted by (3) and (4).

$$CR[kg/s] = CRvol \times \varrho \qquad (3)$$

$$DR[kg/s] = DRvol \times \varrho \qquad (4)$$

The Figure 10 presents the effect of process parameters on deposition (*DR*) and consumption (*CR*) rates for the FS of AA6082-T6 over AA2024-T3. Deposition rates were seen to vary from 200 to 600 mm³/s (0.5-2 g/s), increasing for higher travel speeds. The increase of rotation speed and forging force has a negative effect on *DR*.

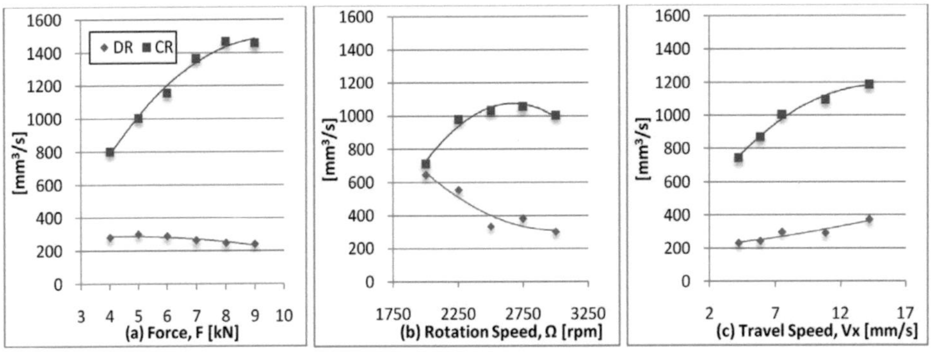

Figure 10. Effect of process parameters on deposition rate (DR) and consumption rate (CR). FS of AA6082-T6, 20 mm diameter consumable rods, over AA2024-T3 plates. Process parameters: a) Ω = 3000 rpm, V = 7.5 mm/s; b) F = 5 kN, V = 7.5 mm/s; c) F = 5 kN, Ω = 3000 rpm.

In order to determine the fraction of consumed material deposited and that is transferred to flash, a deposition efficiency ($\eta_{deposition}$) can be defined as the ratio between DR and CR via (5).

$$\eta_{deposition} = DR/CR \qquad (5)$$

Due to the formation of side unbonded regions, just a part of the deposited material is effectively joined. As such, the joining efficiency ($\eta_{joining}$) is given by the ratio between the bonded width (Wb) and the maximum coating width (Wd) established in (6).

$$\eta_{joining} = W_b / W_d \qquad (6)$$

Thus, the effective coating efficiency ($\eta_{coating}$) reflects the fraction of consumed rod that actually becomes bonded to the substrate and is estimated via (7) by multiplying (5) times (6):

$$\eta_{coating} = \eta_{deposition} \bullet \eta_{joining} = \frac{A_d v}{\pi r^2 V_z} \bullet \frac{W_b}{W_d} \tag{7}$$

Considering the process parameter range depicted in Table 2, in the FS of AA6082-T6 over AA2024-T3, maximum joining efficiency and coating efficiency reached around 82 % and 25 %, respectively. Recently research performed by the authors of the present chapter also shows that flash formation accounts 60 to 70 % of the overall consumable rod consumption.

2.6.2. *Energy consumption*

The mechanical power supplied by the equipment ($\dot{W}_e$) can be divided into three main contributions regarding rod rotation, $\dot{W}_r$, axial plunging,$\dot{W}_z$, and travel, $\dot{W}_x$, as established in (8).

$$\dot{W}_e[J/s] = \dot{W}_r + \dot{W}_z + \dot{W}_x = \frac{2\pi\Omega}{60}(T_1 - T_0) + F_z V_z + F_x V \tag{8}$$

T0 is the torque required to freely rotate the consumable rod without any contact friction, e.g., the torque applied by the machine to impel the prescribed rotation speed and depends on the machine mechanical design, rather than deposition process. When the machine starts to plunge the rod against the plate substrate it raises the torque from T0 to T1.

Hence, for a joining efficiency of 100%, energy consumption per deposited unit of mass (specific energy consumption, EC) is given by (9).

$$EC[J/kg] = \dot{W}_e / DR \tag{9}$$

The Figure 11 depicts the variation of power and specific energy consumption for the tested conditions. Power increases with the forging force and the travel speed, varying between 4-6 kW. The specific energy consumption computed according to equation (9) varies from 8 to 26 J/mm3 (4-10 kJ/g). Specific energy consumption increases with the forging force (Figure 11a). For high rotation speeds, although both the required power and the deposition rate drop, specific energy consumption rises, meaning that the decrease in deposition rate is more significant (Figure 11b). Excessive rotation speeds result in less joining efficiency and increased specific energy consumption per unit of mass. Despite the increase in power, specific energy consumption decreases with travel speed (Figure 11c), given a more significant improvement of deposition rate (Figure 10c). Hence, faster travel speeds allows to improve deposition rates while decreasing specific energy consumption.

Considering the present testing conditions in Table 2, the best processing conditions were achieved for a 500 kN forging force, 2500 rpm rotation speed and a 7.5 mm/s travel speed, which resulted in a specific energy consumption of 12 J/mm3 with a deposition rate of 332 mm3/s. Joining efficiency and coating efficiency was 76 % and 32 %, respectively, while flash accounted for 67 % of consumed material.

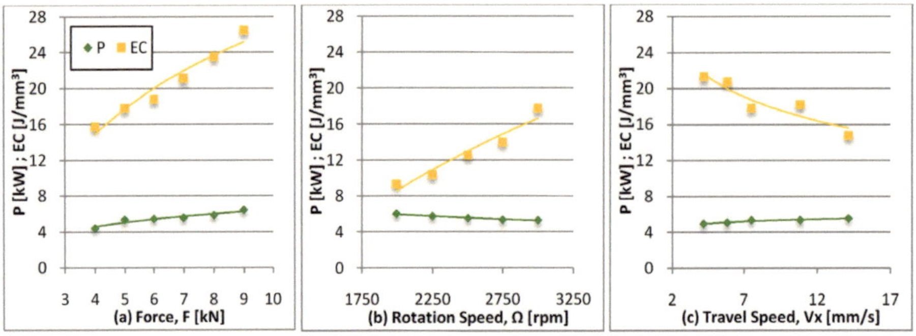

Figure 11. Effect of process parameters on friction surfacing power and specific energy consumption. FS of AA6082-T6, 20 mm diameter consumable rods, over AA2024-T3 plates. Process parameters: a) Ω = 3000 rpm, V = 7.5 mm/s; b) F = 5 kN, V = 7.5 mm/s; c) F = 5 kN, Ω = 3000 rpm.

3. Friction Stir Welding

3.1. Fundaments of the process

The FSW is a process for joining components in the solid phase, using an intermediate non consumable tool, with a suitably profiled shoulder and probe, made of material that is harder than the workpiece material being welded. FSW can be regarded as an autogenous keyhole joining technique, essentially, without the creation of liquid metal.

The rotating tool is plunged into the weld joint and forced to travel along the joint line, heating the components by interfacial and internal friction dissipation, thus producing a weld joint by extruding, forging and stirring the materials from the components in the vicinity of the tool. The basic principles of the process and some nomenclature are represented in Figure 12.

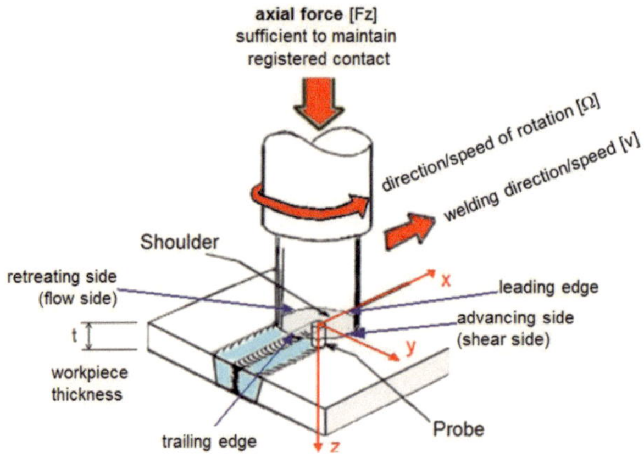

Figure 12. Representation of the main parameters and nomenclature of FSW joints.

The shoulder and the probe thermo mechanically soften and then separate the material being processed by the passage of the probe through the material. The material flows around the probe and is then forge welded together at the trailing edge of the probe. This separation and welding together occur continuously by backfilling from the probe and compaction/containment from the shoulder. This transient separation/rewelding operation happens during and before the trailing edge of the shoulder moves away from the processed/weldtrack. The transient "third-body region" immediately coalesces and forms a solid phase bond as the tool moves away.

The material flow within closed "third-body region" is characterized by [18, 19] :i) on the retreating side of the weld, material is displaced only backward; For any material actually in the path of the probe, the flow reverses and results in material being displaced to a position behind its original position (relative to the welding direction); iii) The material at mid-height in the advancing side near the probe is flowing all around the retreating side being left behind it original position in a kind *swiss roll* movement, that maybe the mechanism of formation of the onion rings in the nugget and iv) Material on the advancing side under the shoulder and near the periphery (but under shoulder influence) is displaced in the forward direction of welding relative to its original position.

3.2. Parameters of the process

The main FSW process parameters are the following:

- Tool geometry;
- Plunge speed and depth of probe in workpieces;
- Tool rotational speed () and direction;
- Travel Speed (V) along the joint line;
- Axial load, Fz;
- Dwell time at start of the weld;
- Clamping system (stiffness, precision and material of the anvil, for easy extraction of workpieces);
- Tilt and side tilt angles;
- Control during plunge, dwell and weld periods: Force control (Fz) *versus* position control;
- Preheating/interpass temperature of workpieces;
- Weld pitch ratio, see equation (10). Varying the weld pitch ratio changes the heat input from the frictional internal and interfacial energy and some effects in the weld region behaviour can be predictedThe weld pitch ratio provides a possible classification for the flow pattern within the third-body zone [1, 20]. For aluminium alloys it is usual to consider the value of weld pitch ratio = 4 as intermediate condition. For weldpitch ratio higher than 4 there is hot condition and for weldpitch ratio smaller than 4 there is cold condition.

Concerning the influence of the hot-to-cold conditions in the metallurgical features of a FSW weld joint, the classification is established in (11), where the HAZ and TMAZ are described in §3.3.

$$\text{weld pitch ratio}\left[\text{rev/mm}\right]=\frac{\Omega\left[\text{rpm}\right]}{\text{v}\left[\text{mm/min}\right]} \tag{10}$$

$$\text{FSW Classification:}\begin{cases} Hot \Leftrightarrow \begin{cases}\uparrow \Omega \\ \downarrow v\end{cases} \Rightarrow \begin{cases}\uparrow HAZ \\ \downarrow TMAZ\end{cases} \\ Cold \Leftrightarrow \begin{cases}\downarrow \Omega \\ \uparrow v\end{cases} \Rightarrow \begin{cases}\downarrow HAZ \\ \uparrow TMAZ\end{cases} \end{cases} \tag{11}$$

3.3. General metallurgical, hardness, mechanical and corrosion resistance features

During the FSW process, the material undergoes intense plastic deformation at elevated temperature, as a rule resulting in the generation of fine and equiaxed re-crystallized grains. This fine microstructure produces good mechanical properties in friction stir welds. Better quality joints are associated with intense three-dimensional material flow. Thus for aluminium alloys is more easy to obtain high levels of quality and reproducibility when welding bigger thicknesses than when welding small thickness, e.g., equal or less than 1 mm, where the material flow tends to be bi-dimensional and the stirring of the materials is consequently poorer [21,22].

The main zones in a FSW joint with distinct metallurgical properties are: i) the thermomechanically affected central zone (TMAZ) that includes the dynamically recrystallized zone (nugget); ii)The heat affected zone (HAZ) and iii) The unaffected parent material or base material (BM).These different zones result from the combined application of mechanical energy and heat energy from frictional dissipation. The typical characteristics of each of these zones for aluminium alloys are the following:

- The BM is the region that was unaffected by the FSW process;
- The HAZ is only affected by the heat energy and presents typically some slight coalescence of grain relatively to the original grain size but is subjected to internal point and linear defects rearrangements. Thus, for the heat treatable wrought aluminium alloys the HAZ may presents some reduction in the distribution of precipitates at grain boundaries;
- The TMAZ grain maintains the characteristics of the HAZ however the grain presents increased deformation as they get close to the interface with the nugget. This fact results from the influence of the material flow prescribed by the movement of the tool and the relatively high maximum temperature reached in this zone;
- The nugget is the region of the TMAZ undergoing dynamic recrystallization with grain size refined and homogenized. The TMAZ/nugget interface enhances the significant dif-

ference between the structure of initial grain and the equiaxial grain resultant of the dynamical recrystallisation process with fine dispersion of the precipitates in the solid solution.

The typical metallurgical structures present in the processed zone of friction stir welds are established and classified in Figure 13 for a heat treatable aluminium alloy (AA2024-T3) and in Figure 14 for a non-heat treatable aluminium alloy (AA5083-H111).

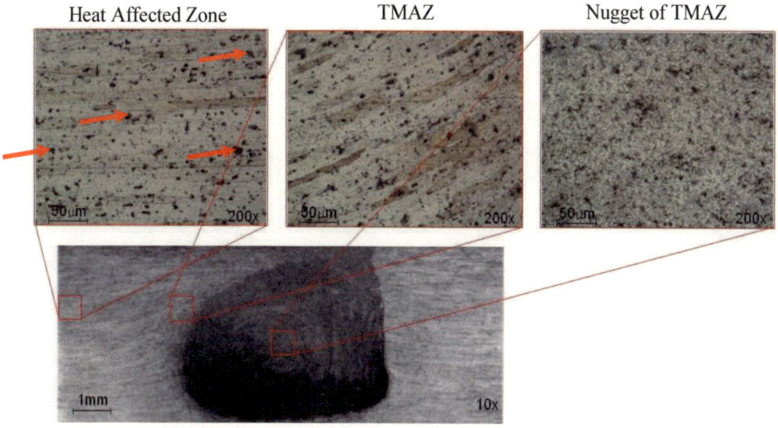

Figure 13. Metallographic analysis of a FSW of AA2024-T3 (thickness = 4.8 mm), Ω = 1120 rpm, v = 320 mm/min. The arrows in the HAZ detail are identifying some of the several precipitates.

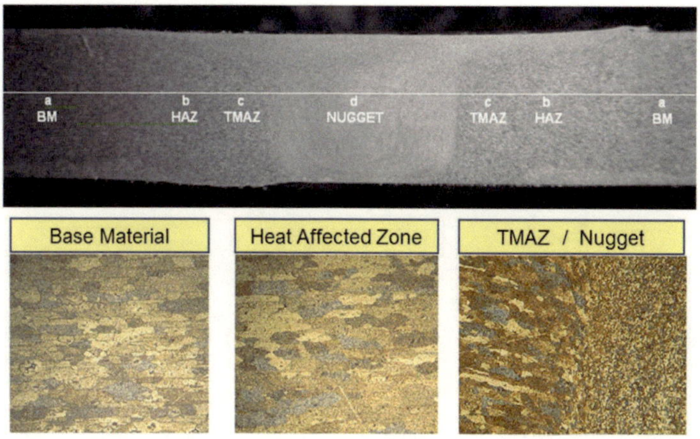

Figure 14. Metallographic analysis of a FSW of AA5083-H111 (thickness = 4.0 mm), Ω = 1120 rpm, v = 320 mm/min.

In Figure 15, it is possible to analyze some relevant characteristics, e.g.: i) relative location of the centre of the nugget; ii) width of the nugget at the centre and iii) width of the nugget at the root of the weld. This last dimension is very relevant in the quality assessment of the

FSW joints because a correct processing of the root of the joint is mandatory to avoid defects located at the root of the bead which are always important [23].

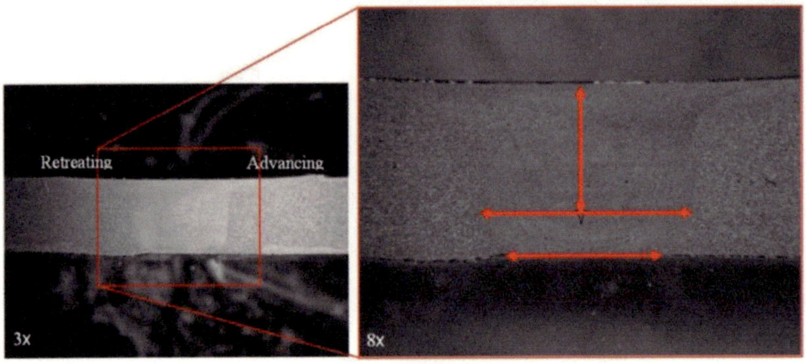

Figure 15. Relevant geometrical characteristics on a metallographic analysis of a FSW cross-section.

The Figure 16 depicts the effective location of the real slip interface which is within the material and not at the top surface in direct contact with the shoulder. In fact, there is a layer (with a thickness 10 μm) of material bellow the shoulder that is dragged to rotate along with the rotating shoulder. This phenomenon is due to the adhesion of the material to the shoulder. Even for very low friction coefficients, the contact pressures are high and the frictional force overcomes the flow stress of the workpieces material. It should also be noticed that the real slip interface is increasingly evident, as it progresses from the center to the periphery of the weld bead, in accordance with the increase of linear speed of the rotating shoulder for higher radius.

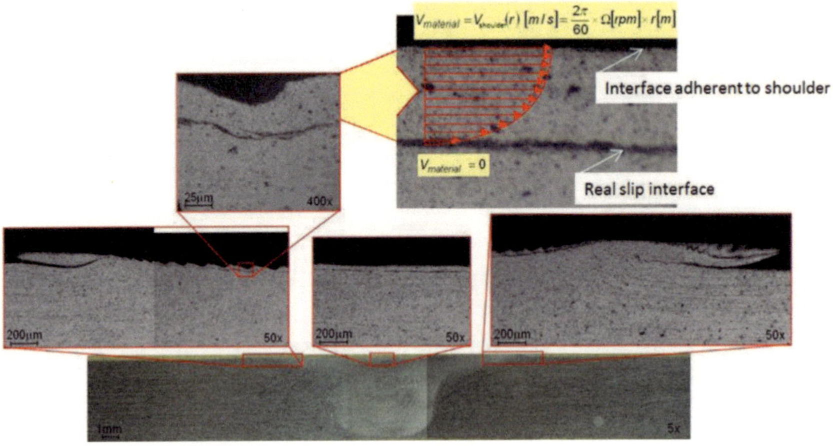

Figure 16. Superficial layer of weld bead material in contact with shoulder that is dragged to rotate along with the shoulder generating a real slip interface about 10μm within the material.

From the Figure 17 showing the behavior of the non-heat treatable alloys it is possible to conclude about the increased hardness of weld bead and heat affected zones when compared with base material. Because these alloys are very sensitive to strain hardening the increase is most significant in all the TMAZ with emphasis for the nugget zone [23,24].

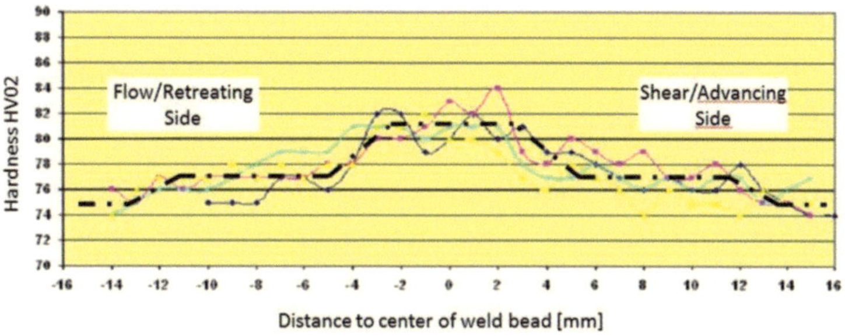

Figure 17. Typical hardness profile for the non-heat treatable wrought aluminium alloys (AA5083-H111; thickness: 4mm).

In Figure 18 it is possible to conclude about the typical location of the global minimum value of the harness field located in the interface between the HAZ and the TMAZ. Along the HAZ there is typically a local minimum due to over ageing. Depending on the pattern of material flow during the FSW... for more cold conditions the minimum value at the flow side (retreating side) is smaller than at the shear side (advancing side). For FSW under more hot conditions the minimum values in both sides of the weld bead have more identical values [1,24].

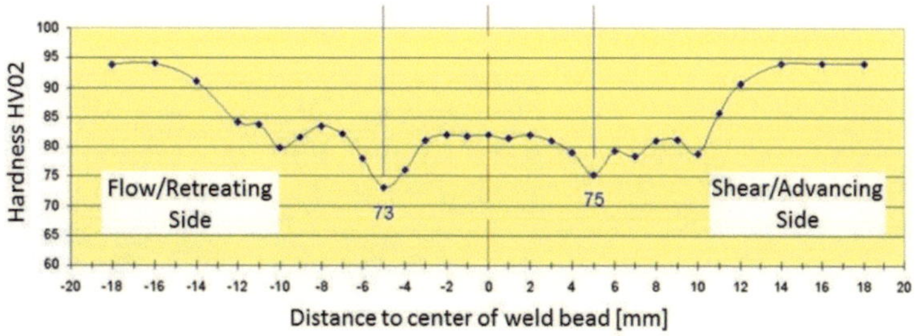

Figure 18. Typical hardness profile for the heat treatable wrought aluminium alloys (AA6061-T4; thickness: 4.8mm).

The hardness profile enables a reliable assessment of the static mechanical resistance of the joints but the fatigue resistance is more dependent on the geometric characteristics of the weld bead at the top and bottom surfaces and eventual internal defects both volumetric or layers/alignment of second phase particles or oxides, e.g., lack of penetration (LOP) root defects [25]. The fracture mechanism of FSW specimens under fatigue load is mainly determined by the size of the defect at the root of weld bead. For the FSW beads in as welded condition the higher critical level of the root defect always resulted in less number of cycles when compared to post weld smoothed root surface condition (Figure 19). The post weld smoothed root surface condition was obtained by mechanical grinding and final polishing of the root surface removing a superficial layer with a thickness of about 0.3mm assuring the removal of root defects. In fact reduced in-depth size of the root defect via surface smoothing increased the fatigue life in about 10X for a stress amplitude of about 80 MPa and R=0.1. Thus it is possible to conclude about the benefit of mechanical resistance when smoothing the root of FSW beads. Moreover FSW beads in smoothed condition shows a fatigue life close to BM results, most significantly for the lowest levels of stress amplitude. The results of FSW specimens in as welded condition are always better than the ones resulting from conventional welding solution: Synergic GMAW. The benefits of the fine equiaxial grain of the FSW nugget in fracture propagation phase plays an important role in this difference [26].

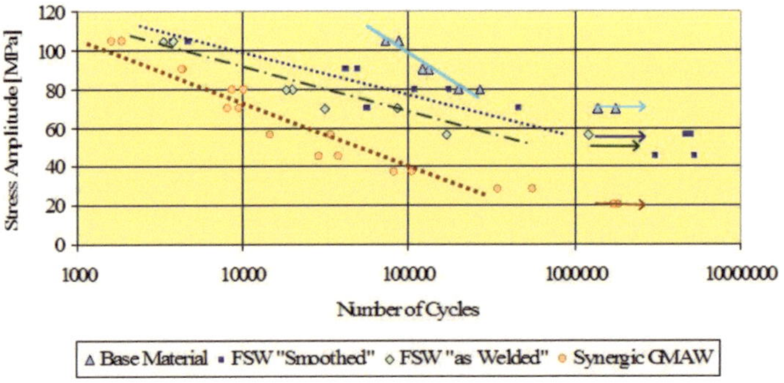

Figure 19. S-N curves from fatigue test trials (R=0.1; f=20Hz) of different testing conditions of AA5083-H111 (thickness = 4 mm).

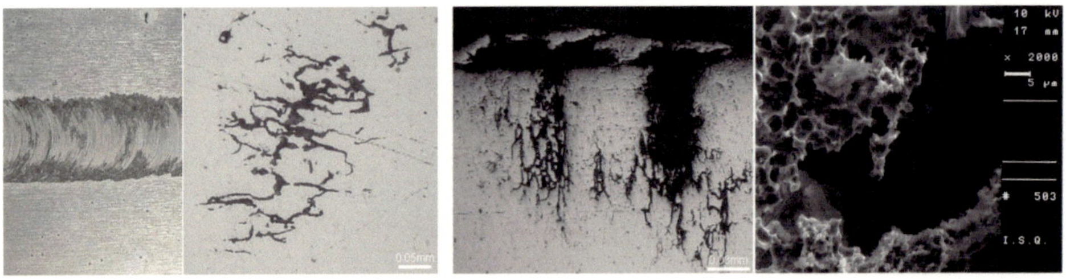

Figure 20. Corrosion developed in base material of AA5083-H111 after 7 days Exposition in Tagus River (20g/l NaCl). Exfoliation of the top surface of a FSW bead and intergranular corrosion along the Al-Mg precipitates resulting in pitting formation mechanism localized under the flash of the weld bead.

Several intergranular corrosion tests were performed for AA5083-H111 focusing the naval applications. These testes were implemented according to ASTM G67: Nitric Acid @ 30 ºC during 24hand also for 7 days of exposition to Tagus river water (20g/L NaCl) as depicted in Figure 20. The results show that the loss mass in the base material samples was much higher

http://dx.doi.org/10.5772/52026

than in the welded samples. This happens because intergranular corrosion mechanism was most susceptible at the intergranular precipitates and these were more abundant in base material. Exfoliation corrosion tests demonstrate that the corrosion resistance of the AA5083-H111 was better when better surface finishing was obtained for the weld bead. Thus, the higher roughness at the top of the weld bead promotes the earlier start and faster development of the corrosive attack [23].The typical geometrical discontinuities left by the shoulder on the top of FSW beads, namely: i) indentation and/or flash at advancing and retreating sides; and ii) semi-circular striates on the processed area under the shoulder, acts as preferential concentration points for the corrosive media. Root defects such as LOP also promote a faster and localized corrosion mechanism at this zone.

3.4. Advantages and limitations

The advantages claimed for the process include:

- Solid-phase nature of the process;
- Capability of welding materials whose structure and properties would be degraded by melting;
- Minimal edge preparation required;
- Machine tool technology, simple to use with good surface appearance;
- Minimal distortion;
- Hot forged microstructure;
- Low residual stress levels, compared with arc welding processes;
- Environmentally friendly with absence of welding fume and excessive noise;
- Suitability for automation;
- Good mechanical properties;
- Welding consumables not required, with exception for inserts that can be used and gas shielding for reactive materials such as titanium and its alloys;
- Not influenced by magnetic forces;
- Continuous - unlimited length;
- Joint can be produced from one side and in all positions.

The current limitations of the FSW process are:

- Backing anvil required (except bobbin stir);
- Keyhole at the end of each weld (except with the introduction of a run-off tab or when a FSW tool with a retractable probe is used);
- Not be able to start the weld joint from the start of the of the joint between the plates to be welded (except when using a run-on tab);

- Workpiece requires rigid clamping (except when the Twin-stir™ variant is used);
- Application not as flexible as certain arc welding processes.

3.5. The challenges in modelling the FSW

The development of computational models can greatly contribute to better understanding any industrial process, particularly FSW. A validated model has the potential to produce reliable information about the deformation and mixing patterns that are important when designing FSW tools and thus should be capable of producing welds free of defects and voids. Further, a model can measure process characteristics that are difficult to observe experimentally such as local strains, strains rates and stresses. These strain and stress fields, together with temperature histories are seen as critical in predicting microstructure evolution. A detailed understanding of microstructural evolution can guide FSW designs by further improving mechanical properties, fatigue strength and corrosion resistance.

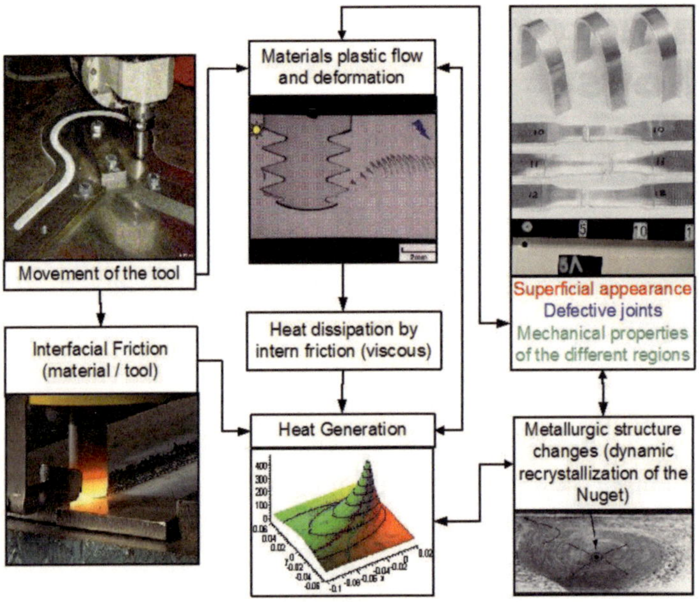

Figure 21. Coupled mechanical/thermal/metallurgical character of FSW process.

While considerable experimental work has been done to improve the knowledge on FSW, there's yet a lot of work needed to create a satisfying global model that can produce consistent results. The main difficulties in modeling FSW are [27]:

- Extensive material deformation in the region containing fully-plasticized material;
- The viscous-plastic flow imposed by the tool rigid surface, into the materials constrained by the interaction with the cold base material, with an essentially elastic behaviour, and the rigid plate (anvil) supporting the joint;

- Heat generated due to the sliding between the surface of the tool and the materials in the joint, depends on an unknown the friction coefficient;
- The correct prediction of the viscous-plastic flow imposed by the tool rigid surface into the materials being welded is also important because the viscous dissipation contributes significantly to the heat development during the performance of the weld bead;
- The materials thermo-mechanical properties vary throughout the FSW process;
- The thermal flow into the tool and support plate, needs to be considered in the models;
- FSW process modeling does not allow geometric simplification because it deals with a complex 3D material flow around the pin;
- The highly rotating tool pin has, typically, a complex geometric profile (e.g. threaded), which is rather difficult to consider for most of the numerical methods available.

The challenge is then to create a model able to fully describe the complex FSW process as illustrated in Figure 21.

4. Friction Stir Channeling

4.1. Fundaments of the process

Friction Stir Channeling (FSC) is an innovative solid-state manufacturing process able to produce continuous internal channels in monolithic plates in a single step. The channels can have any path and variable dimensions along the path. During FSC, a non-consumable rotating tool with a specific shoulder and probe profile is inserted into the solid block, or plate, where the channel is to be opened, and forced to traverse along a predetermined path, creating a fine grain recrystallized microstructure around the new channel. The FSC process results from the application of the correct combination between the direction of tool rotation and the orientation of the probe threads and shoulder scrolls. The following actions will be applied to the visco-plasticized workpiece material: i) an upward action directed to the shoulder, along the threaded probe, combined with ii) an outward/centrifugal action along aspiral scrolled shoulder.

The features of the channels produced with FSC can be controlled by selecting the processing parameters and tool geometry, e.g., during the FSC process, an upward force is generated by rotating a left-hand threaded tool counter-clockwise. The action of the tool produces the "third-body region"and forces part of this viscous material to flow out from the processing zone. Simultaneously, the FSC tool closes the top of the processed zone via the action of the shoulder, enabling the creation of a continuous internal channel. The Figure 22 shows a schematic representation of the FSC process [28].

The FSC process was firstly proposed as a method of manufacturing heat exchanging devices. The applicability of the FSC concept have been discussed and demonstrated in [7,29] to create continuous channels along linear and curved profiles, as well as the possibility of manufacturing mini channel heat exchangers (MCHX). The FSC process was initially based on converting one possible defect in FSW: the formation of internal continuous voids, into a manufacturing technique where all the material extracted from the metal workpiece laid over the processed zone bellow the shoulder [29], within a clearance between the shoulder and the metal workpiece.

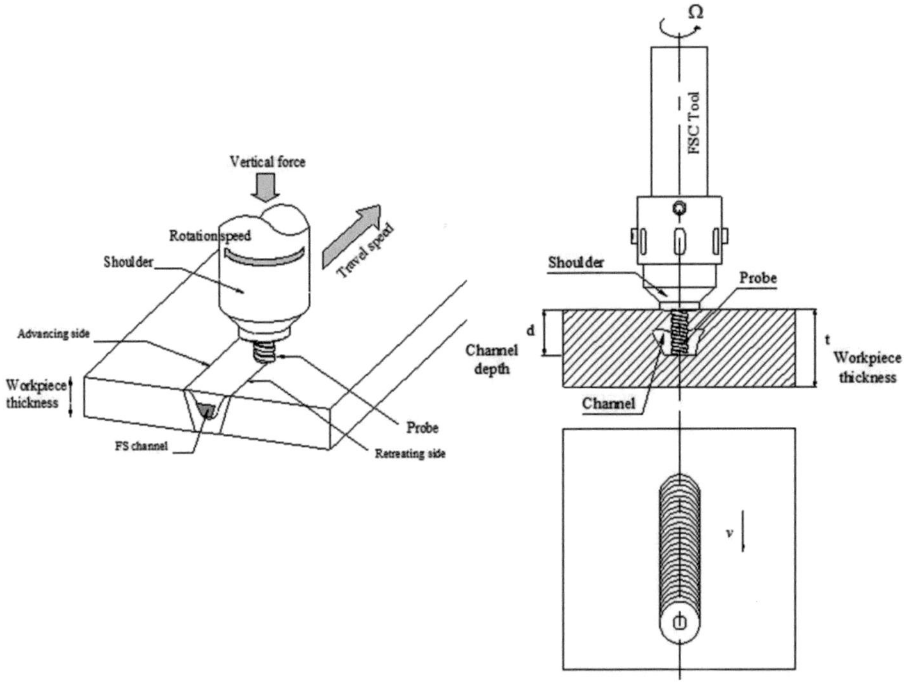

Figure 22. Schematic representation of the FSC process fundaments: parameters and viewsof the cross-section and plan.

There is a study [30] presenting a model based on the flow partition deformational zones for defect formation during FSW. The occurrence of voids in the FSW nugget has been attributed to defective material flow mainly because of the non-optimal processing conditions or geometry of the tool features. The model applies the principle of mass balance to address void formation in the nugget.

A recent development made by the authors of the present chapter and proposed in [31] allows to promote distinct material flow, where a controlled amount of material from the metal workpiece, flow out from the processed zone producing the internal channel. Thus, the material flowing from the interior of the solid metal workpiece is not deposited on the processed surface but flowed outside from the processed zone in the form of flash self-detachable or easy to extract. The position and size of the channels can therefore be controlled and the processed surface can be left at the same initial level. It is also possible to integrate in the tool, a surface finishing feature [31].

http://dx.doi.org/10.5772/52026

The high flexibility and low production costs of this innovative manufacturing process provide this technology a great potential to be successfully introduced in various industries. However, FSC still need to have a considerable development to prove its industrial applicability.Recent advances of tooling for FSC [28] shown to be possible to control during FSC the processed surface finishing and position. The high level of adaptability of FSC makes it possible to apply to many different technical field domains and can bring significant advantages for already existent and new industrial applications.Meanwhile some aluminum alloys have already been subjected to FSC, e.g.: 5 mm thick plates of AA6061-T6 [29], 13 mm thick plates of AA7178-T6 and 15 mm thick plates of AA5083-H11 [32].

The Figure 23 shows a representation of a typical cross section of a channel from FSC with the identification of the advancing and retreating sides and depicting the main microstructural regionsin the vicinity of the channel, namely: i) nugget; ii) TMAZ; iii) HAZ, and iv) BM.

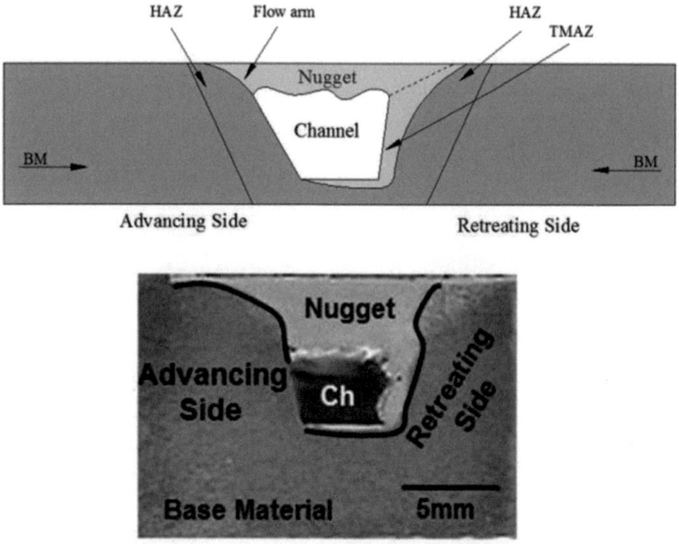

Figure 23. Typical cross section macrograph of a friction stir channel including the main microstructural regions.

4.2. Parameters of the process

Because the FSC is based on the same principles of FSW, the parameters type and the process control in FSC, via position/vertical downward force criteria, are essentially the same of the FSW process. Understanding the material flow, process forces and metallurgy is therefore necessary to control and optimize the channel formation to ensure its industrial successful application.

In order to optimize the friction stirred channels components' performance, and considering that the channel formation is sensitive to FSC parameters variations, it is important to identify, understand and establish interactions between the processing parameters. Some know-how is already possible to establish:

- The probe and shoulder geometry defines the material removed from the interior of the metal workpiece. The material at the "third-body region" is brought against the shoulder by the probe and moved outside by the spiral striate(s) of the shoulder as result of the combination between the i) orientation and geometry of the threads on the probe; ii) orientation and geometry of the shoulder striate(s); iii) direction of tool rotation and iv) remaining process parameters;
- The position of the advancing side and bottom of the channel (channel's depth) are prescribed and well controlled by the shape and position of the probe within the monolithic workpiece material. These surfaces present relative smooth finishing;
- The channel ceiling and retreating side present rough finishing and their position is controlled by the process parameters. Thus, by tuning the parameters it is possible to establish the section of the channel and optionally continuously change this section during the FSC process;
- The shoulder is responsible for constraining the deformed material, closing the ceiling of the channel and its scrolls for driving away a controlled amount of material from the interior of the metal workpiece to outside, in the form of flash;
- If the parameters are correct, the amount of material brought against the shoulder by the probe is equal to the amount of material removed outside by the shoulder;
- If the shoulder is plan and smooth (i.e., not scrolled), the material pushed out by the probe features is deposit underneath the shoulder, at the inherent gap between the shoulder and the workpiece;
- The tool rotation and travel speed control the amount of visco-plastic flow and subsequent heat generation from frictional dissipation during the process. These two parameters, along with plunge depth, if the process is performed via position control, affect the axial force applied on workpiece, which significantly affects the overall channel quality;
- The operating conditions for which the process is more stable and produce better results are cold conditions;
- The FSC parameters are not directly transferable from one aluminum alloy to another, the workpiece material thickness and the thermal conductivity of materials in contact with the workpiece, like anvil and clamping system, influence the cooling rate and the temperature gradients through thickness. Also the heat cooling system of the tool and dissipation through the spindle can influence the process and the tool lifetime.

4.3. Tool features and development

In the FSC tool the probe and shoulder geometry are the most important features that influence the channel formation, shape, localization and properties. The initial FSC studies conducted in [29] used a cylindrical threaded probe with a diameter of 5 mm and 4 mm length and with an initial clearance between the shoulder and the metal workpiece, where the material from the base of the probe is deposited. Since that time, developments have been made by the authors of present chapter [28,31,32], allowing to promote a distinct material flow, discarding the need for a gap between the shoulder and the original surface of the workpiece. At the actual state of the art, with the correct set of parameters there is no need of further finishing operations, but if any is demanded, the actual shoulder design has advanced to geometries capable to integrate a surface finishing feature, and then processed surface can always be left at any required final level, simultaneously with the FSC process, and it is not require to perform a further finishing process. Figure 24 shows the modular FSC tool concept developed [31].

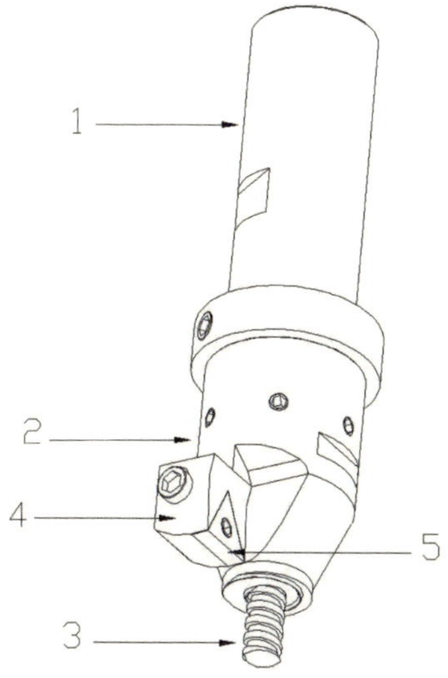

Figure 24. – 3D view of a modular FSC tool. (1) tool body, (2) shoulder, (3) probe, (4) cutting insert clamping, (5) cutting insert.

Resulting from experimental tests, it was established [31] that the number of cutting inserts might have the same that the number of the shoulder spirals striates as shown in Figure 25.

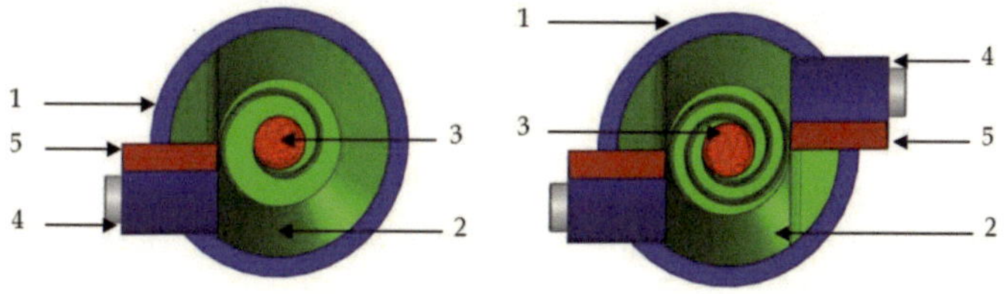

Figure 25. Top view sections showing the equality between the number of scrolls and the number of cutting inserts on the shoulder. (1) tool body, (2) shoulder, (3) probe, (4) cutting insert clamping, (5) cutting insert.

Initially, the FSC tool's components were made of the same material typically applied in FSW: AISI H13, however this material has shown not to be the most appropriate to produce the probes to perform FSC on aluminium alloys at depths greater than 10 mm from the metal workpiece surface. Developments are being made, namely the study of new probe materials and designs, in order to increase the tool's life. Conical probes have long lives comparing to cylinders ones but cylinders probes are more efficient than conical for this technological process which the main objective is to extract material from the metal workpiece.

4.4. Advantages and limitations

The following advantages can be established for the innovative solid-state manufacturing FSC process:

- Able to produce continuous free path internal channels in monolithic plates in a single step;
- The channels can have variable positioning and dimensions along the path;
- The channels have good dimensional stability and repeatability;
- Recent advances of tooling for FSC shown to be possible to control the surface finishing and position of the exterior processed surface, during the execution of the process;
- Capability of producing highly efficient conformal cooling/heating channels. FSC process also has the capability to produce cooling/heating systems in a single component whereat no time is loss joining components;
- The high level of adaptability of FSC makes it possible to apply to many different technical field domains and can bring significant advantages as alternative of already existent technologies, such as, drilling, EDM and milling;
- Environmentally friendly process using a no consumables.

The current limitations of the FSC process are:

- At the actual state of art, the non-consumable tool lifetime is small because of premature fracture under the complex and demanding fatigue loading;
- Sensitivity to change of parameters;
- To produce channels by FSC access from one surface of the metal workpiece is required;
- A residual hole is left open at the end of the channel path.

4.5. Geometric and metallurgical characterization of the channels

The shapes of the channels obtained from FSC are closer to a parallelogram. The channel shape varies with the process parameters. Figure 26 shows the variation of the shape of the channel with changing the FSC process parameters.

The channel geometry can be attributed to the volume of processed material that is displaced out of the "third-body region" by the FSC toolper unit of rotation and also the compacting force that is applied on the channel ceiling during the travel forward movement by the rotating shoulder. The upper surface of the channel (channel ceiling) is rough and wave shape and the channel advancing side (shear side) does not exhibit any significant roughness comparing to the retreating one. The bottom of the channel is relatively smooth and flat due to the flat nature of the tool probe base (Figure 27).

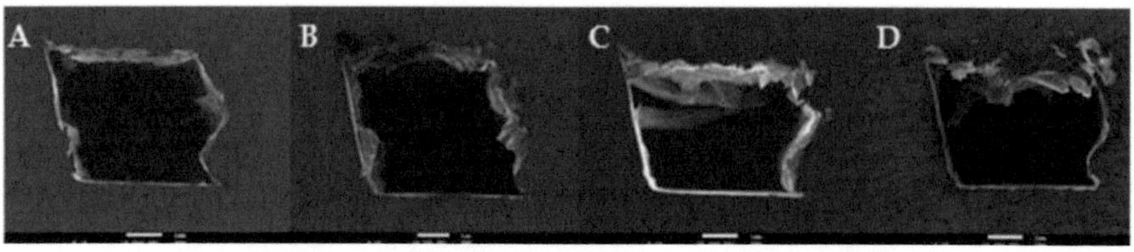

Figure 26. Cross section macrograph showing the channel geometry at different FSC processing parameters for the same FSC tool: a) 600 rpm, 80 mm/min.; b) 600 rpm, 150 mm/min.; c) 800 rpm, 80 mm/min. and d) 800 rpm, 150 mm/min.

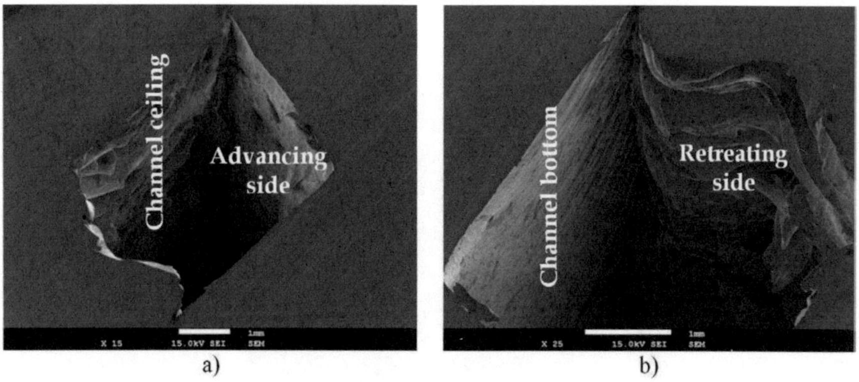

Figure 27. Cross section of a channelshowing the roughness: a) on the ceiling and on the advancing side; b) on the bottom and on the retreating side.

The Figure 28 shows a cross section of a friction stir channel produced with a tool rotation speed of 800 rpm and a tool travel speed of 80 mm/min. In the macrograph depicted in Figure 28 three main regions are visible: i) channel; ii) stir zone (nugget), and the iii) unprocessed base material. The channel nugget (Figure 28b) presents a fine equiaxed recrystallized grain, with a tail heading to the shoulder periphery, at the advancing side (Figure 28c).In details "a)", "c)" and "d)" of Figure 28 it is possible to identify an additional layer surrounding the nugget, referred to as a thermo-mechanically affected zone (TMAZ). Due to tool rotation and linear movement combination, the probe shears the material from the advancing side and flows it around the retreating one, resulting in an asymmetric processed zone. It can be observed in details "e)" and "f)" in Figure 28 that the stir zone (nugget) is more extensive in the retreating side than in the advancing one. The Figure 29, present the hardness field measured in the same cross-section presented in Figure 28.

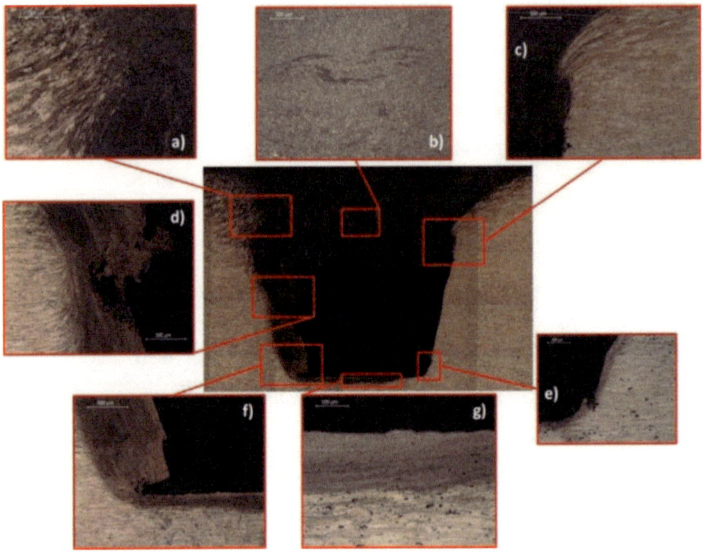

Figure 28. Metallographic results of a cross section at the vicinity of a channel produced with FSC: Ω= 800 rpm and v = 80 mm/min.

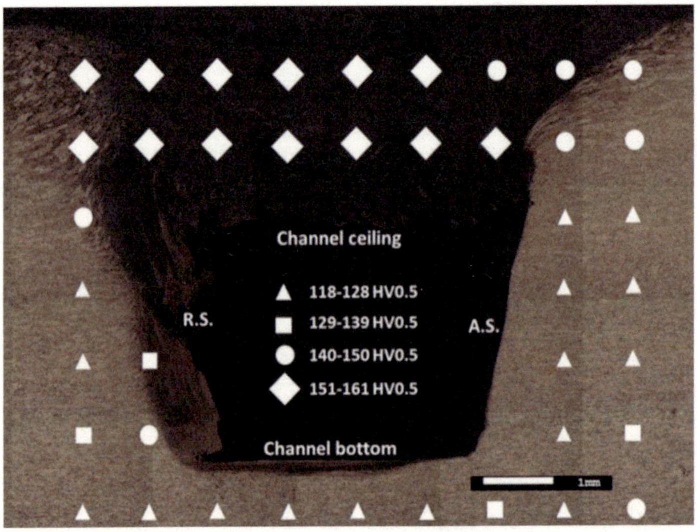

Figure 29. Hardness field in the vicinity of a channel produced with FSC: = 800 rpm and v = 80 mm/min.

In order to standardize the characteristics of the channel shape and to spot possible trends in the channel shape change with the variation of the process parameters, three geometrical parameters of the channel were establish in Figure 30:

- Channel area;

- Closing layer thickness (D);
- Shear angle (α).

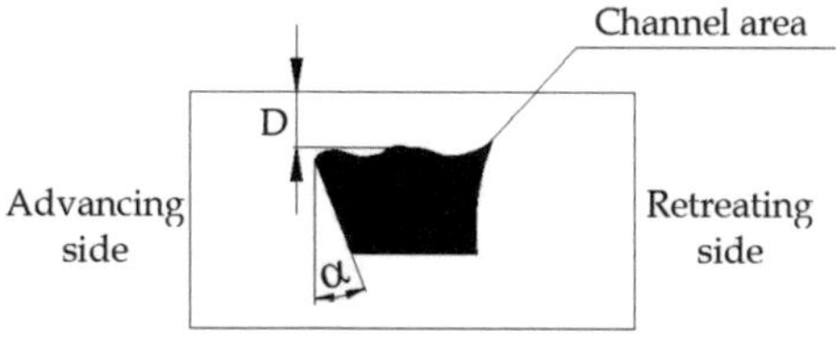

Figure 30. Schematic representation of a cross-section view of a friction stirred channel workpiece showing the geometrical parameters closing layer thickness (D) and shear angle (α).

Results presented in [28] shown that the closing layer thickness values are consistent with those obtained for the channel area, i.e., as the channel area increases, the closing layer thickness decreases. But on the other hand, the shear angle has no relationwith the values of channel areas or even with the closing layer thicknesses for the different FSC conditions.

4.6. New tests for characterization of the channels

Despite the similarities between FSC and FSW processes, some of the testing techniques and criterions typically applied to the inspection and characterization of weld joints properties make no sense to be directly applied to FSC, e.g., direct comparison of mechanical properties of welded specimens with similar base material properties. In order to assess the quality of the channels for different paths, procedures are proposed to assess the quality of FSC results for slop and spiral paths.

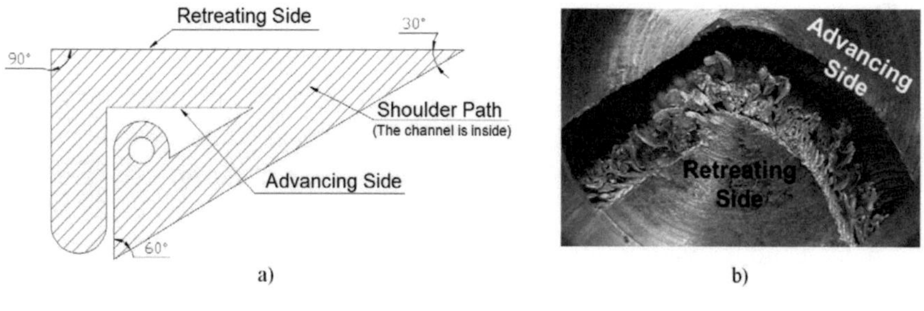

Figure 31. Slope paths. a) linear sloped path tested and b) result of the visual inspection after milling a hole at the 90 degree corner.

For linear sloped paths, namely for the closed ones (Figure 31a), the path should be carried out in order to the retreating side stays outside and the advancing side inside the path. The path presented allows assessing continuous channels with variations of 30, 60 and 90 degrees along its path. Linear sloped paths corners should be inspect. As a procedure, it is proposed milling the FSC plate of the non-processed side to the channel bottom using a milling cutter with the same diameter than the FSC tool shoulder. The result of applying this procedure to the 90 degree corner is shown in Figure 31b.

Such as for linear sloped paths, the spiral paths should be carried out in order to the retreating side stays outside and the advancing side inside the spiral. The minimum allowed gap between spirals is the shoulder radius. To assess the geometrical stability of channels during spiral paths two cuts are recommending: a horizontal cut through the transition points of the arcs and a vertical cut through the outermost points of each arc as shown in Figure 32. Made the cuts, the channels cross sections should be assessed, e.g., according to the geometrical parameters presented in Figure 30.

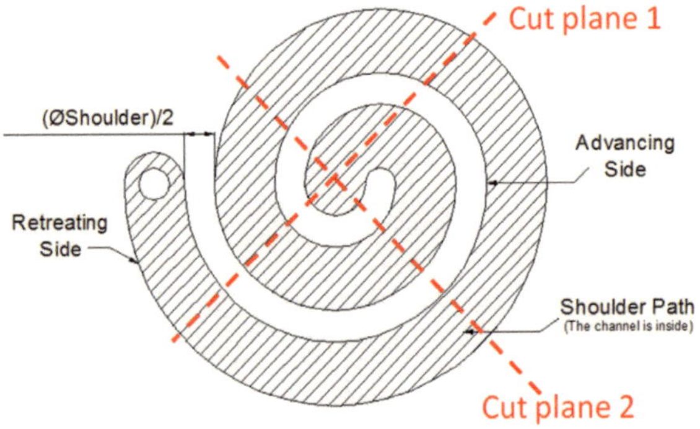

Figure 32. Schematic representation of the spiral path produced showing the minimum allowed gap between spirals (shoulder radius).

In order to evaluate the mechanical strength of the FSC processed zone, a free-bend test is proposed as a FSC performance parameter. The proposal is to analyse the mechanical strength of a large area, the FSC processed zone, rather than a specific and small zone as the welded joints root in FSW. In Figure 33 is shown the 4 points free-bend tests results for four specimens tested which were extracted from the same trajectory of an unstable FSC procedure. From Figure 33 it is possible to verify, that the channel strength is not constant over the path. The friction stir processed zone is more resistance at the beginning of the path than at the end. All specimens tested fractured at the advancing side, namely in the boundary between the nugget and the thermo-mechanically affected zone.

To obtain the indentation resistance of the processed zone with FSC, the authors also propose aMacro indentation resistance test. Thus, specimens with a transversal channel should be produced according to the following dimensions: i) Length ≥7×∅Shoulder; ii) Width= 2×plate thickness; iii) Depth =plate thickness. The specimens should betested according to the apparatus presented in Figure 34. The friction stir processed zone must be in contact with the mandrel. The mandrel diameter must be equal to the probe maximum diameter. It is recommend a mandrel velocity about 1mm/min. The test ends when the maximum force is reached.

http://dx.doi.org/10.5772/52026

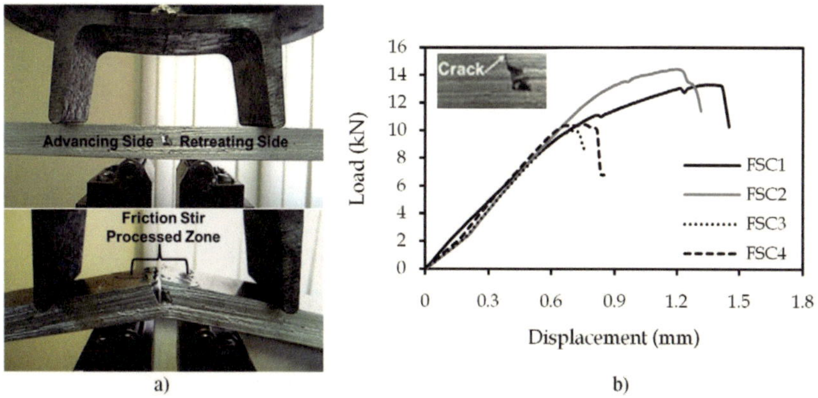

Figure 33. Points free-bend tests for FSC: a) schematic representation of the standard E 290-97a suited for testing FSC specimens 4pt bending resistance and b) experimental results.

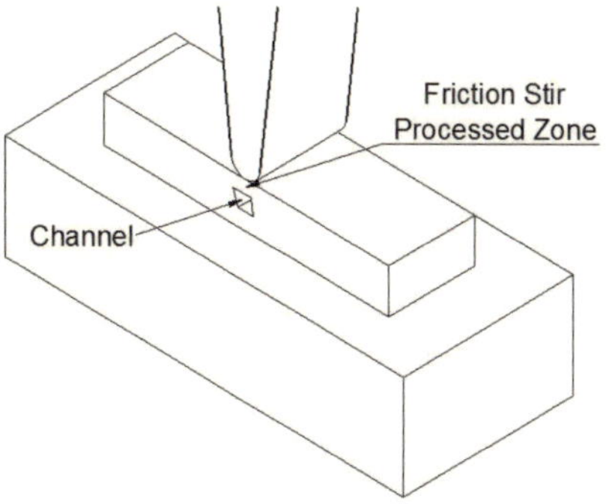

Figure 34. Macro indentation resistance test apparatus.

Because friction stir channels have high potential application in heat exchanger industry and in conformal cooling/heating systems, it is crucial to assess effective air tightness and their mechanical resistance when loaded with internal pressure (in Figure 35, it is proposed an internal hydraulic pressure test). For these tests, it should be produced parallelepiped specimens with the friction stir processed zone centered and according to the following dimensions: i) Length≥5×∅Shoulder; ii) Width =2×∅Shoulder; iii) Depth =plate thickness.

To assess the channels' airtightness should be used pressurised helium (at a minimum of 5 bar) because the helium atom is the smallest among the inert gases and thus, it is possible to ensure that any cracks greater than the helium's atom is detected. In order to see the released gas, the specimen should be immersed in a container with clean water.

Relatively to the internal hydraulic pressure tests, they were carried out using an oil pressing machine and a 100 bar manometer. Themanometerusedshould bewhich ensuresthe FSC system design requirements.

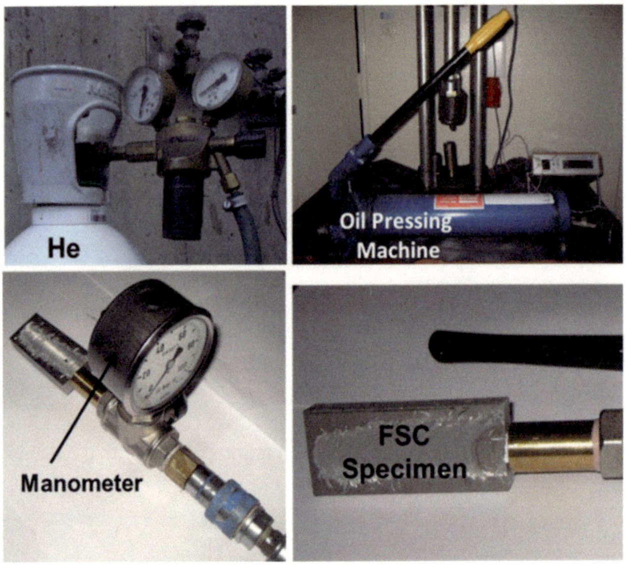

Figure 35. Airtightness and internal hydraulic pressure tests apparatus.

4.7. Establishment of potential applications

The value of a technology is directly related with the amount of solutions and add-value it can provide to the industry. Consequently, if a new technology is being developed it is of major importance to analyse what are the possible industrial applications it has.

The generation of a continuous channel by FSC has the potential to open a wide range of applications, such as, conformal cooling/heating systems, heat exchangers and advanced tailored performance engineering metallic materials.

Conformal cooling is a concepton which the refrigerating channels follow the configuration of the part shape, enabling higher quality and productivity in the cooling/heating procedure. The influence on the cooling time and surface quality is significant and this concept has been growing recently. The FSC process is able to produce these conformal channels.

A prototype was developed for a company producing plastic injection moulds to demonstrate the potential of FSC technology (Figure 36). The prototype was discussed with the company, to determine if it could be an alternative to produce small dimension channels for thermoplastic parts that are produced by mould injection. These moulds have cooling channels behind the part surface that defines the geometry of the part. Owing to the complex geometries of certain components, FSC could be the solution for these channels due to the flexibility of the technology that permits the channels to have any desired path and position within the mould.

The generation of a continuous channel by FSC has the potential to open a wide range of applications also in heat exchanger industry.Compact heat exchangers are generally used in industry, especially in gas-to-gas or liquid-to-gas heat exchangers. For example, vehicular heat exchangers, condensers and evaporators in air-condition and refrigeration industry, aircraft oil-coolers, automotive radiators, and intercoolers or compressors and FSC has the ability to produce the channels for any of these types of heat exchangers, which demonstrates the elevated applicability it has in the various industries [7].

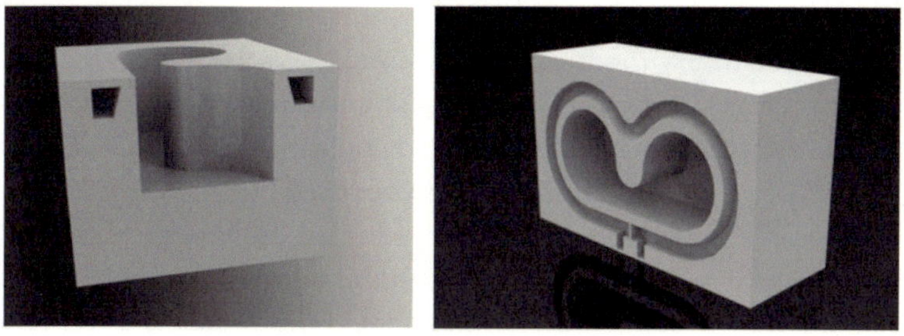

Figure 36. Sample of a prototype developed for a mouldcompany with a conformal cooling channel produced by FSC.

A new application, in development, is the production of Advanced Tailored Performance of Engineering Metallic Materials (ATEM) by FSC. A predetermined friction stirred channel pattern is produced in an aluminum workpiece in order to obtain a component with specific mechanical and metallurgical properties. FS channels can be filled with other materials or be used only to reduce the component structural weight and optimize is stiffness as illustrated in Figure 37. The channels produced with FSC, can also be used for crossing wires within solid components with many potential application for aluminium alloys structures, namely in aeronautics.

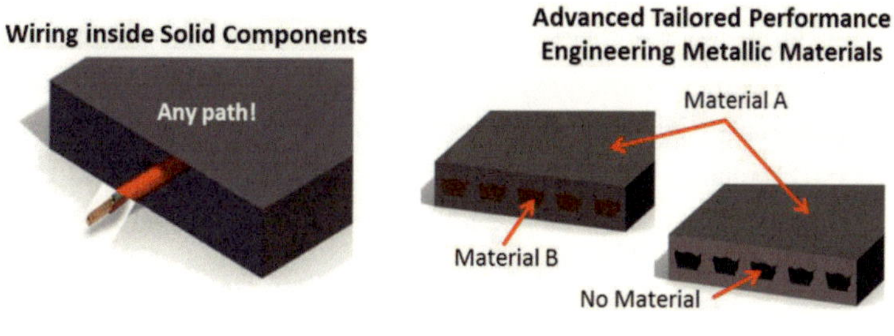

Figure 37. Schematic illustration of two alternative industrial applications with internal channels produced by FSC.

5. Conclusions

From the present work the following conclusions can be drawn:

- Three relevant friction based processing processes covering a wide range of technological applications are presented in it state of the art fundaments and main features. Also, the most significant experimental results are depicted;
- The concept supporting all the three processes addressed in the present chapter, and many others processes on the field of solid state processing, is the material flow within open or closed "third-body region". This concept is introduced in some detail in order to allow readers to be able to further develop the actual solutions and invent new ones;
- The Friction Surfacing (FS) process is a lean coating technology ideal for localized surface engineering applications, requiring the joining of materials with compatibility issues. As a solid state process, there is no melting involved and the coating material is solely provided by a consumable rod. Friction surfacing produces high strength coatings, soundly bonded, with low dilution, no porosity and little part distortion, making FS suitable to process thermal sensitive materials, such as, aluminium alloys. The absence of splashes, toxic fumes and radiation makes friction surfacing a cleaner alternative. However, bonding quality at coating edges need further evolution and post processing is generally required to obtain uniform coating surface and remove the poorly bonded portions;
- The basic fundaments of friction stir welding (FSW) process have been presented. These basic fundamentals have enabled the invention and development of many variants of FSW. FSW is a mature and reliable technology with guidelines in many construction codes, mainly focusing on products made from aluminium and its alloys. The new ISO standard 25239 Friction stir welding - Aluminium alloys, has been published. This ISO standard will help with the implementation of FSW technology in light metal fabrication industries. Moreover, the FSW is a very complex process and difficult to be assessed based on computational modeling;
- The Friction Stir Channeling (FSC) process is a disruptive innovation enabling higher efficiency in energetic applications and advances of structural design of many products. With this chapter the authors proposed new feasible FSC performance parameters based on non-conventional testing techniques, some quantitative and others qualitative. The FSC performance parameters proposed to assess the tightness are based on the amount of material removed, the macro indentation resistance of the FSC processed zone, the bending resistance of the FSC processed surface under tensile stress and the geometrical stability of channels during linear and spiral FSC paths.

Acknowledgements

The authors would like to acknowledge FCT/MCTES funding for the project FRISURF' (PTDC/EME-TME/103543/2008) and PhD grants: SFRH/BD/62963/2009 and SFRH/BD/78539/2011.

http://dx.doi.org/10.5772/52026

Author details

Pedro Vilaça[1,2*], João Gandra[2] and Catarina Vidal[2]

*Address all correspondence to: pedro.vilaca@ist.utl.pt

1 IDMEC, Instituto de Engenharia Mecânica, Lisbon, Portugal

2 IST-UTL, Instituto Superior Técnico, Lisbon, Portugal

References

[1] Vilaça, P. (2003). Fundaments of Friction Stir Welding Process- Experimental Analysis and AnalyticalModeling. *PhD Thesis*, Technical University of Lisbon.

[2] Vilaça, P., Pepe, N., & Quintino, L. (2006). Metallurgical and Corrosion Features of Friction Stir Welding of AA5083H111. Welding in the World. , 50(9/10), 55-64.

[3] Leitão, C., Leal, R. M., Rodrigues, D. M., Loureiro, A., & Vilaça, P. (2009). Mechanical Behaviour of Similar And Dissimilar AA5182H111 And AA6016-T4 Thin Friction Stir Welds. *Materials and Design*, 30, 101-108.

[4] Thomas, W. M. (2009). An Investigation and Study into Friction Stir Welding of Ferrous-Based Material. *PhD thesis*, University of Bolton.

[5] Klopstock, H., & Neelands, A. R. (1941). An Improved Method of Joining or Welding Metals. *Patent application Nº 572789*.

[6] Thomas, W. M., Nicholas, E. D., Needham, J. C., Murch, M. G., Temple-Smith, P., & Dawes, C. J. (1991). Improvements relating to friction stir welding. *US Patent Nº. 5460317*.

[7] Mishra, R. S. (2005). Integral Channel in Metal Components and Fabrication Thereof. *US Patent Nº 6923362*.

[8] Bedford, G. M., Vitanov, V. I., & Voutchkov, I. I. (2001). On The Thermo-Mechanical Events During Friction Surfacing of High Speed Steels. *Surface and Coatings Tecnhology*, 141-34.

[9] Nicholas, E. D., & Thomas, W. M. (1986). Metal Deposition by Friction Welding. *Welding Journal*, 8-17.

[10] Shirzani, A. A., Assadi, H., & Wallach, E. R. (2001). Interface Evolution And Bond Strength When Diffusion Bonding Materials With Stable Oxide Films. *Journal of Surface and Interface Analysis*, 31-609.

[11] Batchelor, A. W., Jana, S., Koh, C. P., & Tan, C. S. (1996). The Effect of Metal Type and Multi-Layering on Friction Surfacing. *Journal of Materials Processing Technology*, 57-172.

[12] Chandrasekaran, M., Batchelor, A. W., & Jana, S. (1997). Friction Surfacing of Metal Coatings on Steel and Aluminum Substrate. *Journal of Materials Processing Technology*, 72-446.

[13] Vitanov, V. I., Voutchkov, I. I., & Bedford, G. M. (2000). Decision Support System To Optimize The Frictec (Friction Surfacing) Process. *Journal of Materials Processing Technology*, 107-236.

[14] Macedo, M. L. K., Pinheiro, G. A., Santos dos, J. F., & Strohaecker, T. R. (2010). Deposit By Friction Surfacing and its Applications. *Welding International*, 24-422.

[15] Nicholas, E. D. (1993). Friction Surfacing. ASM Handbook- Welding Brazing and Soldering. *ASM International*, 6, 321-323.

[16] Rafi, H. K., Ram, G. D. J., Phanikumar, G., & Rao, K. P. (2011). Microstructural Evolution During Friction Surfacing of Tool Steel H13. *Materials and Design*, 32-82.

[17] Gandra, J., Miranda, R. M., & Vilaça, P. (2012). Performance Analysis of Friction Surfacing. *Journal of Materials Processing Technology*, 212-1676.

[18] Seidel, T. U. , & Reynolds, A. P. (2011). Visualization of the Material Flow in AA2195 Friction-Stir Welds Using a Marker Insert Technique. *Metallurgical and Materials Transactions A.*, 32(11), 2879-2884.

[19] London, B., Mahoney, M., Bingel, W., Calabrese, M., & Waldron, D. (2001). Experimental Methods for Determining Material Flow in Friction Stir Welds. Japan. *3rd International Symposium on Friction Stir Welding.*

[20] Vilaça, P., Quintino, L., Jorge, F., & Santos, . (2005). dos. iSTIR- Analytical Thermal Model for Friction Stir Welding. *Journal of Materials Processing Technology*, 169(3), 452-465.

[21] Leitão, C., Leal, R. M., Rodrigues, D. M., Vilaça, P., & Loureiro, A. (2008). Material flow in Friction Stir Welding. *Journal of Microscopy and Microanalysis*, 14(S3), 87-90.

[22] Leal, R. M., Leitão, C., Loureiro, A., Rodrigues, D. M., & Vilaça, P. (2008). Material Flow in Heterogeneous Friction Stir Welding of Thin Aluminium Sheets: Effect of Shoulder Geometry. *International Journal of Materials Science and Engineering A*, 498(1/2), 384-391.

[23] Vilaça, P., Pepe, N., & Quintino, L. (2006). Metallurgical and Corrosion Features of Friction Stir Welding of AA 5083H111. *Welding in the World*, 50(9/10), 55-64.

[24] Vilaça, P., & Quintino, L. (2006). Experimental and Computational Developments of FSW. Welding. *Equipment and Technology*, XVII, 17-24.

[25] Santos, T., Vilaça, P., Reis, L., Quintino, L., & Freitas, M. (2008). Advances in NDT Techniques for Friction Stir Welding Joints of AA2024. New Orleans, USA. *The Minerals, Metals & Materials Society (TMS) 2008 Annual Meeting & Exhibition*, 3, 27-32.

[26] Pépe, N., Vilaça, P., Quintino, L., Reis, L., & Freitas, M. (2006). Fatigue Behavior of Shipbuilding Aluminium Alloy Welded by Friction Stir Welding, Session 3: Welds. Atlanta, USA. *9th International Fatigue Congress*.

[27] Santos, T., Vilaça, P., Quintino, L., Dos, J., & Santos, . (2009). Computational Tools for Modeling FSW and an Improved Tool For NDT. *Welding in the World*, 53(5-6), r99r108.

[28] Vidal, C., Infante, V., & Vilaça, P. (2012). Mechanical Characterization of Friction Stir Channels under Internal Pressure and In-Plane Bending. *Key Engineering Materials*, 105-108.

[29] Balasubramanian, N., Mishra, R. S., & Krishnamurthy, K. (2009). Friction Stir Channelling: Characterization of the Channels. *Journal of Materials Processing Technology*, 209-3696.

[30] Arbegast, W. (2008). A Flow-Partitioned Deformation Zone Model for Defect Formation During Friction Stir Welding. *Scripta Materialia*, 58-372.

[31] Vilaça, P., & Vidal, C. (2011). Modular Adjustable Tool and Correspondent Process for Opening Continuous Internal Channels in Solid Components. *National patent pending Nº 105628 T*.

[32] Vidal, C., Infante, V., & Vilaça, P. (2011). Mechanical Characterization of Friction Stir Channels Under Internal Pressure and In-Plane Bending. Dubrovnik, Croatia. *10th International Conference on Fracture and Damage Mechanics*.

Chapter 8

Dry, Semi-Dry and Wet Machining of 6061-T6 Aluminium Alloy

J. Kouam, V. Songmene, M. Balazinski and
P. Hendrick

Additional information is available at the end of the chapter

http://dx.doi.org/10.5772/51351

1. Introduction

In the industries, several techniques are used to reduce the manufacturing costs and to protect the environment. In machining process, the used of lubrication could play an important role in different parameter such as tool life, cutting temperature, surface finish, chip formation and metallic particle emission.

The use of cutting fluids is between 7 to 17% of manufacturing costs parts [1]. In this point of view dry machining could help to reduce the cost of cutting fluid and to protect the environment. The main problem of using the cutting fluid instead of reducing the cutting zone temperature is the aerosol generation during the machining process [2].

Sutherland et al. [3] showed in their study that the aerosol quantity produces in the lubrication machining could be 12 to 80 times higher compared to the dry machining. Anselmo et al. [4] carried out a study in dry machining of steel 1045 using two different tools. They found that dry machining requires a very hard tool material that is resistant to high temperature. They also found that tool life in dry machining could be similar in lubricated machining if the cutting depth is fewer with high cutting speed.

Although the good performance of dry machining, the major problem is still the tool wear under severe conditions which sometimes need the use of small lubrication fluid. Some researchers [1, 5, 6] developed a method using small lubrication fluid in machining process. Using the absorption characteristics of the ester model and the surface tool analysis some researchers [1, 7] carried out a study on a tribology of cutting performance in semi-dry.

In machining process, Islam et al. [8] in their work studied the turning performance in lubrication condition. They showed that the cutting fluid supply strategies have no influence on the surface finish but the amount of MQL (minimum quantity lubrication) could have an influence. Machado et al. [9] showed in their study on turning of carbon steel that the decreasing of the cutting fluid debit decrease the cutting force and improve the surface finish.

The surface finish is an indicator of the quality of the material following the machining process. Vikram Kumar et al. [10] carried out a study on surface roughness during hard turning in Dry, MQL and wet conditions on AISI 4340 alloy steel. The feed rate range use in this work was from 0.04 to 0.06 mm/rev. The authors showed that in the different lubrication condition the roughness against the feed rate is constant. The feed rate range use in this work is too limited to establish a good relationship between roughness and feed rate and also to observe that the cutting speed has no influence on roughness. At the same feed rate, the roughness is low in the MQL condition compare to wet and dry condition.

Another study was also carried out in a study of surface roughness during hard turning in Dry, MQL and wet conditions on AISI 4340 alloy steel [11]. In this study, the feed rate range was from 0.05 to 0.14 mm/rev and the maximum of cutting speed was 120 m/min. The authors observed that from 0.05 to 0.1 mm/rev feed rate range the roughness is approximately similar and constant at different lubrication condition. From 0.1 to 0.14 mm/rev, the roughness increases with the feed rate and is low in the MQL condition compared to wet and dry condition. In this work again, the feed rate and the cutting speed range larger still limited to establish a good relationship between roughness and feed rate. Ozawa et al. [12] also showed that using MQL gives good results about surface roughness.

Decreasing the chip size improves tool life, surface finish and the energy required for machining. Several research works have analyzed chip formation, including works by Hua et al. [13], Poulachon et al. [14] and Jawahir et al. [15], which have covered different types of chips: continuous chip, segmented chip and elemental chip. Dhar et al. [16] have done some investigations on turning of AISI 1040 steel in Dry, MQL and wet conditions. The authors studied the chip formation by analysing the chip reduction coefficient. In this study, the feed rate range was from 0.1 to 0.2 mm/rev and the cutting speed range was from 60 to 130 m/min. This study was also about turning of AISI 1040 steel. The authors observed that the chip reduction coefficient decreases when the cutting speed increases at different feed rate. In another hand the MQL condition present the lowest values compared to the wet and the dry condition.

Dhar et al. [16] have presented the effect of dry, MQL and wet lubrication on tool wear. This work presents the tool wear after machining during 45 min in the dry, MQL and wet condition in the turning process of steel material. The authors observed that the tool wear is less important in the MQL condition compared to wet and dry condition. In the case of aluminum alloys, the considered criterion for tool life is the tool breakage due to adhesion of the aluminum alloy on the tool as a result of the heat involved in the process or the wear experienced by the cutting tool. Nouri et al. [18], showed in their study in dry drilling of aluminum that the most important causes of tool wear are the diffusion and adhesion of the material, which could degrade the surface finish. Damir et al [19] and Diakodimitris et al. [20] also presented the effect dry, MQL and wet lubrication on tool wear in the milling process. In this work, the authors showed that MQL clearly improves the tool life reducing the tool wear drastically.

Machining processes generate aerosols (in dry, MQL and wet conditions) which are harmful both to the health of operators and to the environment [3, 16, 22, 23, 24, 25, 26]. Sutherland et al. [3], in their machining work, showed that particle emissions is high in wet machining compared to dry machining. They also showed that particle emissions increase with the cutting speed. Khettabi et al. [24] also carried out a study in particle emission during turning of the 6061-T6 aluminium alloy, and 1018 and 4140 steels, using carbide cutting tools. In their work they showed that there is two speed regimes were the particle emissions is minimize. This observation was also confirmed by Kouam et al. [25] in their work on drilling, and by Kouam et al. [26] in their work on friction.

The aim of this research is to investigate the effect of the lubrication (dry, semi-dry and wet) during the turning of 6061-T6 aluminum alloy. The process performance indicators investigated include the surface roughness, tool wear, chip formation and particle emission.

2. Lubrication effect on turning processes of 6061-T6

Turning tests were carried out on 6061-T6. The tool insert used had a TiB_2 PVD (Physical Vapor Deposition) coating over a very deformation-resistant unalloyed substrate with 80° nozzle angle and 11 relief angle.

The experiments were conducted on a lathe turn Darbert Machinery model and the experimental parameters were as follows:

- Cutting speed: 79.40-661.54 m/min.
- Feed rate: 0.0508-0.2845 mm/rev
- Cutting depth: 1 mm
- Lubrication condition: dry, semi-dry (3.06, 1.75 and 0.6 ml/min debit) and wet

The chemical composition of 6061-T6 is in Table 1.

	Si	Fe	Cu	Mn	Mg	Cr	Zn	Ti	Al
6061-T6	0.7	0.5	0.22	0.09	0.93	0.08	0.15	0.08	Balance

Table 1. Chemical composition of 6061-T6 alloys

In semi-dry condition, two lubricants (Mecagreen 550 and Microkut 400) were used and in wet condition, the commercial lubricant at 306 ml/min debit was used.

An original cooling and lubrication method is used here under the form of micro-lubrication or MQL (Minimum Quantity Lubrication) [19]. The main component of this MQL system is an airblast atomizer injector (as the SB202010 shown on Figure 1, with an oil nozzle diameter of 0.25 mm, from System Tecnolub Inc.). The injector operates with pressurized air. The pressurized air arrives to the system and passes through a filter equipped with a dryer; the air then goes through a pressure regulator and reaches the external channel of the atomizer. The oil is transported to the internal channel of the atomizer through a micro volumetric piston pump and various regulators.

Finite element methods were used by the ULB team to simulate and to optimize the obtained spray phenomenon. Computational fluid dynamics (CFD) simulations with FINE/Open 2.11.1 were used to study the air injection in the ambient environment (single phase flow simulations). In Figure 2, a cross section of the annular air channel of the SB-202010 nozzle shows the resulting velocity field for an air flow rate of 31 l/min. As it is observed in Figure 2, blue color characterize the low speed particle (between 0 to 2.5 m/s) and red color the high speed particle (between 17.5 to 20 m/s). The computational domain is made up of around 1 000 000 finite elements.

Figure 1. A typical MQL oil-air atomizer / injector

Experimental characterizations of the oil-air spray were also done at ULB in order to optimize the spray shape and precision and the corresponding injector geometry as well as its working parameters. An example of such a test result is shown on Figure 3 with a PIV (particle image velocimetry) measurement. This experimental testing (also done using PDA and laser diffraction measurements) was used in conjunction with two-phase flow CFD numerical simulations (also with the FINE software) of the oil-air spray.

The oil through the central channel of the atomizer can be either pure lubricating oil or an oil/water mixture (at a ratio like 5/95).

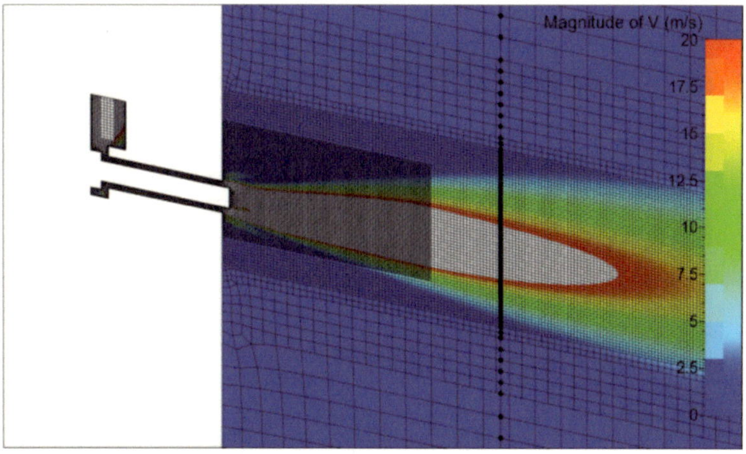

Figure 2. Illustration of the air velocity at the exit of the injector (m_g = 31 l/min)

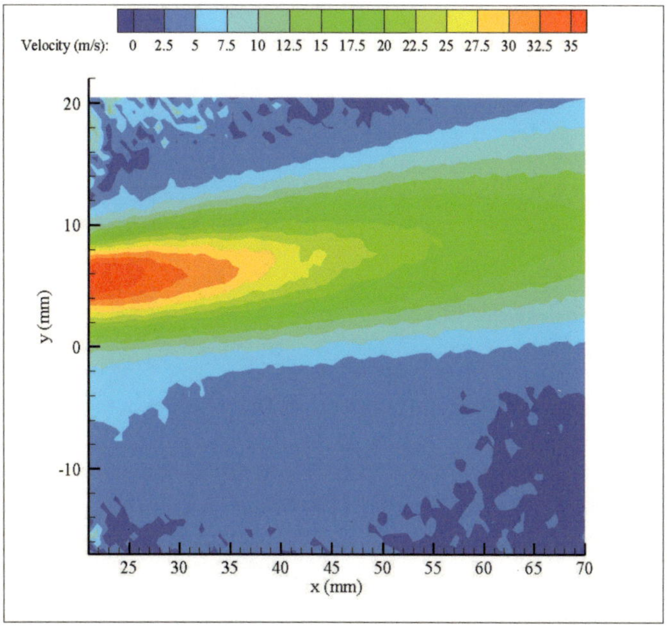

Figure 3. PIV measurement of the oil velocity after the injector

2.1. Effect of lubrication and cutting conditions on part quality

Figure 4 presents the roughness profile against different feed direction of 6061-T6 during turning at 207 m/min cutting speed in dry, semi-dry (Mecagreen 550 at 3.06 ml/min debit) and wet condition at low and high feed rate. The Mitutoyo SJ-400 equipment was used to measure the surface roughness. It is observed that in figure 4a (at low feed rate) the roughness profile is high in the wet condition compared to semi-dry (Mecagreen 550 at 3.06 ml/min debit) and dry conditions. In Figure 4b (at high feed rate) the roughness is high in the wet condition compared to semi-dry (Mecagreen 550 at 3.06 ml/min debit) and dry conditions as in Figure 4a. In this case (Figure 4b) the roughness profile is high the semi-dry (Mecagreen 550 at 3.06 ml/min debit) condition compared to the dry condition, which is the opposite of what is observed in Figure 1a.

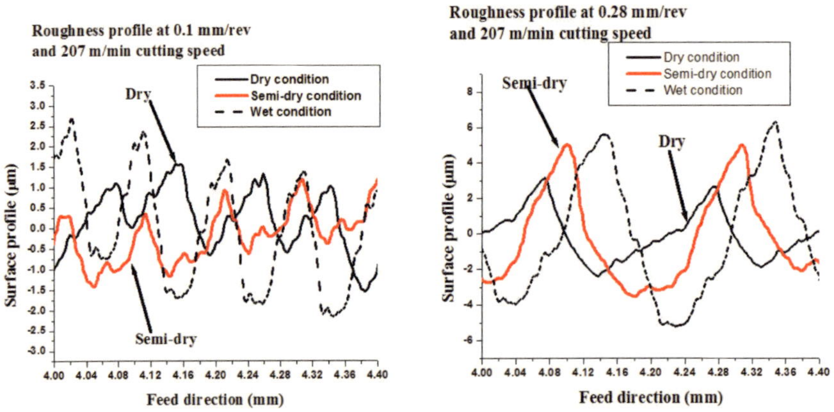

a) at low feed rate

b) at high feed rate

Figure 4. Roughness profile against different feed direction at 207 m/min cutting speed

Figure 5 presents the roughness at different cutting speed and feed rate respectively. It is confirmed on Figure 5 that the cutting speeds does not influence the surface finish which is mostly influenced by the feed rate (Figure 6) and the lubrication conditions.

Figure 6 presents the roughness at different feed rate of 6061-T6 during turning at 207 m/min cutting speed in dry, semi-dry (Mecagreen 550 at 3.06 ml/min debit) and wet condition. It is observed that at low feed rates (0.05 and 0.10 mm/rev) the roughness is high in the wet condition compared to semi-dry (Mecagreen 550 at 3.06 ml/min debit) and dry conditions. The roughness is still high in the dry condition compare to semi-dry (Mecagreen 550 at 3.06 ml/min debit) condition. This observation was confirmed by the work done by Dhar et al. [16] and Kamata et al. [27]. In their work they showed that the cutting performance of semi-dry machining is better than that of dry and conventional wet machining. They have also shown that the dry machining is better than the wet cutting.

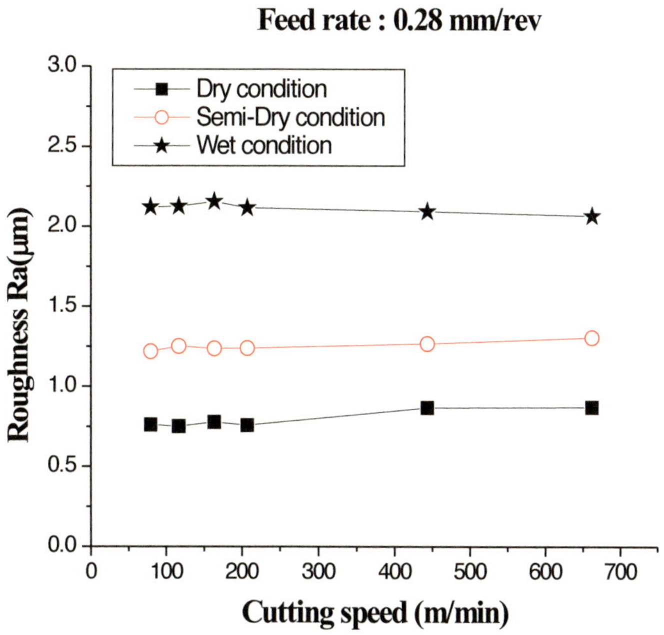

Figure 5. Roughness against cutting speed

In the work done by Vikram Kumar. [10] and Varadarajan et al. [11], they showed that the semi-dry machining is also better than that of dry and conventional wet machining but the wet is better compare to the dry cutting.

In general, the roughness is high in the wet condition compared to semi-dry and dry conditions. For feed rates higher than 0.10 mm/rev, the roughness is high in the semi-dry condition compared to the dry condition.

Figure 7 presents the roughness at different feed rate of 6061-T6 during turning at 207 m/min cutting speed in dry, semi-dry, wet condition and theoretical values. The theoretical roughness R_{ath} was obtained using Boothroyd et al. [28] formula in equation (1) as follow:

$$R_{ath} = 0.0321 \frac{f^2}{r_\varepsilon} \tag{1}$$

where f is the feed rate and r_ε is the tool insert nozzle radius.

It is observed in Figure 7 that at low feed rate (from 0.05 to 0.10 mm/rev feed rate) the theoretical roughness is low compared to dry, semi-dry and wet condition. At high feed rate (after 0.10 mm/rev feed rate) the theoretical roughness still be low compared to semi-dry and wet condition but be high compared to the dry condition which is the opposite of what is observed in the case of low feed rate.

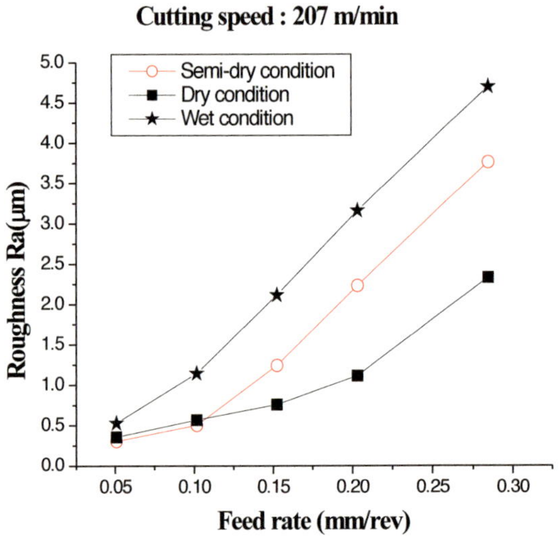

Figure 6. Roughness at different feed rate in in dry, semi-dry and wet condition dry, semi-dry and wet condition

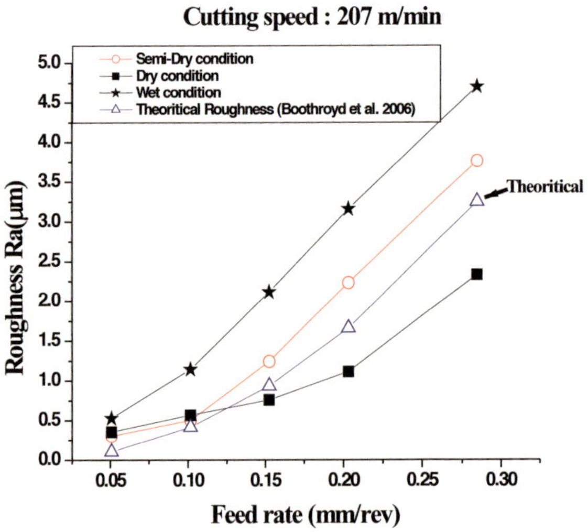

Figure 7. Roughness at different feed rate in dry, semi-dry and wet condition

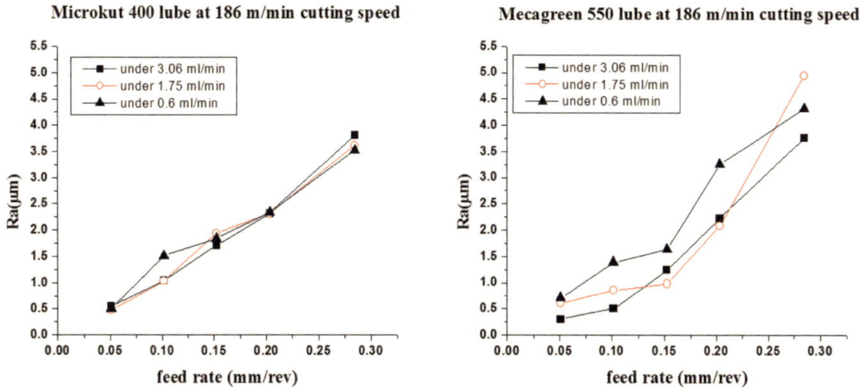

a) Microkut 400 lubricant b) Mecagreen 550 lubricant

Figure 8. Roughness at different feed rate at 186 m/min cutting speed at different lubrication debit using Microkut 400 and Mecagreen 550 lubricant

Figure 8 presents the roughness at different lubrication debit of 6061-T6 during turning at 186 m/min cutting speed using Microkut 400 and Mecagreen 550 lubricants. It is observed that the roughness in turning of 6061-T6 at different feed rate using microkut 400 lubricant (Figure 8a) is similar at different lubrication debit which is different in the case of Mecagreen 550 (Figure 8b). In Figure 8b it is also observed that the roughness is high at 3.06 ml/min debit compared to 1.75 and 0.6 ml/min debit. This observation could be due to the fact that in machining process Mecagreen 550 lubricant is generally used as a cooler of the cutting zone. The average roughness is low using Microkut 400 lubricant compared to Mecagreen 500. This observation confirmed the fact that Microkut 400 lubricant is the recommended one to be used to have a good surface finish in the turning process.

Figure 9 presents lubrication debit comparison on the roughness during turning process of 6061-T6 at the same cutting speed using Microkut 400, Mecagreen 550 and commercial lubricants. The commercial lubricant debit was 306 ml/min and Microkut 400 and Mecagreen 550 lubricant debit was 3.06, 1.75 and 0.6 ml/min.

The general observation (Figure 9a, 9b and 9c) is that the roughness is high when using the commercial lubricant compared to Microkut 400 and Mecagreen 550 lubricants. It is also observed that at 3.06 ml/min lubrication debit (Figure 9a), the roughness using Microkut 400 lubricant and Mecagreen 550 is similar at high feed rate. Using 1.75 ml/min lubrication debit (Figure 9b), the roughness using Microkut 400 lubricant is low compared to Mecagreen 550 high feed rate. Using 0.6 ml/min lubrication debit (Figure 9c), the lower tendency of the roughness using Microkut 400 lubricant compared to Mecagreen 550 high feed rate observed in Figure 9b is more pronounced. This observation confirmed the fact that Microkut 400 lubricant is more useful in the case of having a good surface finish in the turning process.

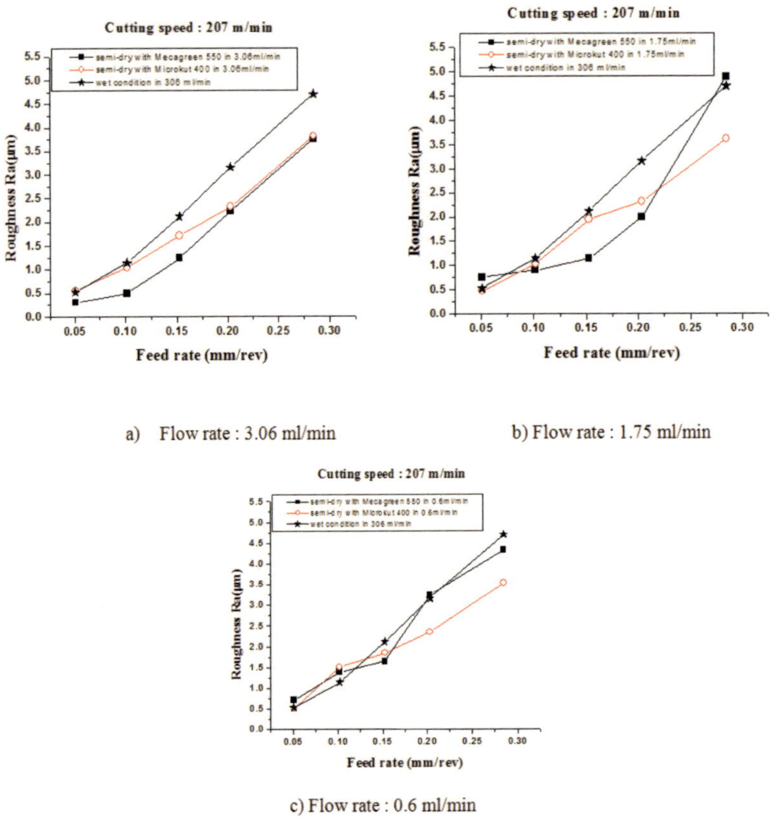

Figure 9. Lubrication debit comparison on the roughness during turning process of 6061-T6 at the same cutting speed using Mecagreen 550 and Microkut 400 lubricant

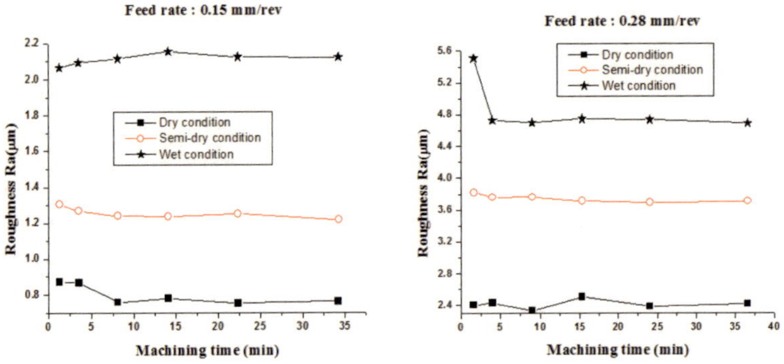

Figure 10. Roughness comparison at different machining time in dry, semi-dry (Mecagreen 550 at 3.06 ml/min debit) and wet conditions at 0.15 and 0.28 mm/rev feed rate

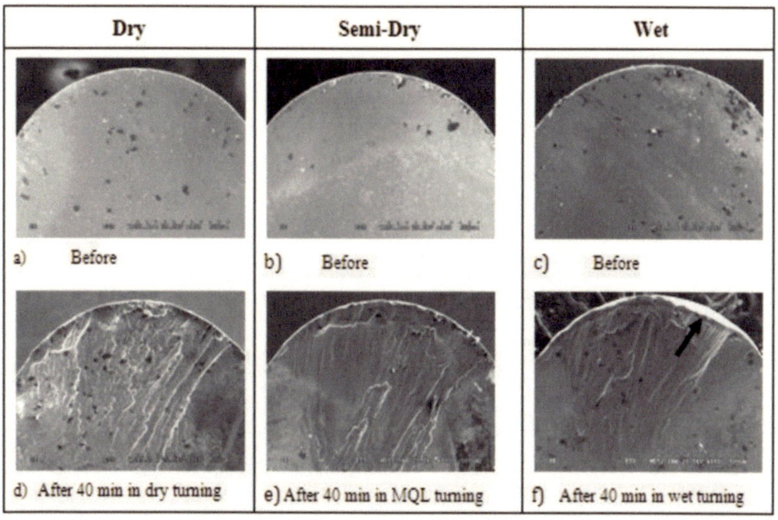

Figure 11. SEM insert tool wear images in dry, Semi-dry (Mecagreen 550 at 3.06 ml/min debit) and Wet conditions: magnification 100x

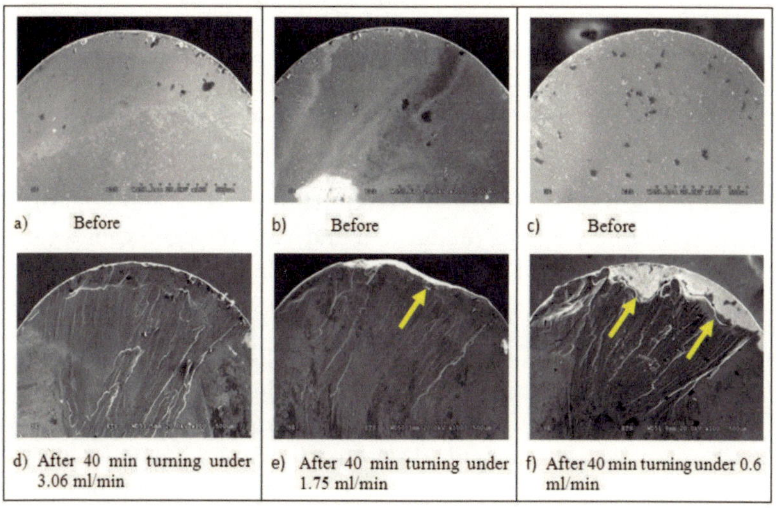

Figure 12. SEM insert tool wear images under Mecagreen 550 lubricant at different lubrication debit: magnification 100x

Figure 10 presents the roughness of 6061-T6 at different machining times in dry, semi-dry (Mecagreen 550 at 3.06 ml/min debit) and wet conditions during turning at 0.15 and 0.28 mm/rev feed rate. It is observed that under different lubrication conditions the surface roughness does not change much with the machining time. This observation is an indication that the tool wear after 40 min machining is not too significant (Figure 9).

2.2. Tool wear

Figure 11 presents SEM images of the cutting tool insert before and after in different lubrication conditions. It was observed that the separate tool insert used to machine the 6061-T6 aluminum alloy exhibited a normal tool insert life under the test conditions used and no premature tool insert wear or breakage occurred during the turning tests.

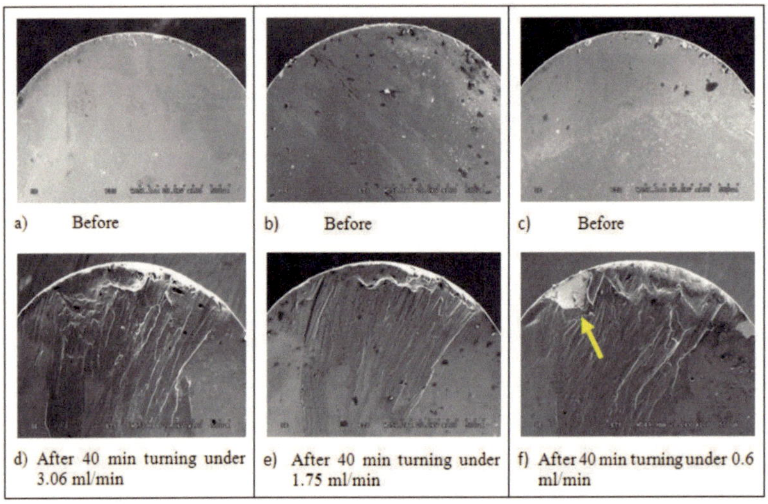

Figure 13. SEM insert tool wear images under Microkut 400 lubricant at different lubrication debit: magnification 100x

After 40 min in turning process, it was observed that in the dry and semi-dry (Mecagreen 550 at 3.06 ml/min debit) conditions (Figure 11d and 11e), no significant wear occurred for the cutting tool insert lip. Figure 11f shows that the cutting tool insert lip became worn and exhibited a small wear in the wet conditions.

Figure 12 presents SEM images of the cutting tool insert before and after at different lubrication debit of 6061-T6 after 40 min turning process using Mecagreen 550 lubricant. As in Figure 11 (a, b and c) it was observed (Figure 12a, 12b and 12c) that the separate tool insert used to machine the 6061-T6 aluminum alloy exhibited a normal tool insert life under the test conditions used and no premature tool insert wear or breakage occurred during the turning tests.

After 40 min in turning process, it was observed that in 3.06 ml/min debit condition (Figure 12d), no significant wear occurred for the cutting tool insert lip. Figure 12e shows that the cutting tool insert lip became worn and exhibited a small wear in 1.75 ml/min debit. In 0.6 ml/min debit (Figure 12f), the cutting tool insert lip became more worn and exhibited wear more pronounced compared to 1.75 ml/min debit (Figure 12e). This observation confirmed the relationship between roughness (Figure 8b) and tool insert wear during turning process.

Figure 13 presents SEM images of the cutting tool insert before and after at different lubrication debit of 6061-T6 after 40 min turning process using Microkut 400 lubricant. As in Figure

11 (a, b and c) and Figure 12 (a, band c) it was observed (Figure 13a, 13b and 13c).It was observed that the separate tool insert used to machine the 6061-T6 aluminum alloy exhibited a normal tool insert life under the test conditions used and no premature tool insert wear or breakage occurred during the turning tests.

After 40 min in turning process, it was observed that in 3.06 and 1.75 ml/min debit condition (Figure 13d and 13e), no significant wear occurred for the cutting tool insert lip. Figure 13f shows that the cutting tool insert lip became worn and exhibited a wear.

In 0.6 ml/min debit the cutting tool insert lip exhibited wear more pronounced using Macagreen 500 lubricant (Figure 12f) compared to Microkut 400 lubricant (Figure 13f). This observation confirmed the relationship between lubricant and tool insert wear during turning process.

2.3. Chip control and chip quality

Figure 14 presents the chip thickness at different cutting speed and different feed rates using Microkut 400 lubrication at two flow rates : 3.06 ml/min and 0.6 ml/min. It is observed that at low feed rate the chip thickness is similar at different cutting speed. But at the high feed rate the chip thickness decrease with the increasing of the cutting speed. The effect of the lubricating flow rate does not seem to affect the chip thickness.

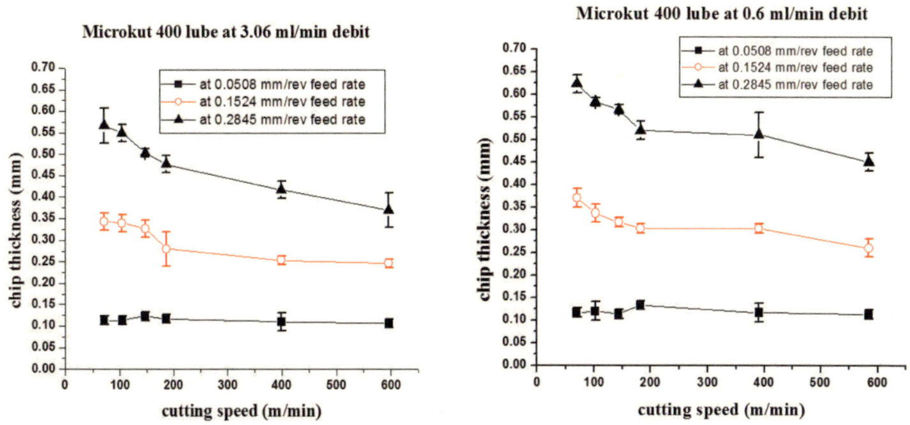

Figure 14. Chip thickness at different cutting speeds and debits

Figure 15 presents the chip reduction coefficient at different feed rate and different lubrication debit using Microkut 400 and Mecagreen 550 lubricant. The chip reduction coefficient (ratio of depth of cut to the chip thickness) permits to have an indication on cutting energy and cutting temperature. The general observation is that the chip reduction decrease when increasing feed rate which confirmed the work done by Dhar et al. [16]. It is also observed that the chip reduction using Microkut 400 lubricant (Figure 15a) decrease with the decreasing of the lubricant debit and inversely it increase with lubricant debit in the case of Mecagreen 550 lubricant (Figure 15b).

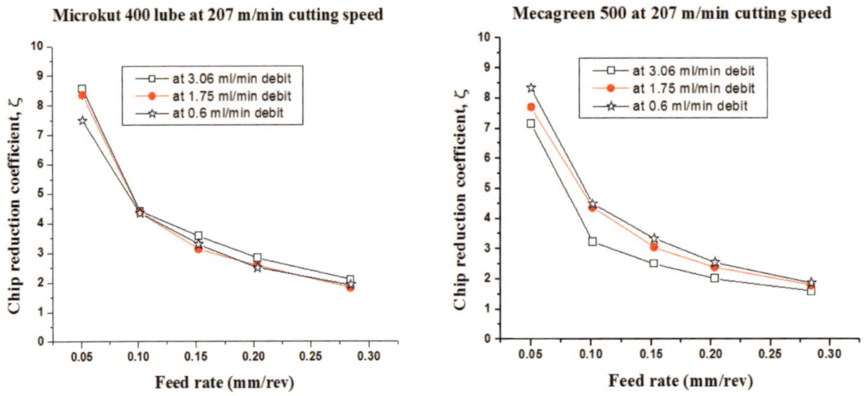

a) Microkut 400 lubricant

b) Mecagreen 550 lubricant

Figure 15. Chip reduction coefficient at different feed rate and different lubrication debit using Microkut 400 and Mecagreen 550 lubricant

Figure 16 presents lubrication debit comparison on the chip reduction coefficient during turning process of 6061-T6 at the same cutting speed using Microkut 400 and Mecagreen 550 lubricants. It is observed that at 3.06 ml/min lubrication debit (Figure 16a), the chip reduction coefficient using Microkut 400 lubricant is high compared to Mecagreen 550 lubricant.

Using 1.75 ml/min lubrication debit (Figure 16b), the chip reduction coefficient using Microkut 400 lubricant and Mecagreen 500 is similar in high feed rate. In low feed rate (0.05 mm/rev) the chip reduction coefficient (Figure 16b) is high using Microkut 400 lubricant compared to Mecagreen 550 lubricant but fewer than using 3.06 ml/min debit (Figure 16a). Figure 16c shows the inversion of the chip reduction coefficient which is low using Microkut 400 lubricant compared to Mecagreen 550 lubricant. This is an indication that Microkut 400 lubricant could be more useful in the reduction of energy consumption compared to Mecagreen 550 in the turning process.

X-ray diffraction analysis (XRD) was used to characterize the chip grain size of each material. To do this, from the X-ray diffraction peaks, the Scherrer relation was used to determine the average size of the chips analyzed.

The X-ray diffraction peaks are characterized by diffraction angles θ_{hkl} and I_{hkl} intensities, depending on the lattice cell and on the wavelength radiation, λ, used. The hkl Miller indexes correspond to the diffracting crystallographic planes (d_{hkl} is the distance between them). These three parameters are connected through the Bragg [29] in equation (2) as follows:

$$2d_{hkl}\sin\theta_{hkl} = n\lambda \tag{2}$$

Figure 17 presents the XRD diagrams of the 6061-T6 material of the diffraction intensity at different diffraction angle and Miller index. It is observed that the (111) Miller present the high intensity peak diffraction value.

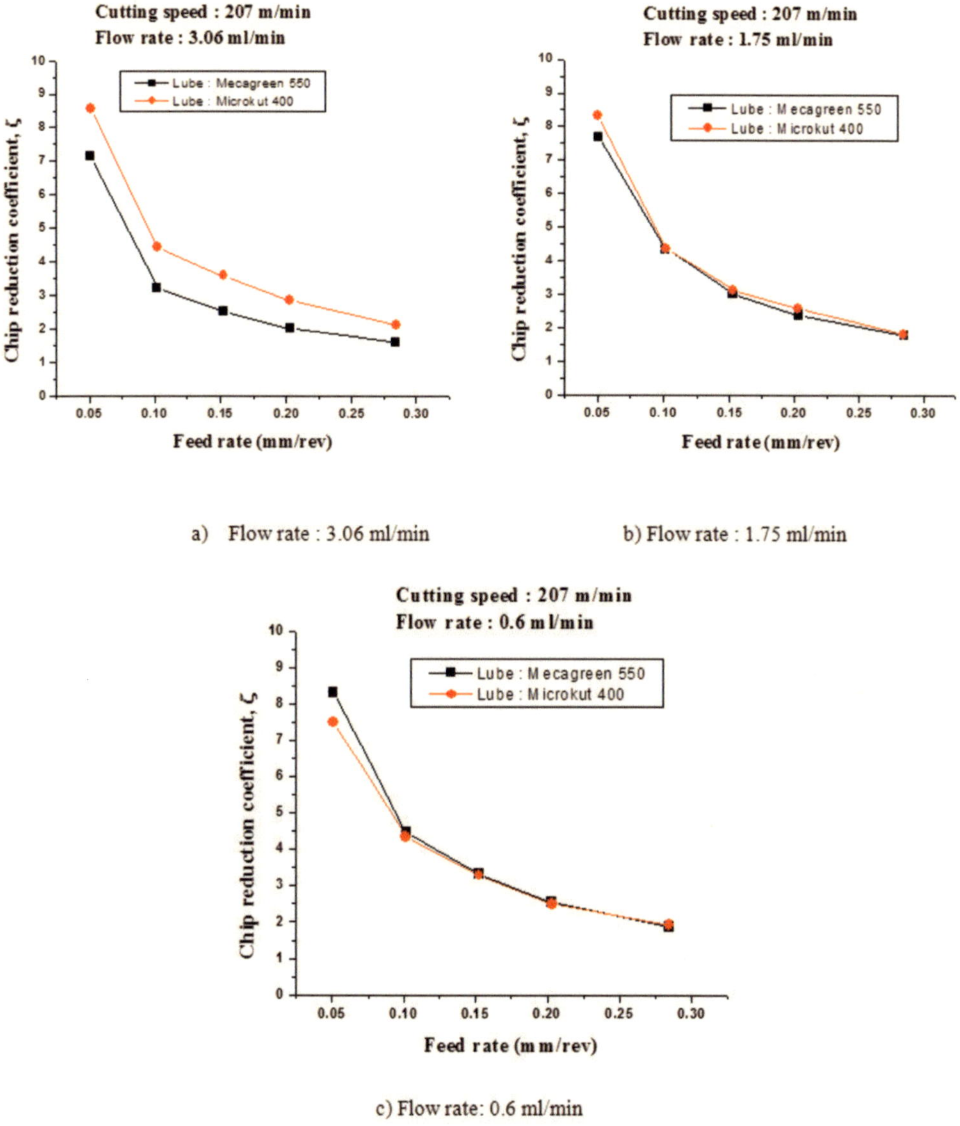

a) Flow rate : 3.06 ml/min

b) Flow rate : 1.75 ml/min

c) Flow rate: 0.6 ml/min

Figure 16. Effect of lube and flow rates on chip reduction coefficient

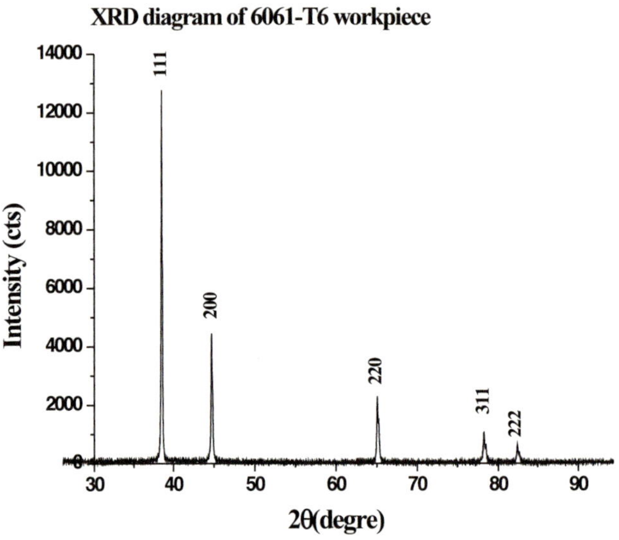

Figure 17. XRD chip diagram and chip grain size at different feed rate

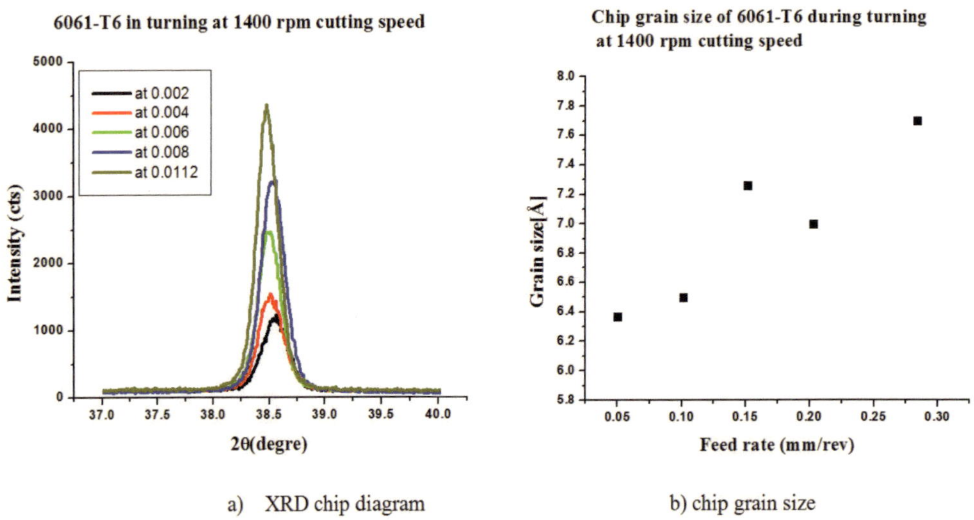

Figure 18. XRD chip diagram and chip grain size at different feed rate during semi-dry machining

Figure 18 presents the XRD (x ray diffraction) and the grain size of the chip thickness different feed rate. Figure 17a presents the XRD chip diagram and Figure 17b presents the chip grain size. From Figure 18a the XRD diagrams were analysed according to the deconvolution of the peak in two Lorentzians corresponding to the two components $\lambda_{K\alpha 1}$ and $\lambda_{K\alpha 2}$ of the wavelength $\lambda_{K\alpha}$ of copper. This procedure yields a measurement of a width for each peak of the full width of the peak at half maximum intensity (FWHM) (parameter ω), corresponding to a well-defined wavelength, and has been applied to the whole diagram. Kouam et al. [30] showed in their work how to decompose these peaks without further decreasing their intensity.

The chip grain size d_m has been obtained from equation (3) using Sherrer formula as follow:

$$d_m = \frac{k.\lambda}{\omega\cos\theta_{hkl}} \qquad (3)$$

where k is a constant close to 0.9; *hkl* are the Miller indexes, θ_{hkl} is the diffraction angle, λ is the wavelength of the incidental radiation, and ω corresponds to FWHM (full width of the peak at half maximum).

It is observed in figure 18a that the peak intensity at the same cutting speed increase with the increasing of the feed rate. This observation is confirmed in Figure 18b in which the grain size increase with the increasing of the feed rate and could be interpreted as owing to crystallographic effects due to the increase in the chip temperature during the drilling process. This change could help determine the chip temperature.

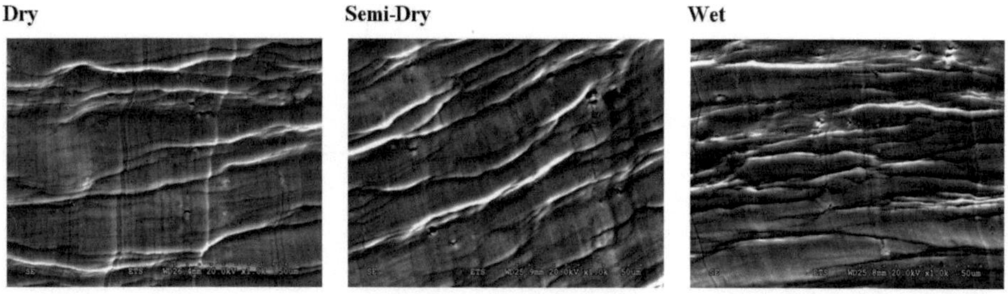

Figure 19. SEM chip segmentation at different lubrication conditions (1000 X magnification).

2.4. Chip formation

Figure 19 presents SEM images of the chip segmentation at different lubrication conditions. The magnification (1000 X) used was enough to show any significant segmentation of the chip at 207 m/min cutting speed and 0.28 mm/rev feed rate.

The segmentation could be defined by the segmentation band density parameter η_s. According to the formulation of Becze and Elbestavi [31], the chip segmentation density parameter η_s can be estimated by the equation (4):

$$\eta_S^{-1}(A + B\exp(CV)) * \left(1 + D\left(\frac{f - f_0}{f_0}\right)\right) \quad (4)$$

where f is the feed rate, V is the cutting speed, and A, B, and C are empirical constants.

Khettabi et al. [32] developed a simple method for determining the chip segmentation density parameter η_s using the distance (l) corresponding to 10 segmentation bands (Equation 5).

$$\eta_S \frac{1}{l_b} = \frac{10}{l} \quad (5)$$

where l_b is the band width.

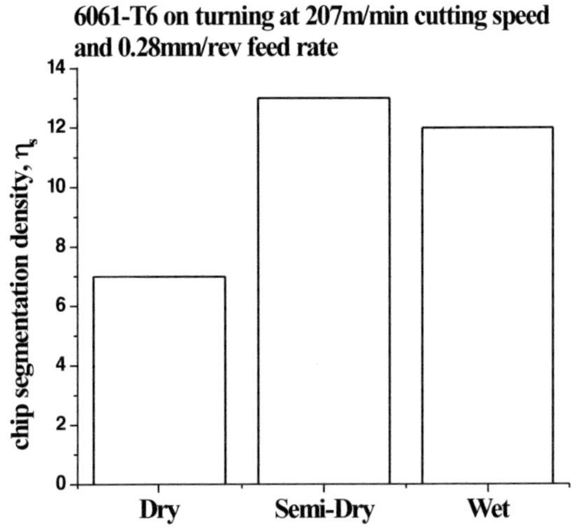

Figure 20. Chip segmentation density η_s for different lubrication conditions

Figure 20 presents the chip segmentation density parameter η_s for different lubrication conditions at 207 m/min cutting speed and 0.28 mm/rev feed rate. Figure 20 is obtained using equation (5). It is observed that the chip segmentation density parameter η_s is low during dry machining compared to the semi-dry (Mecagreen 550 at 3.06 ml/min debit) and wet conditions.

2.5. Metallic particle emission

The machining of metallic components produce also aerosol (wet and dry) which can deteriorate the working shop floor environment. Figure 21 presents the total mass concentration at different cutting speed in dry, semi-dry (Mecagreen 550 at 3.06 ml/min debit) and wet conditions. It is observed that at low cutting speed the total mass concentration increase and de-

http://dx.doi.org/10.5772/51351

crease at high cutting speed. These two speed regimes were observed by [24] in their work on turning, by [25] in their work on drilling, and by [26] in their work on friction.

At the same cutting speed the total mass concentration is low in dry condition compared to semi-dry (Mecagreen 550 at 3.06 ml/min debit) and wet conditions. This phenomenon could be due to the fact that when the chip becomes brittle, dust emission decreases significantly [32], [34]. This observation is confirmed by what has been previously obtained in figure 20.

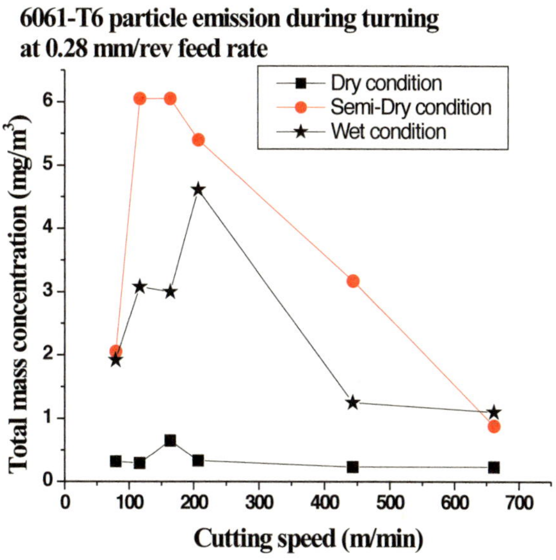

Figure 21. Total mass concentration at different lubrication conditions

3. Conclusion

In this work the effect of the lubrication condition in the turning process was studied. It was found that during this process, the surface roughness, the tool wear, the chip formation and the metallic particle emissions depend significantly on lubrication and cutting conditions.

- At different lubrication conditions, it was found that the surface roughness depends only on the feed rate and not on the cutting speed. At low feed rate the surface roughness is low in the semi-dry lubrication compared to the wet and dry lubrication. In another hand Microkut 400 lubricant is more efficient for a good surface finish in the turning process compared to Mecagreen 550.
- For the tool insert wear the dry and semi-dry (Mecagreen 550 at 3.06 ml/min debit) condition has no significant wear was occurred for the cutting tool insert lip. In the wet condi-

tion the cutting tool insert lip exhibited a small wear after 40 min turning process. At low lubrication debit (0.6 ml/min) the cutting tool insert lip exhibited wear more pronounced using Macagreen 500 lubricant compared to Microkut 400 lubricant. This observation confirmed the relationship between lubricant and tool insert wear during turning process.

- For the chip reduction coefficient it was found that Microkut 400 lubricant could be more useful in the reduction of energy consumption and chip temperature compared to Mecagreen 550 in the turning process.
- From the XRD diagram, the increases in chip grain size with changing cutting conditions confirmed the change of microstructure of chip during the turning process and could for the chip temperature.
- Metallic particle emissions were found to be affected not only by the lubrication condition but also by cutting conditions. At very low speeds, the amount of particles is low, and then increases, reaches a maximum value, and eventually decreases. This observation could help reduce dust emissions, which can have serious consequences on the health of the operator. In the case of cast aluminum alloys, increasing the feed rate led to lower metallic particle emission. In general, the use of semi-dry produce more metallic particle which can be due to the generation of aerosol.

Author details

J. Kouam[1], V. Songmene[1*], M. Balazinski[1] and P. Hendrick[3]

*Address all correspondence to: Victor.Songmene@etsmtl.ca

1 École de Technologie Supérieure (ETS), Montréal, Canada

École Polytechnique de Montréal, Montréal, Canada

2 Université Libre de Bruxelles, Belgique

References

[1] Weinert, K. I., Sutherland, J. W., & Wakabayashi, T. (2004). Dry machining and minimum quantity lubrication. *CIRP Annals- Manufacturing Technology*, 53(2), 511-537.

[2] Bell, D. D., Chou, J., Nowag, L., & Liang, S. Y. (1999). Modeling of the environmental effect of cutting fluid. *Tribology Transactions*, 42(1), 168-173.

[3] Sutherland, J. W., Kulur, V. N., & King, N. C. (2000). Experimental investigation of air quality in wet and dry turning. *CIRP Annals- Manufact. Technology*, 49(1), 61-64.

[4] Diniz, A. E., & De oliveira, A. J. (2004). Optimizing the use of dry cutting in rough turning steel operations. *International Journal of Machine Tools & Manufacture*, 44, 1061-1067.

[5] Suda, S., Yokota, H., Inasaki, I., & Wakabayashi, T. (2002). A synthetic ester as an optimal cutting fluid for minimal quantity lubrication machining. *CIRP*, 51, 95-98.

[6] Mc Cabe, J., & Ostraff, M. A. (2001). Performance experience with near-dry machining of aluminium. *Lubrication Engineering*, 57(12), 22-27.

[7] Wakabayashi, T., Inasaki, I., & Suda, S. (2006). Tribological action and optimal performance: Research activities regarding MQL machining fluids. *Machining Science &Technology*, 10, 59-85.

[8] Islam, M. N., Rafai, N. H., & Heng, B. C. (2012). Effect of cutting fluid supply strategies on surface finish of turned parts. *Journal of Advance Materials research*, 383-390, 4576-4584.

[9] Machado, A. R., & Wallbank, J. (1997). The effect of Extermely Low Lubricant Volumes in Machining. *Wear*, 210, 76-82.

[10] Vikram, K. C. H. R., & Ramamoorthy, B. (2007). Performance of Coated Tools During Hard Turning Under Minimum Fluid Application. *Journal of Materials Processing Technology*, 185, 210-216.

[11] Varadarajan, A. S., Philip, P. K., & Ramamoorthy, B. (2002). Investigations on Hard Turning with Minimal Cutting Fluid Application (HTMF) and its Comparison with Dry and Wet Turning. *International Journal of Machine Tools & Manufacture*, 42, 193-200.

[12] Ozawa, M., Hosokawa, A., Tanaka, R., Furumoto, T., & Ueda, T. (1998). Minimum quantity lubrication turning using tools with oil holes. Kanazawa, Ishakawa, Japan. *ASPE Proceedings, October*, 19.

[13] Hua, J., & Shivpuri, R. (2002). Influence of crack mechanics on the chip segmentation in the machining of titanium alloys. *Proceedings of the Ninth ISPE International Conference on Concurrent Engineering, Cranfield, UK.*, 27-31.

[14] Poulachon, G., Moisan, A. L., & Jawahir, I. S. (2001). Evaluation of Chip Morphology in Hard Turning Using Constitutive Models and Material Property Data. *Proceedings of the ASME Manufacturing Engineering Division*, 12, 179-185.

[15] Jawahir, I. S., & Van Luttervelt, C. A. (1998). Recent Developments in Chip Control Research and Applications. *Working Group on Chip Control, STC-C, CIRP*, 1990-1993.

[16] Dhar, N. R., Islam, M. W., Islam, S., & Mithu, M. A. H. (2006). The influence of minimum quantity lubrication (MQL) on cutting temperature, chip and dimensional accuracy in turning AISI-1040 steel. *J. Material Processing Technology*, 171, 93-99.

[17] Dhar, N. R., Kamruzzaman, M., & Ahmed, M. (2006). Effect of minimum quantity lubrication (MQL) on tool wear and surface roughness in turning AISI-4340 steel. *Journal of Materials Processing Technology*, 172, 299-304.

[18] Nouari, M., List, G., Girot, F.and., & Ge'hin, D. (2005). Effect of machining parameters and coating on wear mechanisms in dry drilling of aluminium alloys. *Machine Tools & Manufacture*, 45, 1436-1442.

[19] Damir, A., Lancereau, S., Attia, H., & Hendrick, P. (2010). On the Performance of Minimum Quantity Lubrication. *Milling Al 6061, CIRP.*

[20] Diakodimitris, C., Hendrick, P., & Iskandar, Y. R. (2011). *Study of Minimum Quantity Cooling (MQC) on the tool temperature in milling operations, CIRP 44th.*

[21] Dhar, N. R., Islam, M. W., Islam, S., & Mithu, M. A. H. (2006a). The Influence of Minimum Quantity of Lubrication (MQL) on Cutting Temperature, Chip and Dimensional Accuracy in Turning AISI-1040 Steel. *Journal of Materials Processing Technology*, 171(1), 93-99.

[22] Songmene, V., Balout, B.and., & Masounave, J. (2008a). Clean Machining: Experimental Investigation on Dust formation Part I: Influence of Machining Parameters and Chip Formation. *Int. J. of Env. Conscious Design & Manufacturing*, 14, 1-16.

[23] Songmene, V., Balout, B.and., & Masounave, J. (2008b). Clean Machining: Experimental Investigation on Dust formation Part II: Influence of Machining Strategies and Drill Condition. *Int. J. of Env. Conscious Design & Manufacturing*, 14, 17-33.

[24] Khettabi, R., Songmene, V.and., & Masounave, J. (2007). Effect of tool lead angle and chip formation mode on dust emission in dry cutting. *Journal of Materials Processing Technology*, 194(1-3), 100-109.

[25] Kouam, J., Masounave, J., Songmene, V., & Gireadeau, A. (2010). Pre-holes Effect on Cutting Forces and Particle Emission During Dry Drilling Machining. *Proc. of 49th Annual Conf. Metallurgists of CIM, Vancouver, Canada*, 253-263.

[26] Kouam, J., Songmene, V., Djebara, A., & Khettabi, R. (2011). Effect of Friction Testing of Metals on Particle Emission. *J. of Materials Engineering and Performance*, s11665-011-9972-6.

[27] Kamata, Y., & Obikawa, T. (2007). High speed MQL finish-turning of Inconel 718 with different coated tools. *Journal of Materials Processing Technology*, 192–193, 281-286.

[28] Boothroyd, G., & Winston, A. K. (2006). *Fundamentals of Machining and Machine Tools. 3rd edition*, Taylor and Francis, 194-195.

[29] Bragg, L. (1949). *The Crystalline State 1.*, G. Bell, London.

[30] Kouam, J., Ait-Ahcene, T., Plaiasu, A. G., Abrudeanu, M., Motoc, A., Beche, E., & Monty, C. (2008). Characterization and properties of ZnO based nanopowders prepared by solar physical vapor deposition (SPVD). *Journal of Solar Energy*, 82, 226-238.

[31] Becze and Elbestawi. (2002). A chip formation based analytic force model for oblique cutting. *International Journal of Machine Tools and Manufacture*, 42(4), 529-538.

[32] Khettabi, R., Songmene, V., & Masounave, J. (2010). Effects of Speeds, Materials and Tool Rake Angles on Metallic Particle Emission during Orthogonal Cutting. *Journal of Materials Engineering and Performance, ASM International*, 16(6), 767-675.

[33] Kelly, J. F., & Correrell, M. G. (2002). Minimal lubrication machining of aluminium alloys. *Journal of Materials Processing Technology*, 120, 327-334.

[34] Attanasio, A., Gelfi, M., Giardini, C., & Remino, C. (2006). Minimal quantity lubrication in turning: Effect on tool wear. *Wear*, 260, 333-338.

Chapter 9

Global Machinability of Al-Mg-Si Extrusions

V. Songmene, J. Kouam, I. Zaghbani, N. Parson and A. Maltais

Additional information is available at the end of the chapter

http://dx.doi.org/10.5772/54021

1. Introduction

Field performance, mechanical properties and workability usually sustain the development of new alloys. As far as the workability is concerned, most extrusions need not only to meet good extrudability, but also good or acceptable machinability as some machining operations (eg. drilling or finishing machining) are usually required. Unfortunately, alloys with higher strength could have better machinability but lower extrudability; Aluminum alloys with excellent extrudability such as AA1060 or AA1100 (Figure 1) often exhibit low machinability, especially due to the chip formation and the workpiece material adhering to the cutting tool leading to the build-up-edge (BUE), modifying the cutting process, leading to tool breakage or deteriorating the surface finish when this BUE is broken. Al-Si-Mg alloys (6XXX series) usually exhibit good machinability and good extrudability; This is one of the reason why about 90% of most extruded aluminum parts are in 6XXX family. The AA6262 which is recognized for its ease chip breakability (a lot of second phase particles which help initiate fracture) generally leads to good machinability but poor extrudability because of its poor formability. Any new aluminum alloy with excellent machinability, extrudability and mechanical properties will therefore lead to considerable advantage compared to existing alloys.

Farmer (1978) proposed a system (based on extrusion ratio, die angle, billet length-to-diameter ratio and statistical analysis) for evaluating the extrudability of aluminum alloys and illustrated the system using AA2011, AA6061, AA6063 and AA 6262 aluminium alloys in their T0-temper condition. The extrusion ratio and the extrusion pressure were shown to able to allow benchmarking the tested alloys according to their extrudability. The AA6063 showed the best performance, followed by the AA6061 and the AA6262 wheras the AA2011 exhibited the poorest performance amongst the criterion tested. The performance of the tested alloys found by Farmer (1978) is in good agreement with the data presented in Figure 1. Both the extrusion ratio and the extrusion pressure gave a good picture of true stress-true strain behavior of the tested alloys. It is however important to notice that the machinability of the alloys was not evaluated.

According to Saha (2012), the maximum extrusion speed is one of the most significant factors influencing the cost and efficiency of the extrusion process. This speed can therefore be used to measure the extrudability of alloys. He (Saha, 2012) compared the extrudability of the AA2XXX, AA6XXX and that of the AA7XXX based on major allowing elements. AA6XXX containing Mg and Si had better relative extrudability (100), while alloys of the same family but containing Cu was rated at 60%. AA2XXX series with Cu, Mg, Mn as major alloying element were rated 15% and AA7XXX with Zn, Mg, Cu, Cr as major alloying elements were rate 10%. This classification is in good agreement with the data presented in Figure 1.

Chena et al. (2009) investigated the dynamic fracture behavior (Charpy test) of extruded AA6XXX and AA7XXX aluminum alloys: AA6060, AA6082, AA7003 and AA7108, all in T6 condition. As expected, they found that the AA7XXX which have higher strength (yield and ultimate tensile stresses) exhibited higher toughness than the AA6XXX but the fracture strain of the AA7XX was lower compared to that of the AA6XXX alloys as a consequence of higher precipitation of grain boundary found in AA7XXX. It is therefore understandable that the AA7XXX show poor extrudability as depicted in Figure 1.

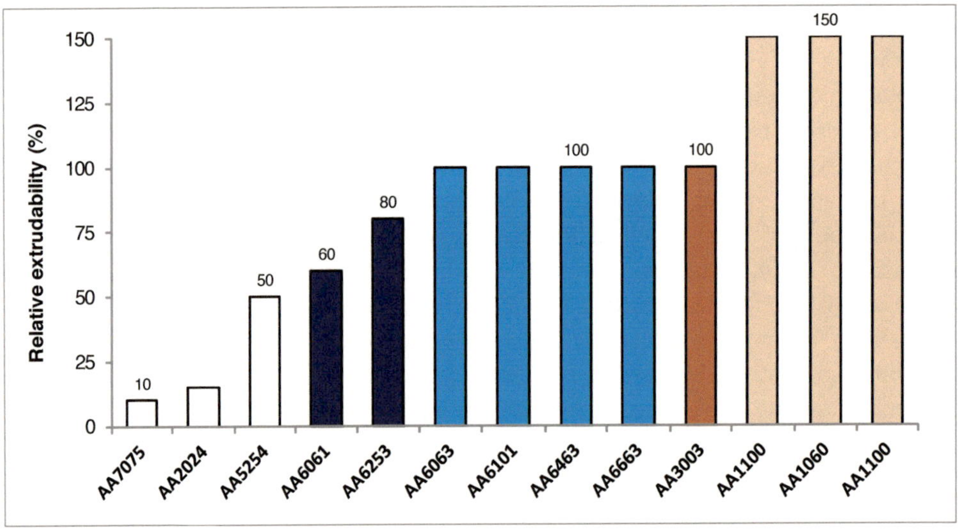

Figure 1. Relative extrudability of some selected aluminum alloys (adapted from Duval, 2012)

Pangborn *et al.* (2012) studied the workability, the mechanical properties of the 6061 and that of the 6005A in T5 and T6 conditions and found that the 6005A has significant advantages over the 6061 in many applications: better toughness, better corrosion resistance, better

http://dx.doi.org/10.5772/54021

quenching sensitivity and better consistency of mechanical properties together with comparable fatigue resistance, machinability rating, formability, weldability, and ease ability to meet minimum mechanical properties with 6061. According to the authors, when initially tried by a customer, <<the 6005A ran at more than twice the extrusion speed, which had resulted in tearing for 6061. The surface finish was excellent and good mechanical properties were obtainer using only air cooling, at the press run-out. No tearing was noted in the 6005A even at highest speed used at the press, indicating the shape could have been pushed even faster>>, Pangborn et al. (2012). After two years period after implementation, the data showed a << net productivity increased over 50 percent in a two year time period>>

Along with the introduction of new material, it has become necessary to characterize the workpiece workability and machinability, Zaghbani et al., (2010). Traditionally, the machinability characterization has been based either on the tool life or on the energy required to shape them, Songmene et al., (1996). In the case of aluminum alloys, the considered criterion for tool life is the tool breakage due to adhesion of the aluminum alloy on the tool as a result of the heat involved in the process or the material ductility. The tool breakage is generally observed for a carbide drill after drilling more than 2000 holes at a depth of around 28.3 mm, Zitoune, (2010). Nouri et al., (2004), while studying the wear mechanism during drilling of aluminium alloys, showed in their study that some of the most important causes of tool wear are the diffusion and adhesion of the material, which could degrade the surface finish. These results were confirmed by List et al., (2005).

In the drilling process, the cutting forces can be a good indicator for a comparison of different alloys under the same machining conditions, Zaghbani and Songmene, (2009). Some authors, such as Shaw (1989), Subramanian et al., (1977), Kouam et al., (2010) and Kouam et al., (2012), have shown that the thrust force increases with the feed rate. In the case of isotropic materials, the cutting forces follow a linear behavior when the feed is varied, Altintas, (2000). For new materials, it is important to verify this behavior. If the behavior is non-linear, this can indicate a high non-homogeneity within the material.

The machinability of an alloy, while not being a standard property, defines its ability to be machined or shaped using a cutting tool (Sandvik, 1994), and can be evaluated using tool wear, tool life, productivity, part quality, cutting forces, or chip form. Songmene et al. (1996) established a procedure for testing the tool life and the machinability of materials and he also defined a global machinability rating taking into account the tool life, the cutting forces, the surface finish and the chip formation. Decreasing the chip size can improve the tool life, the surface finish and lower the energy required for machining. The surface finish is an indicator of the quality of the material following the machining process. One other main difficulty encountered in the machining process is burr formation. Its removal is costly and is considered a non-productive operation. The burr morphology depends on the cutting conditions and the mechanical properties of the workpiece material and on the tool used. Rivero et al., (2006) showed in their work that burr formation could have an influence on power consumption and on the tool temperature. In his work, Gillespie (1973) identified some mechanisms responsible of burr formation: material lateral deformation, chip bending and chip tearing. These mechanisms are of course dependent on workpiece materials, cutting tool geometry and machining strategies. Later, Gillespie et al., (1989) linked burr formation mechanisms with deburring processes and techniques. Deburring is however a non-productive and costly finishing process that must be minimized or avoided. Any material leading to limited burr formation is therefore advantageous.

In general, aluminium alloys generally have good machinability (higher tool life; low cutting forces; higher cutting speeds can be used) but some issues might arise with chip control and management, build-up edge: material adhering on the cutting tool causing poor surface finish and burr formation. Burr removal is costly and is considered as a non-productive although it is a finishing operation. High silicon content aluminum alloys (hyper-eutectic) are usually more abrasive and wear out quicker the cutting tool, especially carbide tooling. Diamond tools should be considered. These tools are expensive but are able to produce a large number of parts before getting dull. The addition or the change in additive elements content can impact the mechanical and the machinability properties, therefore could make a product more or less competitive. In fact, economical and maximum productivity machining conditions (speeds and feeds) depend on the machinability of the workpiece material. Therefore, it is important to develop materials with excellent mechanical and field properties but also with good machinability and it could influence the competitiveness and the acceptability of a given product.

The main objective of this work was to compare the machinability of four Al-Mg-Si aluminum alloys manufactured by Rio Tinto Alcan: AA6262-T6 (37.3 HRA); AA6061-T6HS (39.6 HRA); AA6061-T6 (29.7 HRA) and AA4XXX-T6 (36.6 HRA). The evaluation included tool wear, cutting force, surface finish, chip form and burr height. As the current tendency in most machining shops is to eliminate the lubricant for machining costs reduction purposes and to respect environmental and occupational safety regulations, the dry drilling is used in this work to evaluate machinability.

2. Global machinability model

The machinability can be defined as the relatively ease or difficulty to shape a workpiece material using a cutting tool. Several factor affect the machinability; They include the cutting data and conditions (cutting speed, feed, depth of cut and type of operation- lubricated or not), the tool data (tool material and coating, tool geometry including the rake angle, the lead angle, the clearance angle, the nose radius, the edge preparation); but also the operation (continuous or interrupted) and of course the material. Amongst the workpiece material data affecting the machinability, one can list: the mechanical properties (strength, ductility, toughness, and hardness), thermal conditions, inclusions, work hardening, microstructure, conditions, and chemical composition.

The machinability of a material can be evaluated using one or a combination of the following criteria: cutting tool wear/life, energy required for machining or specific cutting forces, part quality including surface texture, burr formation and chip formation. The global machinability rating developed by Songmene et al., (1996) and refined by Zaghbani et al., (2010)

http://dx.doi.org/10.5772/54021

was further improved by adding the sticking tendency and the material ability to form burr. Using specific ratios related to tool life (R_{TL}), sticking tendency (R_{ST}), cutting force (R_{CF}), surface finish R_{SF}, burr height R_{BH} and chip length R_{CL}) the global machinability (GMR) of the tested alloys was assessed:

$$\begin{aligned} Global \quad Machinability = \lambda_1 \times \cdot Tool\_life_{ratio} + \frac{\lambda_2}{Sticking\_tendency_{ratio}} + \frac{\lambda_3}{Forces_{ratio}} \\ + \frac{\lambda_4}{Burr\_height_{ratio}} + \frac{\lambda_5}{Chip\_length_{ratio}} + \frac{\lambda_6}{Surface\_finish_{ratio}} \end{aligned} \tag{1}$$

$$GMR = \lambda_1 \times \cdot R_{TL} + \frac{\lambda_2}{R_{ST}} + \frac{\lambda_3}{R_{CF}} + \frac{\lambda_4}{R_{BH}} + \frac{\lambda_5}{R_{CL}} + \frac{\lambda_6}{R_{SF}} \tag{2}$$

Where λi (i = 1 to 6) are the specific weights: $\lambda_1+\lambda_2+\lambda_3+\lambda_4+\lambda_5+\lambda_6 = 1$; Each ratio (Tool life ratio R_{TL}; sticking tendency ratio R_{ST}; cutting force ratio R_{CF}; surface finish R_{SF}; burr height ratio R_{BH} and chip length ratio R_{CL}) is computed as the ratio of the performance index for the tested material to that of the reference material.

- **Tool life** can be based on a number of holes drilled before the drill is dull. The tool life criteria can be set upon a surface finish value, an increase in cutting forces or on a given drill flank wear value (ie. VB = 0.3 mm).

$$R_{TL} = \frac{Tool\_Life\_M}{Tool\_life\_AA6262} \tag{3}$$

where Tool_life_M is the thrust force of material M and Tool_life_ $_{AA6262}$ is the tool life for the reference AA6262-T6 alloy.

- **Cutting forces:** The ratio of thrust force was used :

$$R_{CF} = \frac{F_z\_M}{F_z\_AA6262} \tag{4}$$

where $F_{z\text{-}M}$ is the thrust force of material M and $F_{z\text{-}AA6262}$ is the thrust force for the reference AA6262-T6 alloy.

- **Material sticking:** One of the issues that may interest a machinist dealing with aluminum alloys is sticking. It is known that some aluminum alloys exhibit a high sticking tendency, which is due mainly to their relatively high thermal conductivity. This causes thermal softening, which leads to the sticking of the work piece material to the tool. During machining, this problem can be observed on different regions of the tool, with the greatest concentration being on the chisel edge where the cutting speed is at its lowest. However, it is difficult to arrive at a conclusion as to which material sticks more than the others based on these images. Normally, the material that sticks the most is the one that will accumulate the highest quantity of metal on the chisel edge. However, as the adhered material can be removed from one hole to another, comparing the accumulated quantity of materials for each alloy is complex. For these reasons, a new criterion for sticking is proposed. This criterion is based on a variation of the axial cutting force. It is known that a new tool has no adhered material on it, and so the cutting force generated using such a tool can be considered as a reference force. As material stick to a new tool, its geometry changes. If a second hole is drilled with the same tool (now with adhered material on its chisel edge) the "new tool geometry" will cause a variation of the cutting force, with this variation being proportional to the adhered material. This criterion which characterizes the sticking tendency (ST) can be defined as the variation of the cutting forces knowing that no flank wear will occur when drilling one hole on aluminum :

$$Sticking\_Criterion = ST = \frac{\Delta F}{F} = 1 - \frac{Max\_F_z\_hole\_1}{Max\_F_z\_hole\_2} \tag{5}$$

The higher is the value of ST (Criterion of sticking) the higher is the tendency of the material to adhere to the tool. Using the proposed sticking criterion, ST, it is possible to rationally compare each alloy tested to the reference AA6262-T6 alloy in terms of sticking. From the sticking tendency of each alloy (Eq. 5), the sticking tendency ratio R_{ST} was defined as:

$$R_{ST} = \frac{ST\_M}{ST\_AA6262} \tag{6}$$

- **Burr formation:** Burrs produced during machining affects the quality of part, especially for precision component, and can easily jeopardize the assembly and the functionality and the life of the components and machines. They cause misfits in precision assembly, blockage in internal ducts during operation (when detached) and early failure of the component, Olvera and Barrow (1995). Deburring and edge finishing in aerospace industry can easily represent 30% of part cost (Gillespie, 1973), or 10-12 % of total machining time. Deburring, like inspection, is a non-productive operation and, as such, should be eliminated or minimized to the greatest extent possible, Dornfeld (2004). Burr control is necessary to improve workers safety, reduce manufacturing and assembly costs and to improve the productivity.

The burr form and height are dependent on the material properties and cutting conditions. A material generating fewer burrs is therefore more interesting than a one generating more. The ability of a given material to generate burr can therefore be a measure of its machinability. To do so, the burr type or the burr size could be used to compare the machinability. Let's R_{BH} be the ratio of the burr size for a tested material compared to the AA6262-T6 reference material.

$$R_{BH} = \frac{BH\_M}{BH\_AA6262} \tag{7}$$

http://dx.doi.org/10.5772/54021

where BH_{-M} is the burr height of material M and $BH_{-AA6262}$ the burr height of the AA6262-T6 alloy.

- **Chip formation:** The success of a machining operation can also be determined by the chip formation which is strongly associated with the properties of the workpiece material and the cutting conditions. Short and discontinuous chips are usually preferred for chip management purpose and also because of their beneficial indirect effects on tool life, surface finish, cutting forces/energies and on metallic particle emission (Songmene et al., 2011). The chip length ratio (R_{CL}) was computed as follow:

$$R_{CL} = \frac{CL\_M}{CL\_AA6262} \tag{8}$$

where CL_{-M} is the measured chip length of material M and $CL_{-AA6262}$ that of the reference AA6262-T6 alloy.

- **Surface finish:** For the surface finish ratio, the arithmetic average roughness (Ra) was used.

$$R_{SF} = \frac{Ra\_M}{Ra\_AA6262} \tag{9}$$

For the evaluation of the global machininability of the Al-Mg-Si alloys (Equation 2), the weights (coefficients of Equation 2) were varied according to potential interest that could be encountered in metal working industries.

3. Al-Mg-Si alloys: Microstructures and manufacturing conditions

The tested materials were manufactured by Rio Tinto Alcan: AA6262-T6 (37.3 HRA); AA6061-T6HS (39.6 HRA); AA6061-T6 (29.7 HRA) and AA4XXX-T6 (36.6 HRA). All four alloys were DC cast and subjected to commercial homogenization cycles. The AA6061 was a typical commercial variant of AA6061 suited for high speed extrusion into thinner wall shapes. The AA6061 High Strength (AA6061HS) contained higher levels of the major alloy addition to promote increased age hardening response. The AA6262 and AA4XXX compositions were based on AA6061 but contained additions of 0.5 wt% Pb -0.4 wt% Bi and 8% Si respectively. The AA6061 variants were DC cast directly as 101mm diameter billet whereas the AA4XXX and AA6262 were commercially cast as 300mm dia. The billets were extruded on the 780 tonne, 101mm dia. RTA extrusion press into a 3 x 41mm strip with an extrusion ratio of 70/1. The section was water quenched and aged for 8hrs/170°C to give the T6 temper. Extrusion exit temperatures were in excess of 510°C to promote good press solutionising. Typical extruded microstructures and associated second phase particle distributions are shown in Figure 2.

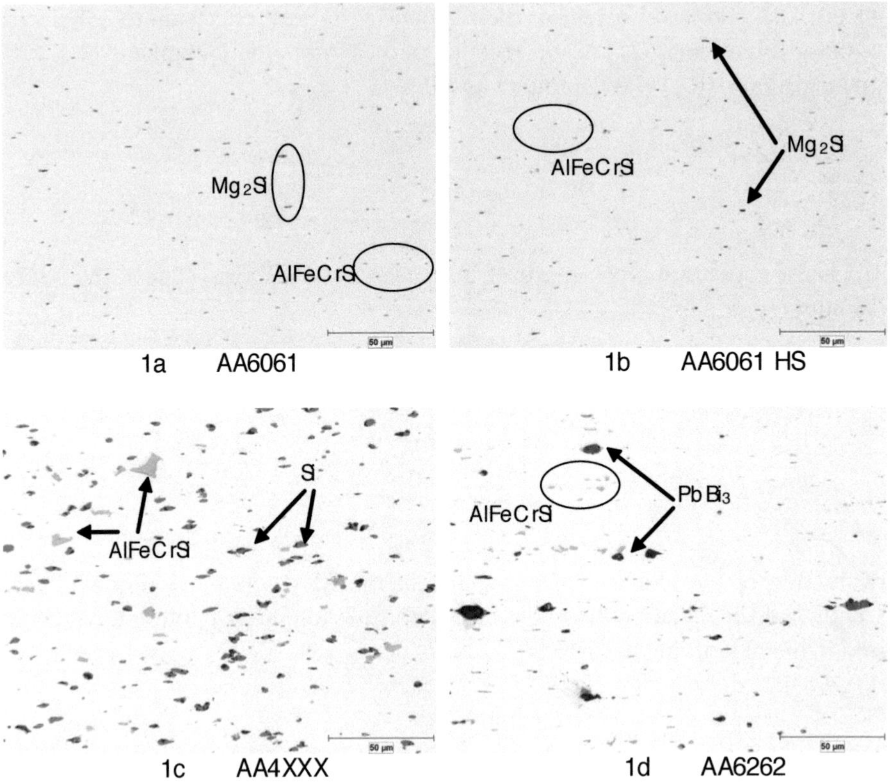

Figure 2. Extruded microstructures (longitudinal, as polished)

The AA6061 variants contained fine Al-Fe-Si intermetallics (grey) and a small volume fraction of undissolved Mg_2Si which is normal for AA6061. As expected, the AA4XXX material with 8% Si contained a much higher volume fraction of second phase consisting of Si particles (dark grey) ~ 5 microns in size and Al-Fe-Si (light grey). The latter was coarser than in the AA6061 variants due to the increased ingot diameter. The main feature of the microstructure in the AA6262 was the globular low melting point $BiPb_3$ particles ~ 2-12 microns in size. The extruded grain structures are shown in Figure 3. The AA6061, AA4XXX and AA6262 extrusions were all fully recrystallised but with varying grain size. The AA6061 HS exhibited a mixed fibrous/partially recrystallised structure.

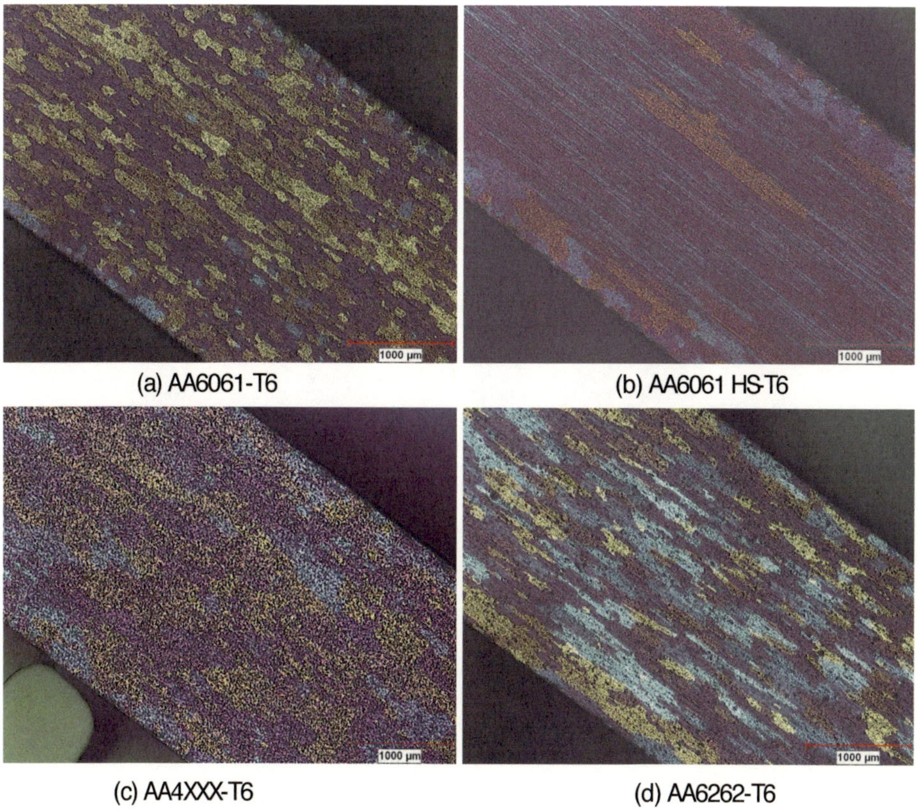

Figure 3. Extruded Grain Structures (Barkers Reagent)

The tensile properties are summarized in Figure 4, with the alloys ranked in terms of increasing strength. The yield strength matches the hardness data presented in section 4 except for the AA4XXX-T6 which gave the lowest yield strength but the second highest hardness. This discrepancy is probably due to the high volume fraction of silicon particles in the microstructure.

All alloys satisfied AA6061-T6 minimum properties but the overall range of yield strength was ~ 60MPa. Toughness was measured using the Kahn Tear test and results are presented in Figure 5. The AA6061-T6 variants exhibited the highest toughness. Normally toughness is expected to fall with increasing yield strength but the AA4XXX-T6 and AA6262-T6 materials did not follow this trend and gave lower toughness for the same or lower yield strength, probably due to the increased volume fraction of coarse second phase particles.

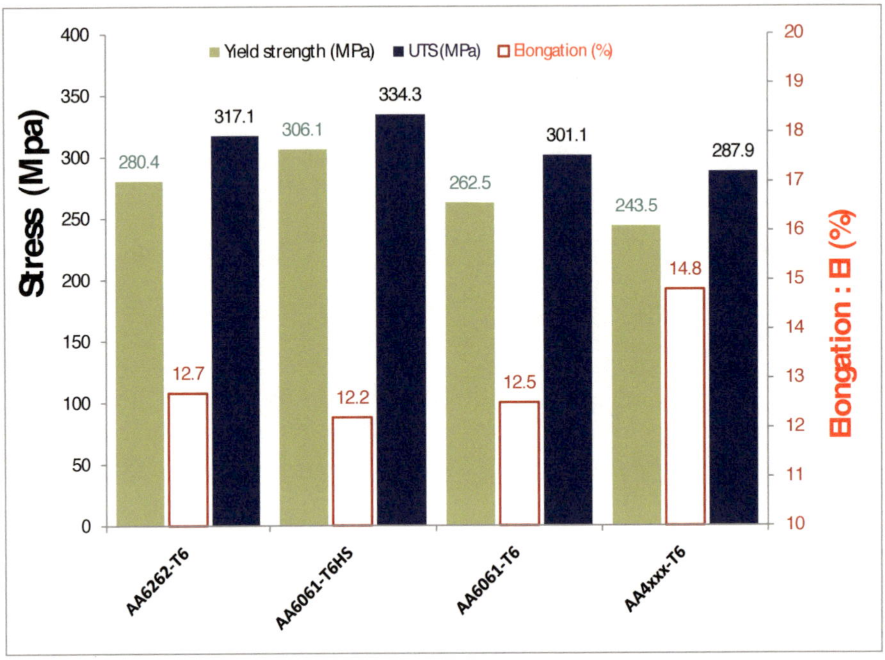

Figure 4. Tensile properties of the tested materials

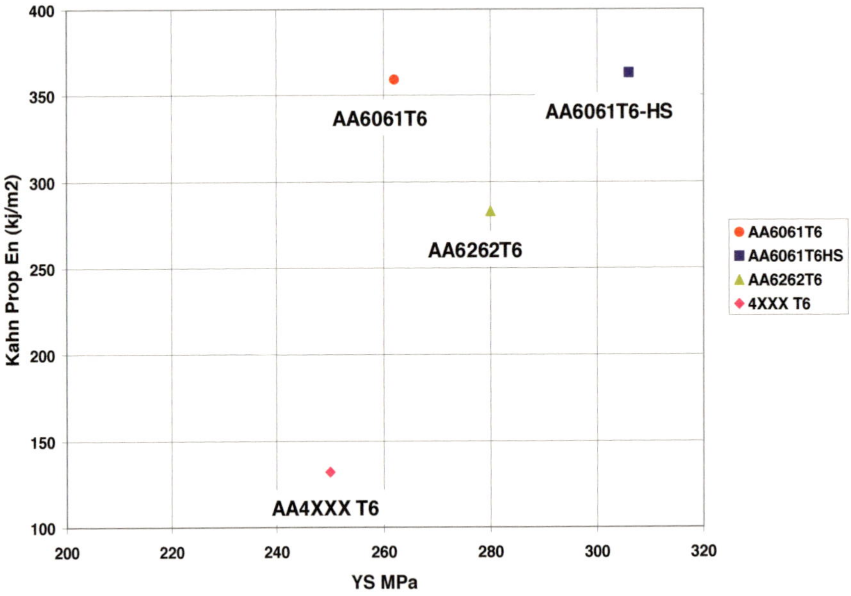

Figure 5. Toughness properties of the tested materials

4. Experimental procedure

The procedure used in this study for machinability testing consisted of drilling parts and recording the following machinability indicators:

- The tool life during drilling operations;
- The surface finish of drilled holes;
- The peak and average values of the cutting force when drilling holes;
- The exit burr height of drilled holes;
- Chip form and chip formation mechanism during drilling and milling.

All these information are necessary for better machinability characterization as presented in section 2- Global machinability model.

The machine-tool and instruments used in the present study are displayed in Figure 6. They consisted of:

- A vertical CNC milling machine-tool (Mazak NEXUS 410 A: 10,000 rpm, 20 hp) which was used to drill the samples or to perform milling operations necessary for chip formation study; Such an industrial scale machine-tool was necessary to reach speeds and conditions currently used in industries. The workpieces were mounted on a 3-axis table dynamometer (Kistler 9255B) to allow the measurement of axial cutting forces during the drilling operations.
- A surface measuring instrument (Mitutoyo SUFTEST SV 600) was used to measure the hole surface finish after the tests;
- An optical microscope (tool maker microscope) equipped with a digital camera, Clemex captive, was used for tool wear inspection and measurement;
- A scanning electron microscope (Hitachi) was used for tool wear mechanisms and patterns studies and for chip form analysis;
- A high speed camera (120 000 fps) was used to study the chip formation during the milling process conducted using the machine-tool presented above.

The following drilling parameters and conditions were used:

- Tool :uncoated high speed steel (HSS); diameter: 9.525 mm (3/8 inch); helix angle:30°; Taper angle: 110°
- Feed per tooth: 0.1016 mm/rev to 0.2286 mm/rev
- Cutting speeds: 45 m/min to 274 m/min
- Each test was repeated 10 times except for the tool wear test which was not repeated because of the large number of holes required (about 1500 holes).
- Lubrication: none

The microstructures, the composition, the temper designations, the manufacturing conditions and the mechanical properties of the alloys tested were described in section 3: These alloys can be summarized as:

- AA6262-T6 (37.3 HRA)
- AA6061-T6HS (39.6 HRA)
- AA6061-T6 (29.7 HRA)
- AA4XXX-T6 (36.6 HRA)

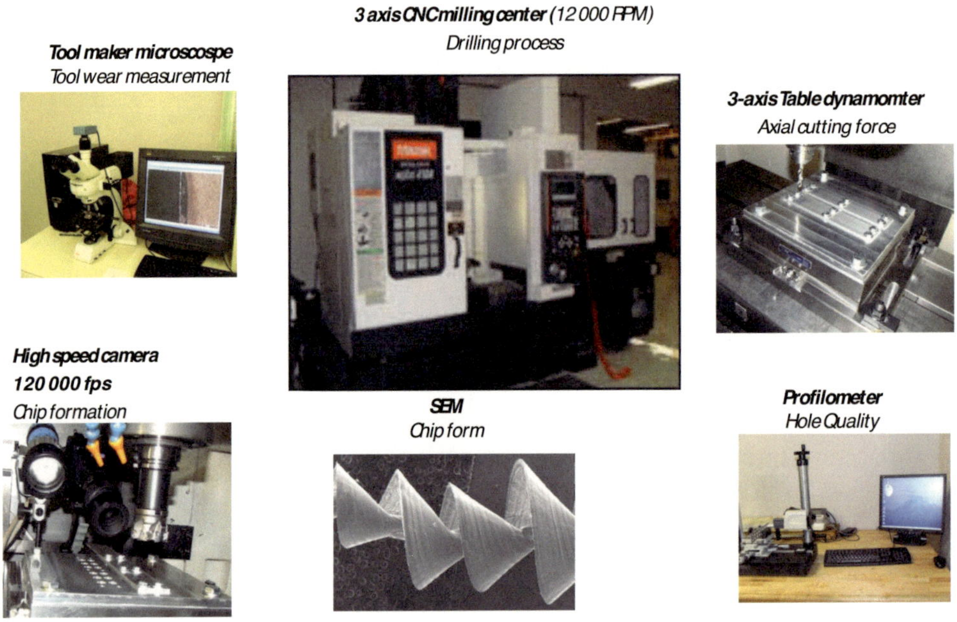

Figure 6. Images of equipment used

5. Results and discussions

5.1. Cutting forces

Figure 7 presents a typical axial force-time data for the four alloys acquired at a sampling frequency of 48 kHz. This high sampling frequency permitted accurate tracking of cutting force variations at the tool tip contact until the retreat of the tool. The similarity of the cutting force profiles for different alloys allows a reasonable comparison of their amplitude. A typical drilling force profile presents different process stages: drill engagement (P1-P2); cutting (P2-P3); deformation of the uncut material leading to burr formation (P4) and drill exit (P4-P5). Following these stages, the drilling process performance can be assessed. Generally, the average or the peak cutting force evaluated when the drill is in full action (between points P2 and P3) can be used to compare different alloys in terms of amplitude of the axial cutting forces. This force can then be plotted against the materials, the workpiece conditions or the cutting parameters.

The variation of the cutting force as a function of feed rate was established in terms of maximum force, and the results are presented in Figure 8. For all the tested alloys, the cutting force followed a linear trend with feed rate. This observation confirms the work by other authors, such as Shaw (1989), Subramanian et al. (1977), Songmene et al. (2011) and Kouam et al. (2012). The linearity was confirmed by correlation coefficients higher than 99%.

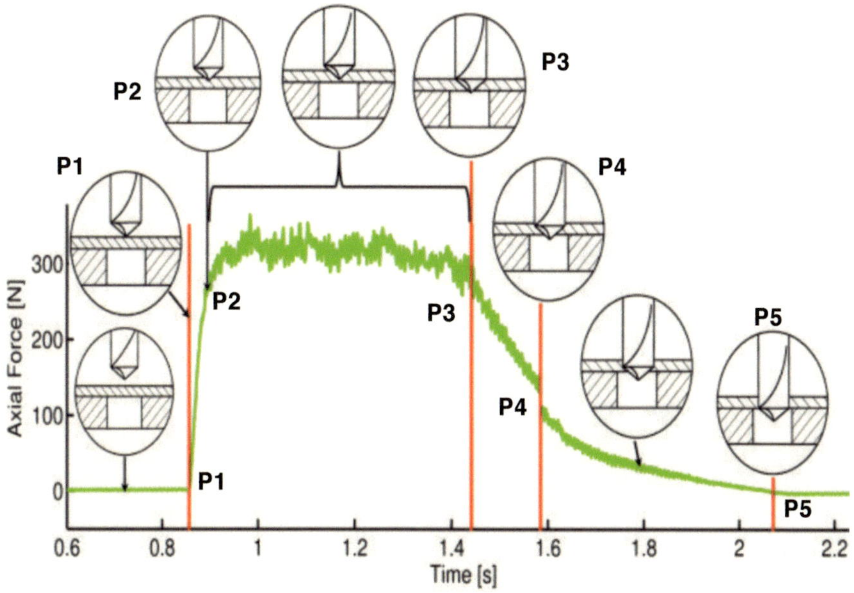

Figure 7. Typical axial force profile during drilling of AA6262-T6

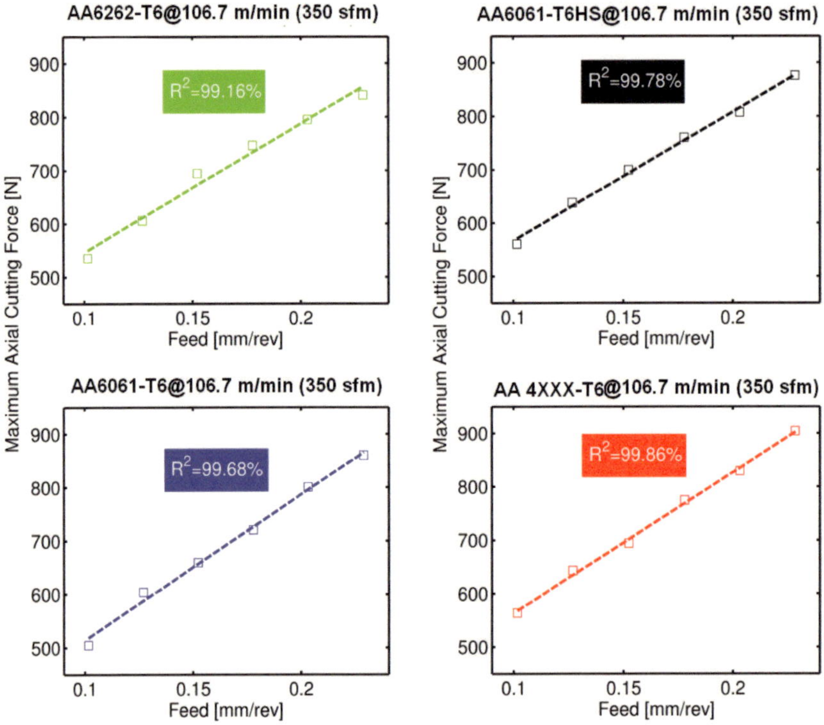

Figure 8. Variation of the maximum axial force of different alloys at different feed rates

Taking into account the decomposition of the axial force into shearing and indentation components (as suggested by Kouam et al. 2010 and 2012), the results in Figure 8 allow the maximum cutting force to be written as a function of the feed and tool radius as follow:

$$F_z = K_d(f \cdot r) + F_{indentation} \tag{10}$$

where : F_z (N) is the maximum cutting force for a given alloy.

- K_d (N/mm^2) is mathematically the slope of the line, and physically, represents the resistance of the material to deformation or extrusion during the drilling process.
- r (mm) the tool radius and f (mm/rev) the feed rate;
- $F_{indentation}$ is mathematically the coordinate at the origin, and physically, represents the indentation effect ($f \approx 0$) which describes the resistance of the material to penetration. The indentation force is proportional to the material hardness as shown in Figure 9a.

The plot in Figure 9b shows that both the resistance to deformation (K_d) and the indentation force are needed to better estimate the axial force as no correlation appears between the indentation force and the resistance K_d. The 6061-T6 alloy presented the higher K_d-value, meaning that for a given feed rate, the increase in cutting axial force for this alloy is higher and compensate for the low indentation force.

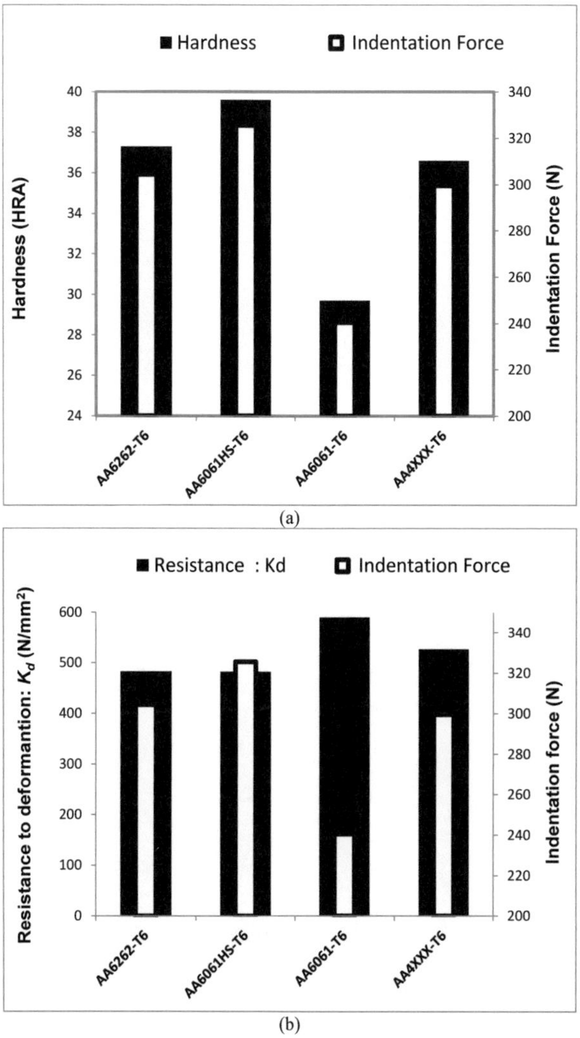

Figure 9. Comparison of components of maximum axial forces (Eq. 10): a) Correlation between the indentation force and hardness; b) comparison between indentation force and the resistance to deformation

Statistical analysis of the cutting forces revealed that there is a significant difference in the behaviour of the four tested alloys in terms of force (average force and maximum peak force). The AA6262-T6 alloy required the lowest force while the AA4XXX-T6 required the highest force (Figure 10). This can be related to the presence of second phases within the alloys (See Figure 2). The 6061-T6 and the 6061-T6HS alloys which had comparable ductility (Figure 4) and toughness (Figure 5) showed comparable cutting force.

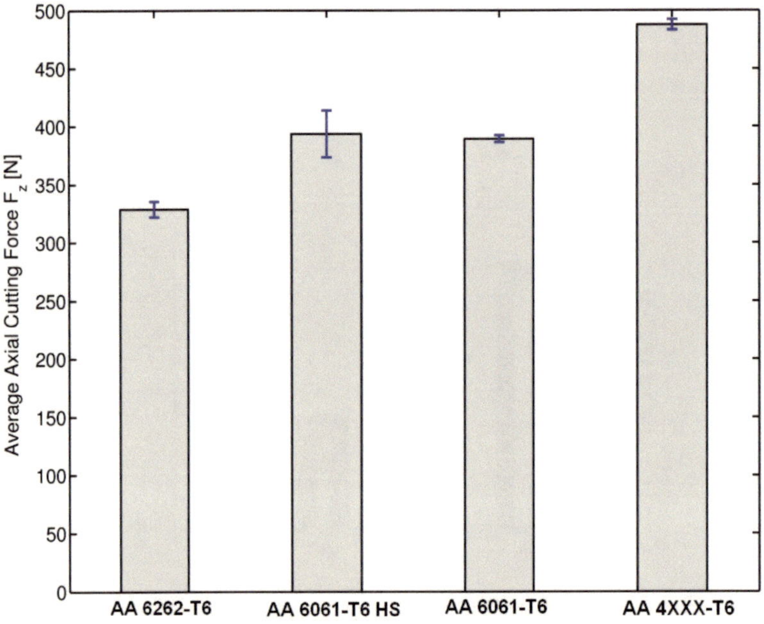

Figure 10. Average axial cutting forces comparison

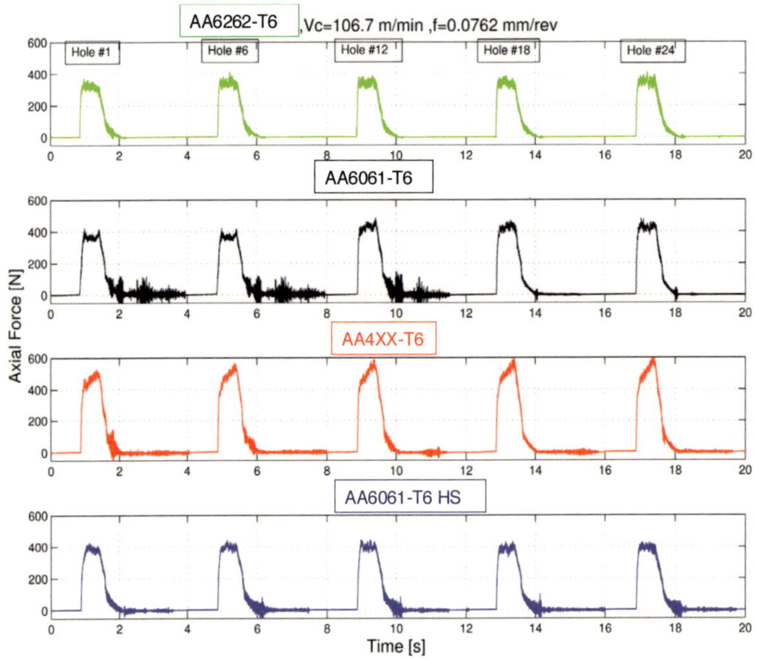

Figure 11. Effect of possible tool wear or material adhesion on cutting force profiles

The force profiles were compared to check for possible tool deterioration (wear, material adhesion, etc.) that could affect the cutting forces, Figure 11. The force profiles recorded for each material repeated themselves well for the first 24 holes drilled, except for the AA6061-T6 for which the first, the sixth and the hole number twelve were different. This difference can be attributed to a possible adhesion of the material on the cutting tool, which might have modified its geometry and led to a different burr formation when the drill exited. Based on the figure 11, it can be reinforced that for the first 24 holes, the cutting did not experienced a wear susceptible of altering the cutting forces.

5.2. Tool wear and tool life

Figure 12 presents SEM images of the cutting lips after drilling many holes. It was observed that the separate drills used to machine the four alloys exhibited a normal tool life under the test conditions used and no premature tool wear or breakage occurred during the drilling tests. The criterion for tool life that is generally considered for aluminum alloys is tool breakage. Each tool was examined by SEM after drilling a given number of holes (432, 856 and 1439 holes), to detect any significant wear on the cutting lips or on the chisel edge.

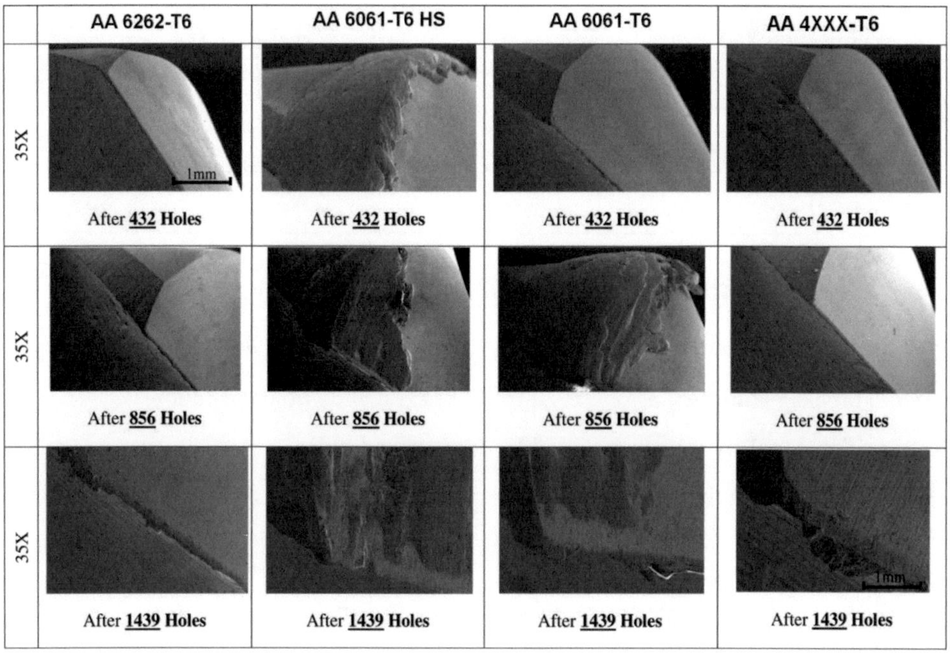

Figure 12. SEM images of the chisel edge after drilling many holes (magnification 35X)

When the number of holes was approximately doubled (856 holes), the cutting lips again did not show any significant wear, as can be seen in the second row of Figure 13. However, when the number of holes reached 1439, the AA6061-T6 HS alloy (second column-third line) began to exhibit significant wear. The arrow added to the graph points to the region of cutting lip wear. The latter was observed at higher magnification (200 X). The SEM images of the cutting lip of the drill used for machining the AA6061-T6HS are presented in Figure 13.

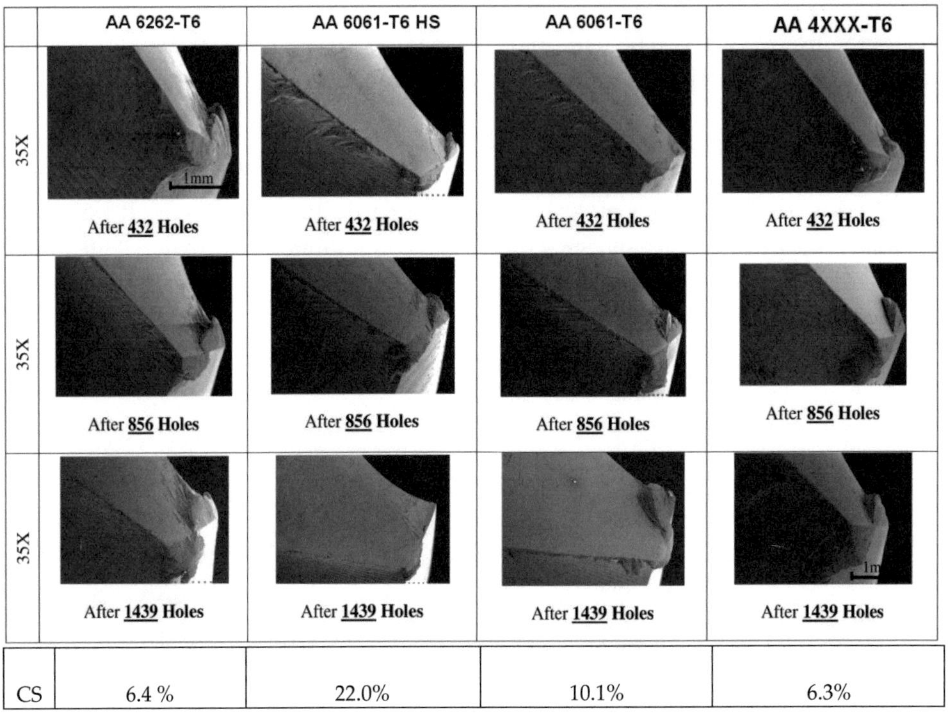

Figure 13. Validation of the sticking criterion (Cs, Eq. 5) with microscopic observations

In general, the four tested aluminum alloys exhibited normal tool life:

- No tool breakage failure was observed for the 1439 drilled holes per alloy,
- No significant tool wear was observed for 1286 drilled holes,
- The aluminum alloy that caused the highest tool wear was the AA6061-T6HS; This can be explained by high mechanical resistance.

It can be considered that the two alloys AA6061-T6 and AA4XXX-T6 have the same tool life as the reference alloy AA6262-T6. While a tool life index (Equation 3) of 94% can be assigned to the AA6061-T6HS. The number of drilled holes with no significant wear was 1286 holes; at 1439 holes there was a significant wear. It can be assumed that the tool wear appeared between 1286 and 1439 holes around 1362 holes.

The sticking tendency of the alloys was evaluated using Equation 5 and the cutting forces profiles. The sticking tendency was confirmed with SEM observations of the drill tips (Figures 12 and 13) and good correlations were found between the sticking tendency and the tool life (Figure 14). Materials with low sticking tendency led to higher tool life (case of the AA6262-T6, AA6061-T6HS and AA4XXX-T6), while the one with high sticking tendency led to lower tool life, Figure 14. When the workpiece material adheres to the cutting tool tip, it modifies the tool geometry, thus increasing the forces required to cut the metal. The surface finish of the machined part is also deteriorated in presence of a built-up-edge; the modification of the tool geometry changes the shearing direction and when the BUE is evacuated, it move to the tool-workpiece interface and contribute to the 3-body wear.

The higher is the value of CS (Criterion of sticking) the higher is the tendency of the material to adhere to the tool. Using this CS, it was possible to rationally compare the four alloys in terms of sticking as shown in Figure 14.

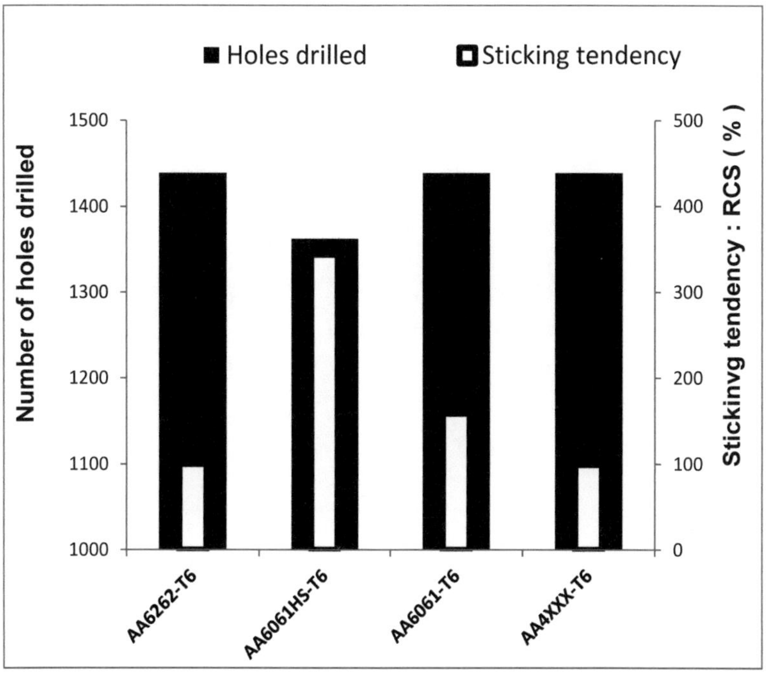

Figure 14. Tool performance and sticking tendency comparison

5.3. Surface quality

Figure 15 presents the average roughness (Ra) and the quadratic roughness (Rq) of holes produced when drilling the different materials. A statistical analysis confirmed that the roughnesses of the four tested alloys are statistically different. The best surface roughness (Ra and Rq) is obtained for the AA6061-T6HS which also exhibited the highest yield strength while the higher values of Ra and Rq were obtained for the AA6262-T6 material. The performance of the AA6262-T6 could be related to sticking of workpiece material onto the cutting edge of the tool (see Figure 12, sticking tendency), forming the build-up-edge

(BUE). It is known that the BUE is usually responsible for the deterioration of the machined part surface finish. However, the recorded values of surface roughness for all the four tested alloys are within acceptable ranges. For a drilling operation, a value between 6.3μm and 1.6μm is considered acceptable for general applications. For more demanding applications of Ra value between 1.6μm and 0.8μm is desirable. For each of the four materials, the ratios of Rq to Ra values were between the ASME recommended brackets.

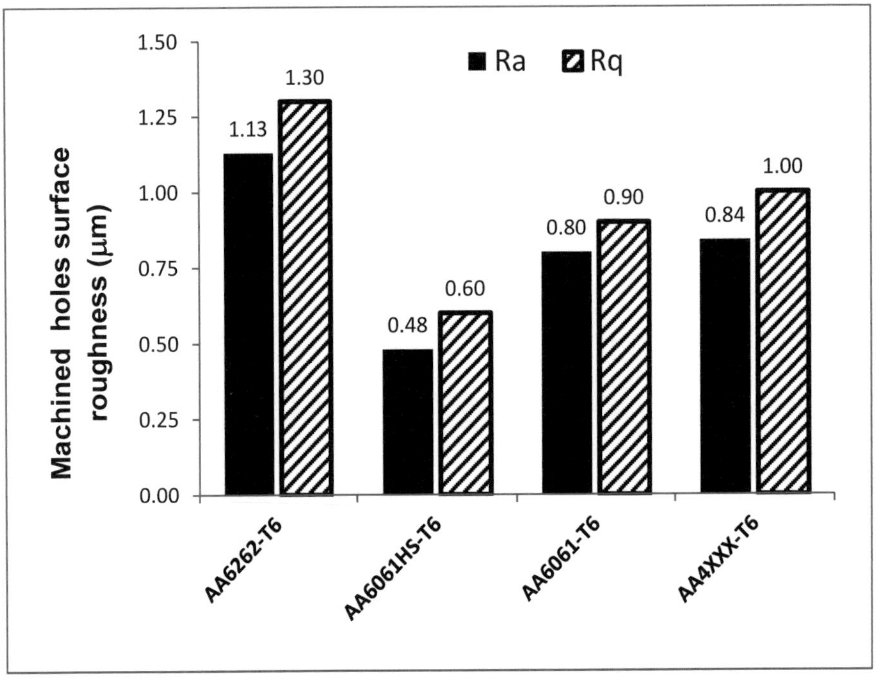

Figure 15. Arithmetic average roughness (Ra) and quadratic roughness (Rq) of holes obtained on different materials

5.4. Burr formation

One other main difficulty encountered during machining of ductile materials is burr formation. Its removal is costly and is considered a non-productive operation. The burr morphology depends on the cutting conditions and the mechanical properties of the workpiece material and on the tool used, Hashimura et al 1999); Rivero et al., (2006) showed that burr formation could have an influence on power consumption and on the tool temperature. Gillespie et al., (1989) linked burr formation mechanisms with deburring processes and techniques.

Burr removal is a non-value added process (Aurich et al. 2009) and might represent as much as 30 percent of the cost of finished parts (Gillespie, 1999). Niknam and Songmene (2012), while modeling and studying the burr formation during milling of AA6061-T6 and AA2024-T321 found that the burr thickness, which control the deburring difficulties and the deburring cycle time, is highly sensitive to material mechanical properties such as yield strength and to the cutting force. As deburring is non-productive and costly finishing process, it should be minimized or avoided. Any material leading to limited burr formation is therefore advantageous.

For assembly purposes, it is important to have holes which are burr-free. For general applications, the hole must be burr-free at a magnification of 5X, while for more critical applications, the magnification can go up to 30 X. The burr form and height are dependent on the material properties and cutting conditions. Images showing typical exit hole appearance are presented in Figure 16 as a function of feed rate. The worst case for burr was obtained for the AA4XXX-T6 (Figure 16) alloy which is the most ductile one. The burr observed was a transient burr type. The other alloys exhibited a uniform burr (type I) or crown burr (type II), Costa (2009):

- The AA6262-T6 and AA6061-T6HS alloys produced only uniform burrs (type II).
- The AA4XXX-T6 and AA6061-T6 alloys produced both uniform burrs (Type II) and transient or crown burrs (type I). The latter are generally difficult to remove.
- The AA4XXX-T6 was problematic in terms of exit burr height.

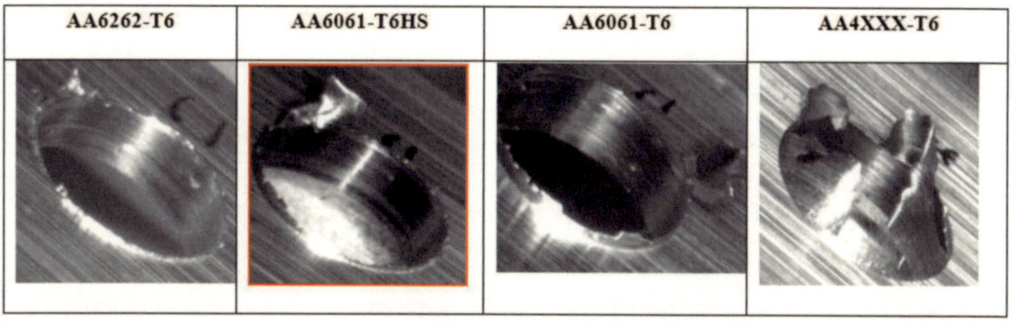

Figure 16. Optical microscopy images of exit burrs observed on drilled holes as a function of feed rate (cutting speed: 45.7 m/min; Feed rate: 0.0508 mm/rev)

In general, the burr form and height was found to be dependent on feed rate, exception of the AA 4XXX-T6 alloy. The lower the feed rate, the higher the burr height obtained, Figure 17. The AA4XXX-T6 produced most of the times high size burrs and only in very limited cases, the burr size was comparable to others alloys tested. At lower speeds, the burr size observed was higher compared the one obtained at high cutting speed; this denotes a possible interaction of the feed rate and the cutting speed on burr formation.

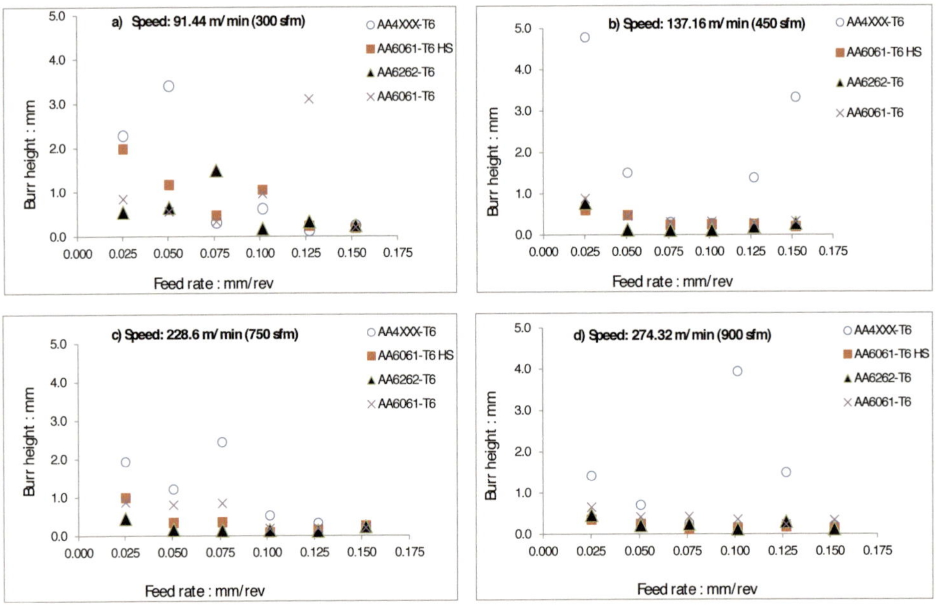

Figure 17. Burr height progression as a function of feed rate, cutting speed and workpiece materials

5.5. Chip formation

The success of an alloy depends also on the chip obtainable during machining of the alloy. A bad chip formation can shorten the tool life, slow down the production, deteriorate the machined part surface finish, increase the machining costs and increase the emission of metallic particles. The AA6262-T6 for example is often preferred for its ability to deliver short and broken chips. Figure 18 displays samples of the chips collected during the drilling of the tested alloys. Under the used cutting conditions, the four tested materials all generated long and continuous chips, but some longer than others (Figure 20).

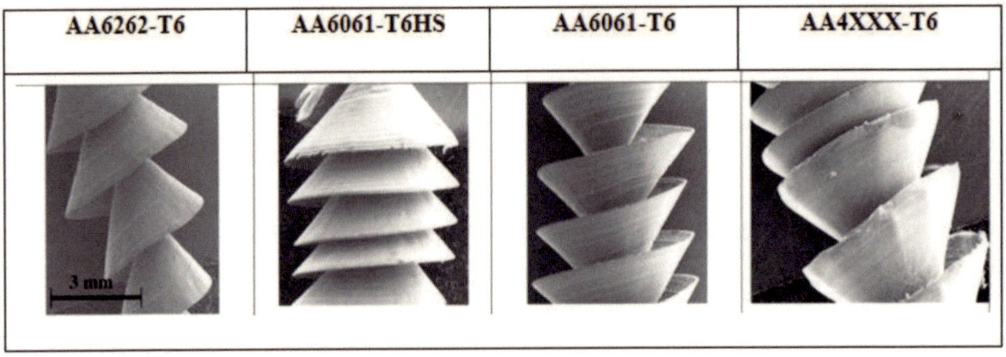

Figure 18. SEM images of the drilled chips (magnification 12 X)

http://dx.doi.org/10.5772/54021

However, during a milling test, a difference was found for example in chip formation for the AA6262-T6 and the AA6061-T6 (Figure 19). The chip formation was recorded using a high speed camera (4 000 fps). In Figure 19, the milling tool progresses for a1 to a3 for the AA6061-T6 and from b1 to b3 for the AA6262-T6. The following observations were made:

- In milling, the chips collected are continuous (Figure 19) and not conical helical as it has been seen in the case of drilling. In figure 19, it can be observed that the produced chips are longer and have more tendencies to adhere to the rake face of the tool. Which may confirm the AA6061 alloy is more adhering than the AA6262, and consequently addition causing more tool wear.

- The chip produced when machining the AA6262-T6 reference material is more curved, leading to high possibility of chip breaking when it comes into contact with workpiece. The difference of chip forms for the four alloys may be explained by the mechanical and thermal properties of each alloy.

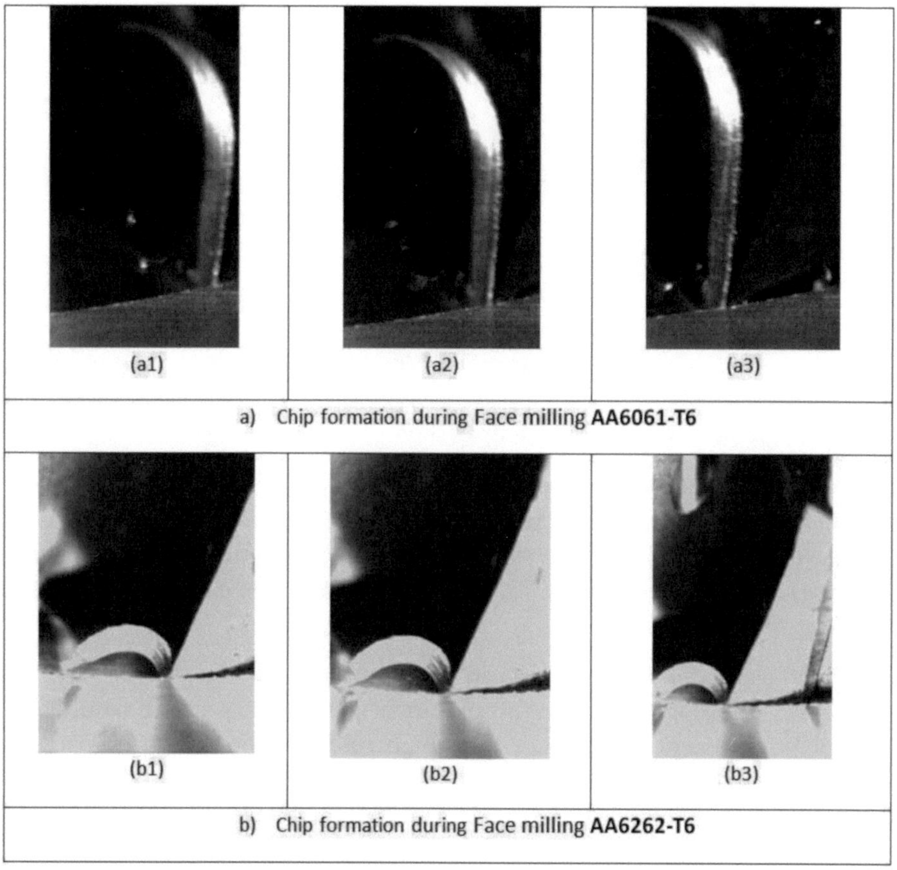

Figure 19. Chip formation during Face milling: feed=0.03 mm/tooth, speed= 191.5 m/min, DOC=1mm; High speed Camera image (4000 fps).

Figure 20 displays the effects of the feed rate and cutting speeds on drilling chip length for each the material tested. It can be observed that the three alloys, AA6262-T6, AA4XXX-T6 and AA6061-T6, behaved similarly in terms of chip length characteristics, which decreased with increased cutting speed and feed rate. For the AA4XXX-T6, however, the data scattering showed some minima at specific combinations of speed and feed.

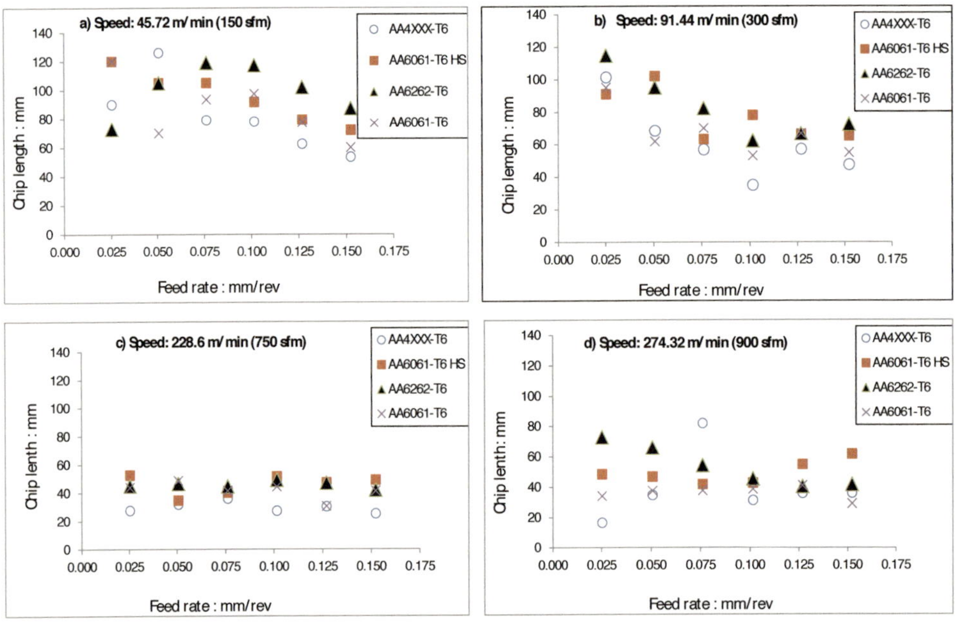

Figure 20. Variation of drilled chip length as a function of feed rate and cutting speeds for the tested alloys

5.6. Partial and global machinability comparisons

In order to compare the performance of the four alloys globally in terms of the various measures of machinability, specific ratios for the main factors of interest were defined. Figure 21 presents a summary of these ratios including, thrust force tendency (Figure 21-a), sticking tendency (Figure 21-b), burr tendency (Figure 21-c) and chip length tendency (Figure 21-d) relative to AA6262-T6.

A higher coefficient value corresponds to a lower machinability. It was observed that AA4XXX-T6 was the worst case, in terms of burr height and thrust force requirement. The latter was probably due to the presence of the high volume faction of second phase consisting of Si particles which also raises the hardness of the AA4XXX-T6. This alloy also had the lowest yield strength and highest elongation which may be related to the poor burr height performance. In contrast, in terms of chip length and sticking tendency the AA4XXX-T6 was equivalent or superior to AA6262-T6. The use of AA6061-T6 HS vs. standard AA6061-T6 gave inferior performance in terms of sticking and chip length which is an interesting result as in often in the industry the trend is to move in this direction to solve machining problems. However, the high strength version was superior in terms of surface roughness and burr height. As expected the AA6262-T6 performed well in most categories but surprisingly in these tests was the worst in terms of chip length.

Figure 22 displays the global machinability of the tested alloys as a function of the weights of the different component of the machinability (λ_1: tool life, λ_2: sticking tendency, $\lambda_{3:}$ cutting force, $\lambda_{4:}$ burr height, $\lambda_{5:}$ chip length and λ_5 surface finish). These weights (λ_1, λ_2, λ_3, λ_4 and λ_5) must be set according to the application, the machine-tool limitations and the manufacturer preference.

It appears for Figure 22 that amongst the tested alloys, only the AA6061-T6 HS performed better (in spite of its high resistance) that the reference material (AA6262-T6) at all the evaluated combinations. The global performances of the AA6061-T6 and the AA4XXX-T6 are comparable but remains lower that of the AA6262-T6.

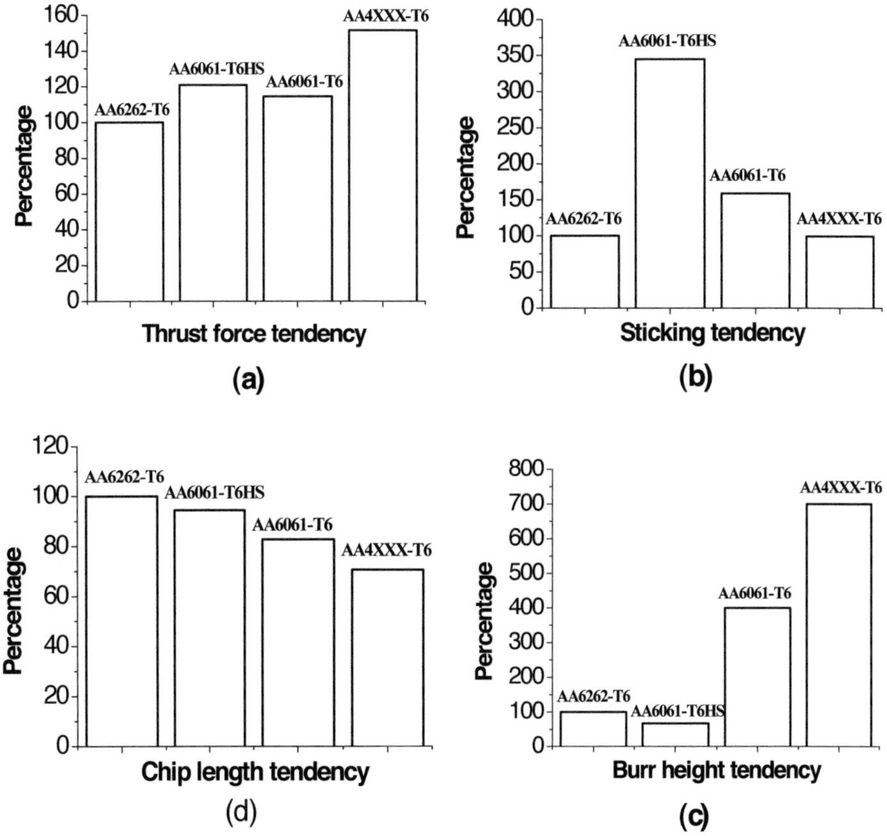

Figure 21. Summary of the force, the sticking, burr and chip length tendency for different tested materials

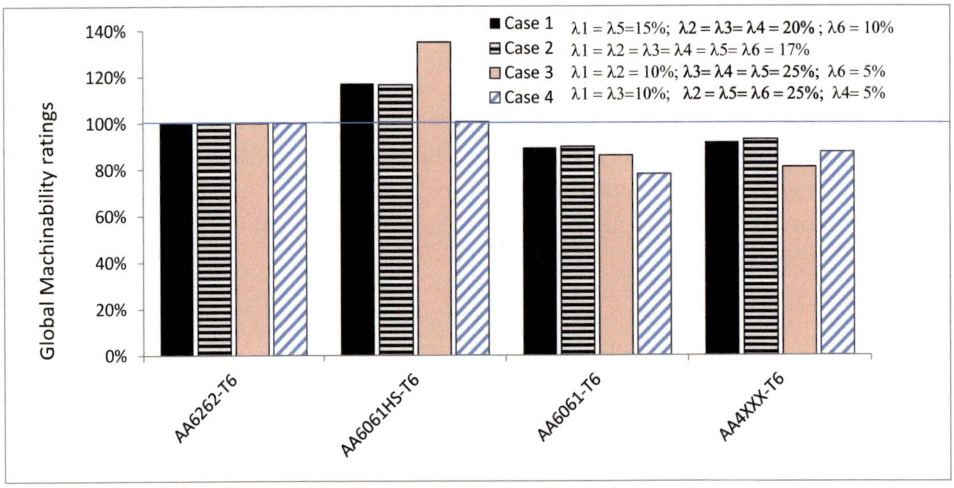

Figure 22. Computed Global Machinability (Eq. 2) for different materials as compared to AA6262-T6

6. Conclusion

The development of aluminum alloys is often conditioned by aeronautical requirements, but aluminum is very interesting for several applications in other sectors. Depending on the nuances, the composition, the treatments and the cutting conditions of these alloys, the material can be classified according to its extrudability, machinability, recyclability, etc. In this work, the machinability performance (tool life, force, surface finish, chip form and burr size) of four commercially available Al-Mg-Si alloys was investigated. It can be concluded that for a non lubricated drilling operation using typical conditions (3/8 inches diameter drill at a cutting speed of 106 m/min (350 sfm)):

- When the global machinability (tool life, material sticking, cutting force, surface finish, chip form and burr size) is concerned, only the AA6061-T6HS) outperformed the benchmark AA6262-T6 while the two other alloys (AA6061-T6 and AA4XXX-T6) showed low machinability compared to the AA6262-T6.

- All the materials exhibited insignificant tool wear after drilling more than 1000 holes. However, when the number of holes reached 1439, the AA6061-T6 HS alloy began to exhibit noticeable wear which may be related to the fact that it had the highest strength compared to others materials. This wear could be reduced by selecting appropriate cutting tool materials or coatings.

- In terms of cutting force, the AA6061-T6HS and the AA6061-T6 were comparable but inferior to the AA6262-T6, whereas the AA4XXX material required the highest force. This may be due to the high volume fraction of Si particles in the microstructure. In a situation

http://dx.doi.org/10.5772/54021

where the machine-tool is powerful enough to accommodate the higher cutting force and the burr could be controlled, the AA4XXX-T6 could become a very interesting material.

- The chip forms obtained were similar for all materials tested: Long or short chips could be obtained depending on the machining conditions. Regardless of the material type, the chip form and the chip management could be controlled by selecting appropriate feeds and speeds.
- In terms of hole quality, the surface finish produced on AA6262-T6 was poor compared to the others alloys tested (AA6061-T6HS, AA6061-T6 and AA-4XXX-T6). The AA6061-T6HS produced a lower burr height which is beneficial in reducing deburring costs.

Author details

V. Songmene[1], J. Kouam[1], I. Zaghbani[1], N. Parson[2] and A. Maltais[2]

1 École de Technologie Supérieure (ÉTS), Montreal, QC, Canada

2 Rio Tinto Alcan, Jonquière, QC, Canada

References

[1] Altintas, Y. (2000), Manufacturing Automation, Cambridge University Press.

[2] Aurich J, Dornfeld D, Arrazola P, Franke V, Leitz L, Min S (2009) Burrs-Analysis, control and removal. CIRP Annals-Manufacturing Technology 58 (2), 519-542

[3] Chena Y., Pedersenb K.O.; Clausena A.H., and O.S. Hopperstada, (2009), An experimental study on the dynamic fracture of extruded AA6xxx and AA7xxx aluminium alloys, Materials Science and Engineering A 523 (2009), 253–262.

[4] Costa E. S., da Silva M. B. and Machado A. R., Burr Produced on the Drilling Process as a Function of Tool Wear and Lubricant-Coolant Conditions, J. of the Braz. Soc. of Mech. Sci. & Eng. Vol. XXXI, No. 1, 2009, 57-63.

[5] Dornfeld, DA., 2004, Strategies for Preventing and Minimizing Burr Formation, Consortium on Deburring and Edge Finishing, (University of California, Berkeley), internal report No Dornfeld-1-04, 19 pages.

[6] Duval, M., (2012), L'Aluminium et sa transformation, notes de cours, ÉTS, Montréal, Canada, 6 juin 2012 (In french).

[7] Farmer, L. E., (1978), a system for evaluating the cold extrudability of Aluminium alloys, Journal of Mechanical Working Technology, 1 (1978), 371-386.

[8] Gillespie, L. K., Deburring and edge finishing, Handbook, 1999.

[9] Gillespie, L.K., 1973, Formation and properties of Machining Burr, M.S. Thesis, Utah State University, Logan, UT.

[10] Hashimura M, Hassamontr J, Dornfeld D (1999) Effect of in-plane exit angle and rake angles on burr height and thickness in face milling operation. Journal of manufacturing science and engineering, 121:13

[11] Kouam J., Masounave J. and Songmene V., (2010) Pre-holes Effect on Cutting Forces and Particle Emission During Dry Drilling Machining, In Proc. of 49th Annual Conf. Metallurgists of CIM, Vancouver, BC, Canada, 2010, 253-263.

[12] Kouam J., Zaghbani I., Songmene V., Parson N. and Maltais A., (2012), Factors affecting the machinability of Al-Mg-Si extrusions, Proc. 10th Int. Aluminum Extrusion Technology Conf., Miami Florida USA, 15-18 May 2012, 443-461.

[13] List G., Nouari M., Ge´hin D., Gomez S., Manaud, J.P. , Le Petitcorps Y. and Girot F., Wear behaviour of cemented carbide tools in dry machining of aluminium alloy, Wear 259 (2005), 1177-1189.

[14] Nouari M., List G., Girot F. and Ge´hin D., Effect of machining parameters and coating on wear mechanisms in dry drilling of aluminium alloys, Machine Tools & Manufacture 45 (2005), 1436-1442.

[15] Niknam, S. A. and Songmene, V., (2012), Modeling of burr thickness in milling of ductile materials, International Journal of Advanced Manufacturing Technology, DOI 10.1007/s00170-012-4479-3

[16] Olvera O., Barrow, G., 1996, An experimental study of burr formation in square shoulder face milling, J. of Mach. Tools Manufacture., vol. 36, No 9, 1005-1020.

[17] Pangborn J.; Hoff E.; Dickson R.; Devadas C., (2012), 6005A: Practical Evidence of Advantages for Extruders and Their Customers, Proc. Tenth international Aluminum Extrusion Technology Seminar & Exposition, May 15-18, 2012, Fontainebleau, Miami Beach Resort, Miami, Florida, USA, 49-54.

[18] Rivero A., Aramendi G., Herranz S. and L´opez de Lacalle L.N., An experimental investigation of the effect of coatings and cutting parameters on the dry drilling performance of aluminum alloys, Advanced Manufacturing Technology (2006) 28, 1-11.

[19] Saha P.K., (2012), Economics of Aluminum Extrusion for Aerospace Applications, Proc. 10th Int. Aluminum Extrusion Technology Conf., Miami Florida USA, 15-18 May 2012, 357-366.

[20] Sandvik C., Modern Metal Cutting - A Practical Handbook, Sweden, 1994.

[21] Shaw M.C., Metal Cutting Principle, Oxford Science, 1989.

[22] Songmene V., Khettabi R., Zaghbani I., Kouam J., and Djebara A., (2011), Machining and Machinability of Aluminum Alloys, Aluminum Alloys, in Theory and Applications", Tibor Kvackaj (Ed.), ISBN: 978-953-307-244-9, InTech, 2011, 377-400.

http://dx.doi.org/10.5772/54021

[23] Songmene V., Stefan, I., Stefan, M., Yan, D., Hirholzer, J., Tool Life & Machinability Testing - Testing Procedure & Database, Report of project, Industrial Research & Development Institute, IRDI, Midland, ON, Canada, Nov. 1996.

[24] Subramanian K. and Cook N.H., Sensing of Drill Wear and Prediction of Drill Life, J. of Engineering for Industry, Trans of ASME, Vol. 99, Series B, 1977, 295-301.

[25] Zaghbani I., Songmene V., A force-temperature model including a constitutive law for Dry High Speed Milling of aluminium alloys, Materials processing technology 209 (2009), 2532-2544.

[26] Zaghbani I., Songmene V., Kientzy G. and Lehuy H., Evaluation of Sustainability of Mould Steels Based on Machinability Data, International Journal Machining and Machinability , Vol. 7, Nos. 1/2, 2010, 58-81.

[27] Zitoune R., Krishnaraj V. and Collombet F., Study of drilling of composite material and aluminium stack, Composite Structures 92 (2010), 1246-1255.

Section 3

Heat Treatment and Welding

Chapter 10

Pure 7000 Alloys: Microstructure, Heat Treatments and Hot Working

P. Leo and E. Cerri

Additional information is available at the end of the chapter

http://dx.doi.org/10.5772/3354

1. Introduction

7000 alloys are used above all in automotive industry and architectural applications. These materials exhibit medium strength and ductility at room temperature and can be strengthened by aging treatment. Moreover they are characterized by low quench sensitivity, good corrosion resistance (due to the absence of Cu addition) and good extrudability (higher than 6061 alloy) [1-5].

Because of their commercial importance, much effort has been spent on investigation of the precipitation process in Al–Zn–Mg alloys [6-10]. The high strength exhibited in the hardened state is due to a fine distributions of precipitates, notably of the metastable η′ phase $MgZn_2$, produced by artificial aging from a supersaturated solid solution. The temperature of artificial aging influences the kinetics and the sequence of precipitation and if heterogeneous nucleation of the equilibrium phase appears, a less efficient hardening is obtained. In this study the response to artificial aging with and without previous solution treatment has been analyzed in the range of 130°C-210°C in order to evaluate which effect on hardening is due to the absence of supersaturation of vacancy rich cluster (VRC) and alloying elements coming from a solution heat treatment and rapid quenching.

There is strong academic and industrial interest in recrystallization driven by the need to understand and control this phenomenon in order to optimize properties through the careful control of thermomechanical processing schedules [11]. In this paper, the effects of different heat treatments and Zr content on rate of recrystallization induced by annealing heat treatment after RT deformation and on further deformation in terms of strain hardening rate (SHR= $d\sigma/d\varepsilon$), have been analyzed. Recrystallization due to hot deformation by torsion and tension test at 200°C-500°C and $10^{-5}s^{-1}$-$10^{-3}s^{-1}$ has been investigated too. During hot working the Al-Zn-Mg alloys exhibit lower flow stress and higher ductility than Al-Mg alloy (for example 5182 and 5083) [12,13]. Generally dynamic recovery (DRV) is the sole restoration mechanism in Al alloys [14-17] developing a subgrain structure inside elongated grain. As a consequence, the flow curves (stress vs strain; σ vs ε) exhibit SH to a steady state regime, although adiabatic heating may cause a peak and a gradual softening particularly at high strain rate ($\dot{\varepsilon}$) and low temperature (T). Ductility is usually high because DRV softened grains allow accommodation of differential grain boundary (GB) sliding, slowing crack formation [18-22]. Solutes, in the form of atmospheres hinder dislocation glide reducing DRV and ductility and raising the flow stress [21-23]. Moreover fine dispersoids pin dislocations and reduce DRV [24,25]. Precipitation hardening alloys may present varied behaviours as a result of changes in precipitate morphology. Growth of precipitates during hot working leads to good ductility and lower stress as shown for Al-Mg-Si [26,27], Al-Cu-Mg [28,29] and Al-Zn-Mg-Cu [30,31]. Solution treated alloy can exhibit high peak stress and dislocation density due to dynamic precipitation (DPN), followed by rapid softening as particles coalesce [29-33]. In this paper the microstructure of hot deformed Al-Zn-Mg samples (even modified with Zr) both by torsion and tension test have been analyzed by SEM and optical microscopy in order to justify the stress-strain curve shape.

Hot working of many engineering alloys is often accompanied by the formation of internal cavities [34-38]. The cavitation process depends strongly on alloy composition and microstructure as well as on the imposed processing condition [3 4 36 37 39]. Particularly large particles and inclusions, notably on GB, introduce new sources of fissure nucleation lowering ductility; solidification segregation and low melting constituents, especially if they spread along the GB, create severe problems [39]. Such cavitation may lead to premature failure (i.e. failure at strains lower than those expected based on material properties such as the strain rate sensitivity index and the strain hardening exponent) or result in a finished part with degraded mechanical properties. The cavitation process comprises three distinct stages, which in most cases occur simultaneously, i.e., (i) cavity nucleation, (ii) cavity growth, and (iii) cavity coalescence. Cavities, which usually nucleate preferentially at GB, triple points, or second-phase particles, grow by either plasticity- or diffusion-controlled mechanisms, or a combination of the two [35,37,39]. For a given material, the particular mechanism varies with the imposed deformation conditions.

2. Experimental procedures

The compositions of the alloys studied in this investigation are reported in Table1.

In order to distinguish easily the two materials with regard to Zr content they have been designated respectively as 7000 and 7000Zr. The materials were supplied in the form of DC cast billet of 20 cm in diameter and 40 cm in length. Cylindrical samples with gage length of 13 mm and 5mm diameter were cut parallel to the longitudinal axis of the billet for tensile and torsion tests. From the same billet, cube samples of 10 mm edges were cut for heat treatments. Artificial aging has been carried out at 130°C, 160°C, 190°C and 210°C up to 432h on the as-cast samples and 48h and solutionized ones (2h at 490°C). The effects of heat treatments were analyzed by hardness (HRF) and electrical conductivity curves.

	Zn	Mg	Fe	Si	Ti	Zr
7000	5.5	1.2	0.07	0.03	0.01	
7000Zr	5.6	1.2	0.07	0.03	0.01	0.16

Table 1. Composition of the alloys (wt%)

The microstructure of as-received alloys has been investigated by optical microscope (Nikon Epiphot 200) and Scanning Electron Microscope - Focused Ion Beam (SEM FIB) ZEISS 1540. The chemical composition of the matrix and particles was investigated by Energy-dispersive X-ray spectroscopy (EDS) analysis. For polarized light observation the samples were ground according to standard methods, electropolished (80ml perchloric acid, 120ml distilled water, 800ml ethanol, 20V) and anodized (Barker's reagent). The average grain size has been evaluated on a population of at least 200 grains by using the NIS software for imaging analysis.

RT tensile tests were performed on as-cast, solutionized (490°C-2h) and peak aged (490°C-2h + 160°C-24h) samples and SHR plotted versus (σ-σy). One half of each fractured sample coming from an RT tensile test was ground parallel to longitudinal axis up to the middle plane and annealed at 500°C for 3h. After 1,5 h of annealing, the sample was water quenched and the average grain size calculated in a fixed area close to the fracture by using the LUCIA G software. Then a second step of annealing at the same temperature and time (total 3h) was applied to each sample in order to follow the recrystallization behaviour.

Hot tensile tests have been performed on as-cast alloys in the range 250°C-400°C and $10^{-5}$to $10^{-3}s^{-1}$. The temperature was measured by two independent thermocouples placed close to the sample. The true stress-true strain curves were calculated from recorded load-displacement data according to the usual formula. Hot torsion tests have been performed in the range 250°C-500°C and $10^{-2}$to $5s^{-1}$. The torque and surface strain have been transformed into equivalent stress and strain by the traditional means. One half of each fractured sample coming from hot deformation tests were ground parallel to longitudinal axis up to the mid- plane in order to investigate on both recrystallization (RX) and cavitation phenomena by optical and SEM analysis. Particularly, for cavitation analysis, micrographs at 20× have been taken along the length of each metallographic section (up to 4mm by fracture mid line) and collected into a montage [Fig.15 a,b,c]. The area of cavities inside the area of metallographic section (4mm by fracture mid line) and the same area of metallographic section has been evaluated using NIS software. Cavity area fractions Cs (%) (area of cavities divided by the area of metallographic section) were determined. Assuming that cavities are randomly distributed inside the specimen, it has been shown that the area fraction is equal to volume fraction Cv (%) [30].

3. Results and discussion

The microstructures of as received alloys exhibit interdendritic segregation (Fig.1). In 7000 alloy (Fig 1a) the addition of small amount of Ti produces grain refining because the Al_3Ti particles act as nucleation sites and moreover lead to smaller precipitate free zones (PFZ) and finer grain boundary precipitation [43,44]. In 7000 Zr alloy (Fig 1b) the average grain size is higher compared to that of 7000 alloy (210±60 μm vs 145±40 μm). It is suggested that Zr reacts with Al_3Ti complex to make it a less potent nucleation site [44]. From SEM/ EDS analysis (Fig.2) hard insoluble brittle particles $FeAl_3/FeAl_6$ type have been detected along grain boundaries (Fig.2a) and $MgZn_2$ or $Mg_3Zn_3Al_3$ both along and grain boundaries and inside grains (Fig.2b).

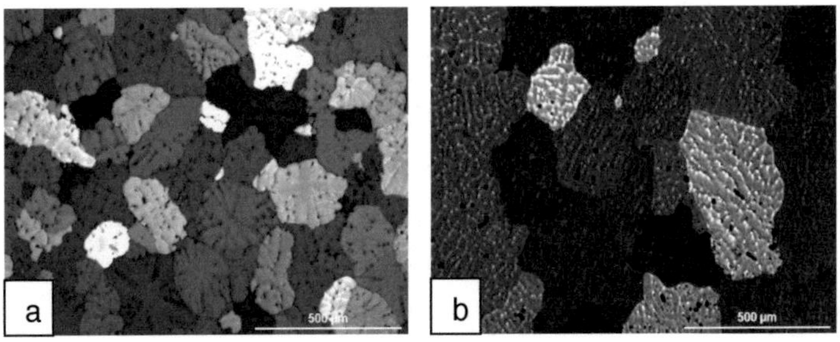

Figure 1. Optical micrographs of 7000 (a) and 7000Zr (b) alloys (5X) showing that the microstructure of both alloys is characterized by dendritic microsegregation. Different grain size is evident comparing (a) and (b)

The average values of hardness in the as-cast and solutionized state (490°C-2h) are slightly higher for 7000Zr alloy (Table 2) despite its grain size being higher. In contrast, the as-cast 7000Zr electrical conductivity is lower (22Ms/m vs 23,5Ms/m) for the higher amount of alloying. As shown in Fig.3 solution heat treatment (490°C-2h) reduce microsegregation and through dissolving hardening particles the hardness is reduced too (Table 2). EDS analysis did not find Al-Zn-Mg particles in solutionized alloys while some undissolved $FeAl_3/FeAl_6$ type particles were found.

http://dx.doi.org/10.5772/3354

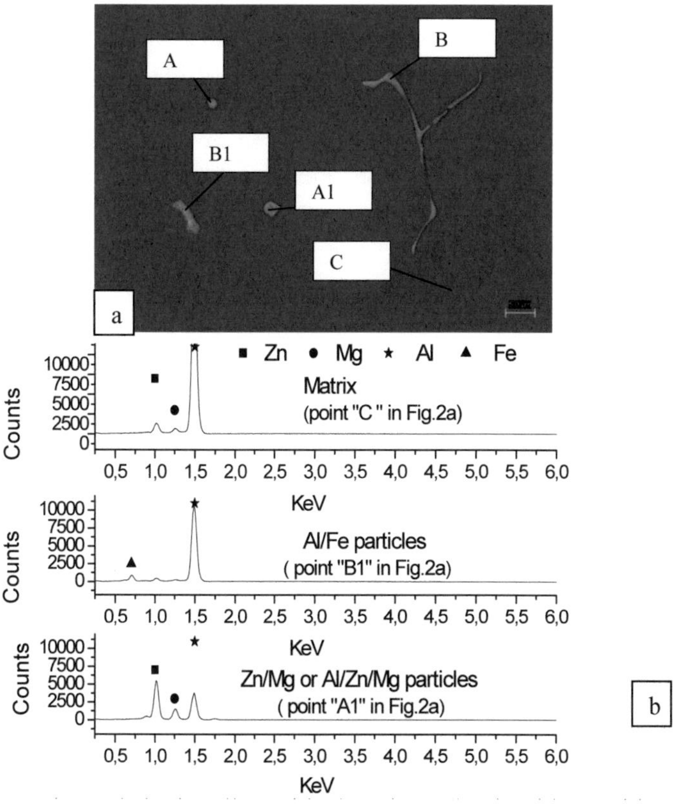

Figure 2. SEM micrograph showing Al/Fe particles (named B, B1) and particles containing Zn and Mg (named A,A1) (a); EDS spectrum of elements content (b) into the matrix (named C in Fig.2a), and B1 and A1 particles.

The hardness and electrical conductivity ageing curves performed on solution treated samples (T6 type) and on the as cast samples at 130-210°C are shown in Fig. 4 and Fig. 5 respectively. The aging treatments (Fig. 4) of solution treated (490°C-2h) samples lead to 15 and 13 point increments of hardness respectively for 7000 and 7000Zr at the two lowest temperatures of aging. At the higher temperatures, the precipitation kinetics are faster but the hardening is less efficient due to heterogeneous nucleations and overaging starts before the peak is reached. Moreover as the VRC are not retained at high temperature of aging, a lower density of hardening precipitates is expected. Comparing the behaviour of the two alloys, the response to T6 heat treatment is better at the higher temperature (190°C and 220°C) for the alloy containing Zr while it is similar for both alloys at the lower temperature of treatment. This behaviour could be due to Al_3Zr compounds that don't dissolve at high temperature of treatment and act as nucleation sites for hardening precipitate η' phase [45-49].The electrical conductivity increases with temperature of aging and time because of the draining of solute from the matrix as the precipitation process proceeds. The aging curves of the as cast samples (Fig. 5) do not show an increment of hardness with respect to the starting state. The absence of super saturation of VRC and alloying elements coming from a solution heat treatment and rapid quenching, substantially reduces the nucleation of hardening precipitates. Even for this heat treatment the response of 7000Zr alloy is better at the higher temperature (190°C and 210°C) compared with the behavior of the alloy without Zr while it is similar for both alloys at the lower temperature of treatment. The electrical conductivity increases with temperature of aging and time but it is always lower that in the case of T6 heat treatment because the lower supersaturation of the matrix.

	HRF As received	HRF Solutionized (490°C-2h)
7000	93	86
7000Zr	95	88

Table 2. HRF of as-cast and solutionized 7000 and 7000Zr alloy

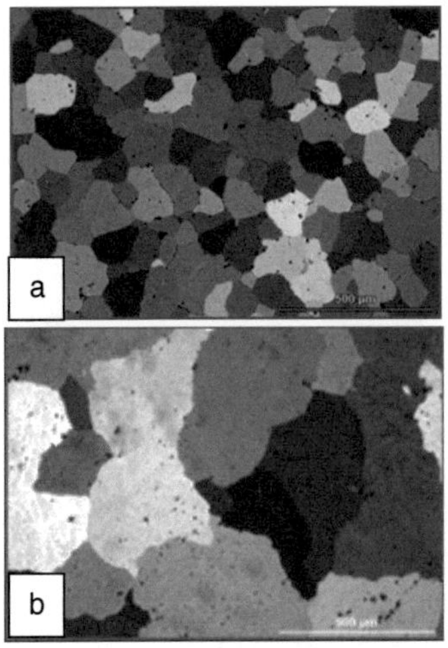

Figure 3. 7000 (a) and 7000Zr(b) alloys after solution heat treatment (490°C-2h). The dissolution of interdendritic segregation is evident compared to Fig.1

RT tensile tests on as-cast and solutionized samples indicate that solution heat treatment lead to low peak stress and high ductility (Fig.6) due to dissolution of both brittle phases and hardening particles. Moreover, the RT ductility is always higher for 7000Zr alloy. In term of SHR (Fig.7), it is always higher for the alloy in the as-cast state compared to solutionized because of large uncuttable particles cause Orowan hardening. Moreover 7000Zr alloy exhibits higher SHR probably due to the interaction of Al_3Zr particles with dislocations.

One half of each fractured samples was ground parallel to the longitudinal axis down to the mid plane and annealed at 500°C. After 1,5 h of annealing, the samples were water quenched and analyzed by optical microscopy for checking any recrystallization phenomena. Then, a second step of annealing of 1,5h was applied to each sample (total 3h). Fig. 8 an Fig.9 illustrate the anodized microstructure of respectively 7000 and 7000Zr specimen as tensile tested (first line) and 3 hours annealed (second line). The first column presents pictures from the as cast sample, the second from the solution treated. Annealing treatment lead to recrystallization rate that is faster on as- cast alloys compared to the solutionized. This result can be clarified by considering that dissolved atoms and fine precipitates formed in the matrix limit the movement of dislocation during annealing and delay the nucleation and growth of new grains. The as cast alloy exhibits both the highest strain hardening rate and low dissolved atoms; both these aspects lead to a shorter recrystallization time. However as shown in Fig.9 the recrystallization of 7000Zr alloy is incomplete and not homogeneous even after three hours of annealing because of additions of Zr.

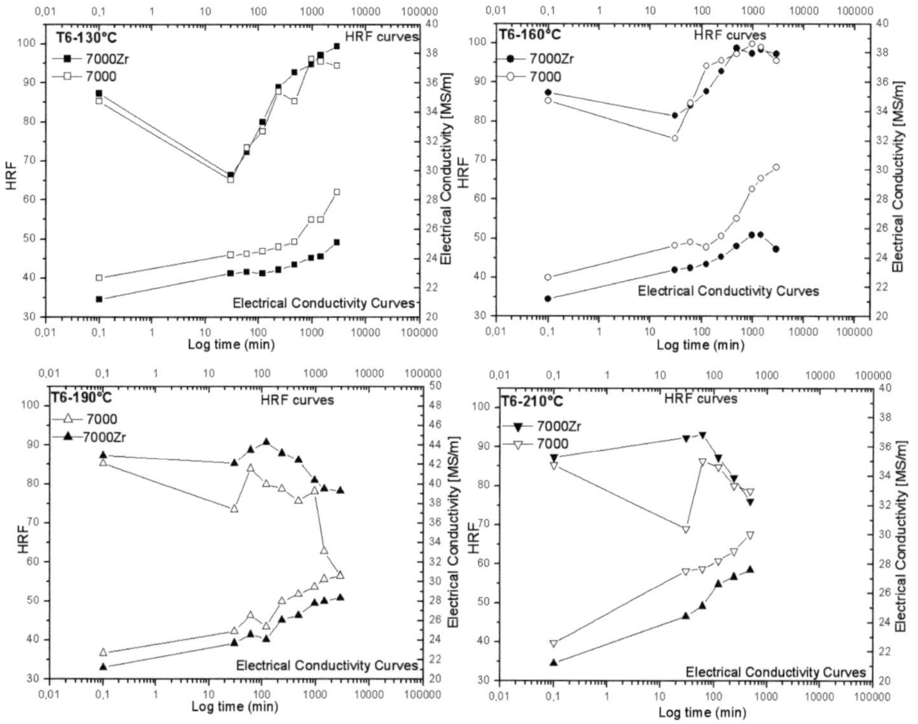

Figure 4. Hardness and electrical conductivity of 7000 and 7000Zr alloys during aging at 130°C, 160°C, 190°C and 210°C after solution treatment at 490°C-2h (initial value at 0,1min).

The peak stress σ_p decreases with increasing T at constant $\acute{\varepsilon}$; moreover, it decreases with decreasing $\acute{\varepsilon}$ at a fixed temperature [Fig. 10]. For each fixed temperature T, the ductility decreases as $\acute{\varepsilon}$ increases for 7000 alloy deformed by torsion and tension test while rises with $\acute{\varepsilon}$ for 7000 Zr tensile samples. This behavior being more evident with increasing temperature is common in creep [51,52]. In contrast for hot working as $\acute{\varepsilon}$ decreases, ductilities increases since the improved DRV mitigates stress concentration and nucleation of voids (usually at triple junctions). Moreover, at fixed $\dot{\varepsilon}$, as T increases, recovery is improved and therefore ductility increases. Ductility in torsion is always higher than in tension because the low normal to shear stress ratio enhances the role of DRV in inhibiting cracking [39]. The peak values are always higher in torsion because the higher $\dot{\varepsilon}$ involved. Constitutive analysis for torsion test gave a Q value of 161Kj/mol [1]. The value of Q close to that of pure Al is due to precipitated particles that are inefficient in interacting with dislocations as confirmed by the low declines in the flow curves and by the very low value of average n (1,5)[1]. In fact the microstructure of as-cast samples hot torsioned at temperature higher than 300°C exhibits significant precipitated particles. Their number decrease as T increases do to overaging or cooperative growth of particles and/or to their dissolution while at fixed T increases with strain rate due to strain hardening effect on enhancing precipitation kinetics (Fig. 11). Optical analysis of torsioned samples on longitudinal plane close to fracture surface after chemical etchant (Keller) shows that only the samples deformed at 500°C exhibit recrystallized grains from SRX (Fig. 12). At 400°C the microstructure is characterized mainly by subgrain as is evident from anodized longitudinal plane of Fig.13. Subgrain have been observed in grains at 300°C 0.1s^{-1} and 0.01 s^{-1} too. Even the samples deformed by tensile at 350°C-400°C show some SRX (Fig.14). Static recrystallization is much more evident in the alloy without Zr, close to the fracture surface and near grain boundaries and due to stress localization (Fig.14).

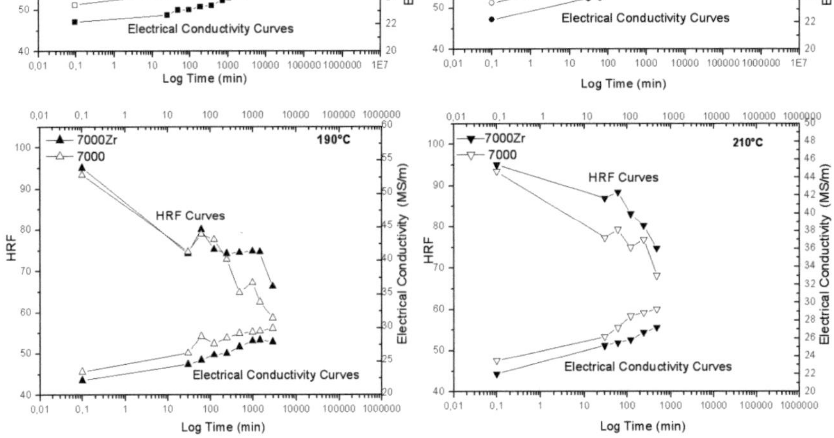

Figure 5. Hardness and electrical conductivity of as-cast 7000 and 7000Zr alloys during aging at 130°C, 160°C, 190°C and 210°C (initial value at 0,1min).

http://dx.doi.org/10.5772/3354

The microstructure of tensioned samples is characterized by cavitation (Fig.15) phenomena. Cavitation is much more evident in the alloy without Zr (Fig.15 b) and it is strongly reduced only if the alloy is solutionized (490°C-2h) before tensile test (Fig.15c). In fact this phenomena is mainly due to both segregation and stress concentration at particles (Fig.15d). Solution heat treatment reduce microsegregation (Fig.3) and dissolve soluble Mg-Zn particles, leading therefore to a reduction of the both hardness and cavitation. Some brittle particles (for example $FeAl_3$/ $FeAl_6$ in Fig.2) are not dissolved from solution heat treatment and continue to act as cavity nucleation point.

Cavitation (Cs%) Vs strain rate is shown in Fig.16 for as the cast 7000 sample deformed at 250°C and 400°C. Cavitation increases with T and, at fixed T it increases as strain rate decreases. The distribution of cavities along the longitudinal sample surfaces at 250°C and 400°C is shown in Fig. 17. At the lowest T, cavitation values decrease as distance from fracture increases while at 400°C the cavities are randomly distributed along the longitudinal area of the samples. So it can be assumed a more active role of grain boundaries sliding (GBS) on cavitation nucleation and growth at the highest T. Theoretical modelling on cavitation in general takes into account the three distinct stages of damage generation, i.e. nucleation, growth and coalescence [40], but, among these, the growth appears the critical phase [41].

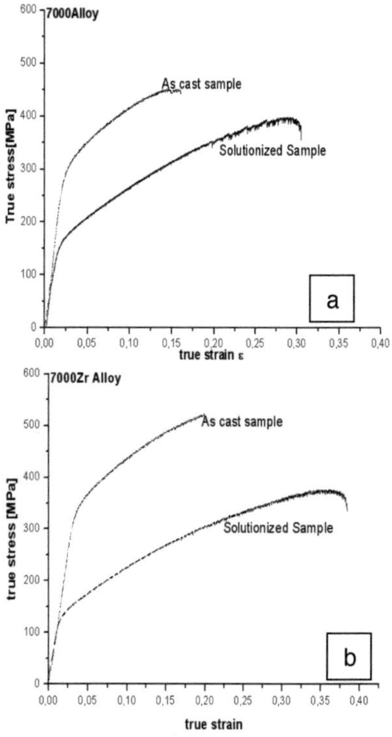

Figure 6. Room temperature tensile curves for 7000 (a) and 7000Zr samples (b) in the following conditions: as cast and solution treated 490°C-2h

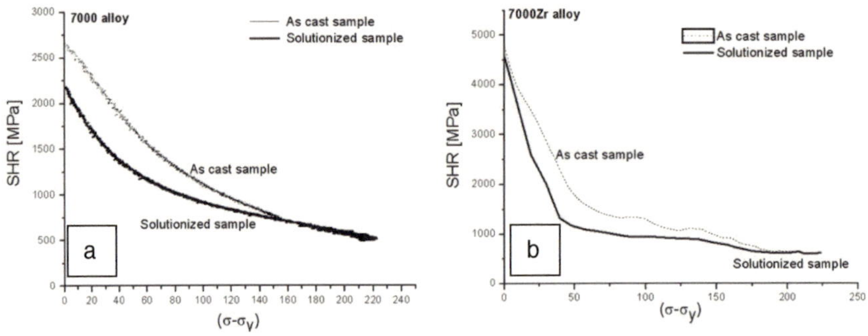

Figure 7. Strain hardening rate curves for the RT tensile curves of 7000 (a) and 7000Zr (b) alloy in the as-cast and solutionized state.

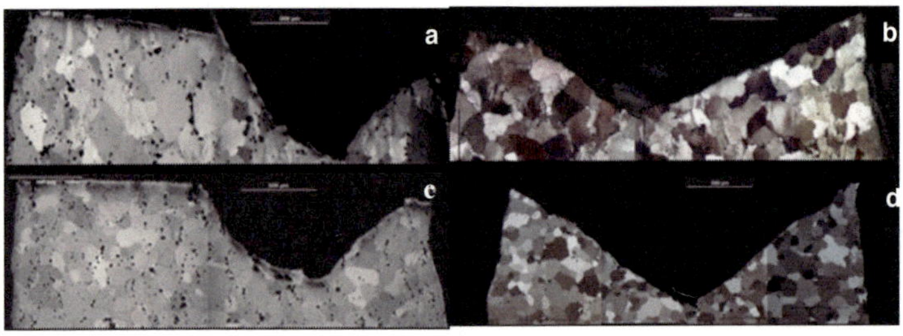

Figure 8. Optical micrographs of 7000 tensioned specimens before (a,b) and after 3h of annealing at 500°C (c,d) where (a,c) are as cast, (b,d) are solution treated

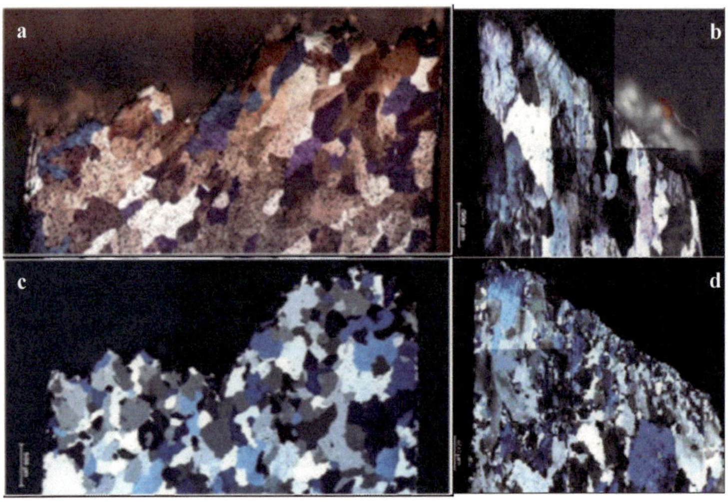

Figure 9. Optical micrographs of 7000Zr tensioned specimens before (a,b) and after 3h of annealing at 500°C (c,d) where (a,c) are as-cast, (b,d) are solution treated.

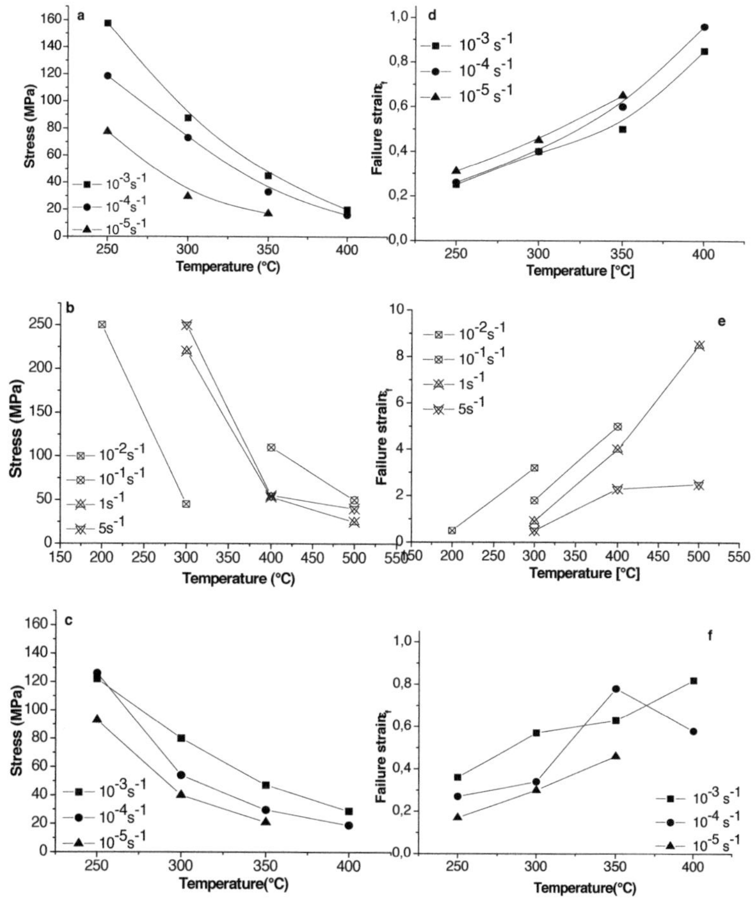

Figure 10. Peak stress variations (a,b,c) and failure stress (d,e,f) as a function of T for the as-cast 7000 alloy hot deformed by tension (a,d) and by torsion (b,e) and for 7000 Zr alloy hot deformed by tension (c,f).

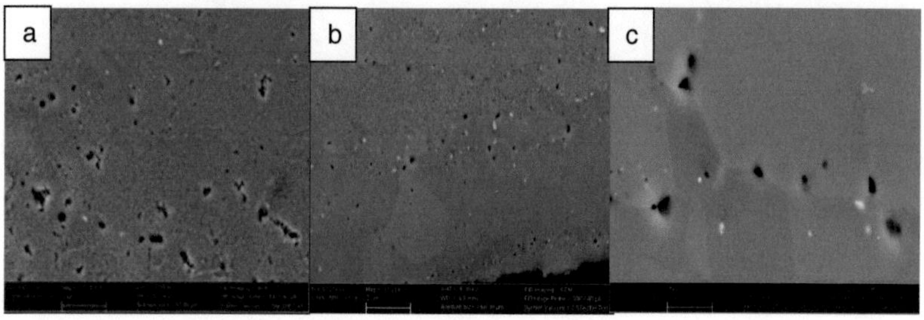

Figure 11. Precipitated particles in hot torsioned samples at 300°C-0,1s^{-1}(a), 300°C-0,01s^{-1} (b) and 400°C-0,1s^{-1}(c): at fixed T (a,b) the number of particles increases with strain rate due to enhanced precipitation on dislocations, while as T increases the size of particles increase due to overaging (a,c).

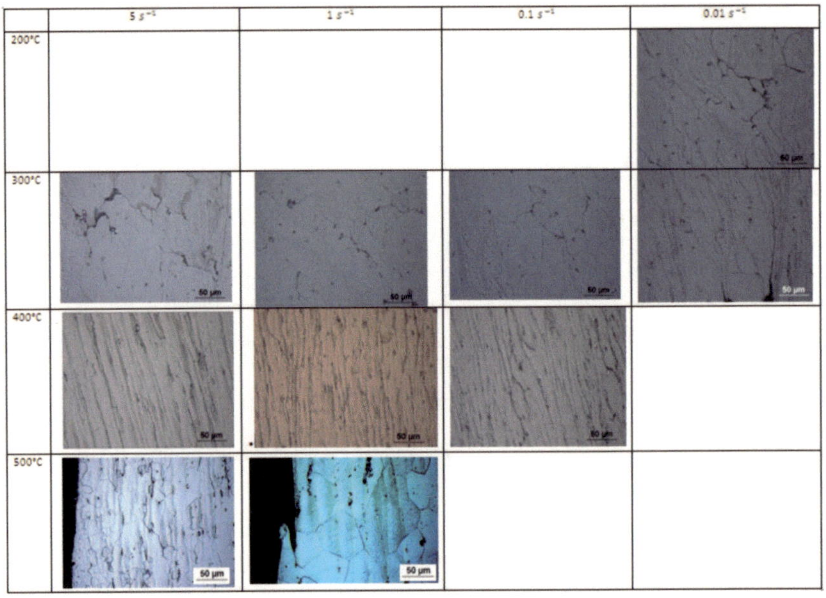

Figure 12. Optical micrographs (50X) of torsioned samples on longitudinal planes close to fracture surface after Keller etchant showing that only the samples deformed at the highest T exhibit SRX.

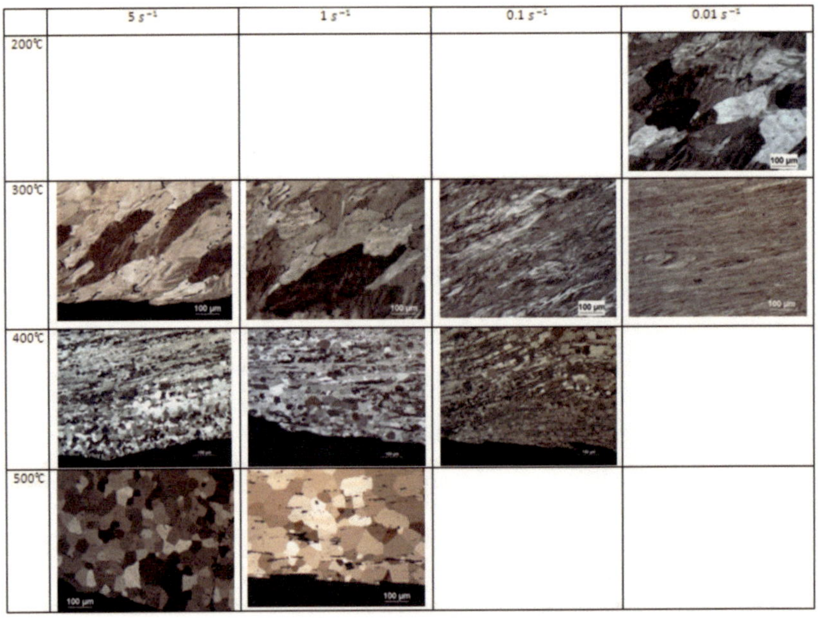

Figure 13. Optical micrographs (20X) of torsioned samples on longitudinal planes close to fracture surface after anodizing showing that the microstructure at 400°C and is characterized mainly by subgrains. Subgrains have been observed in some grains at 300°C and 0.1 s^{-1}-0.01s^{-1}

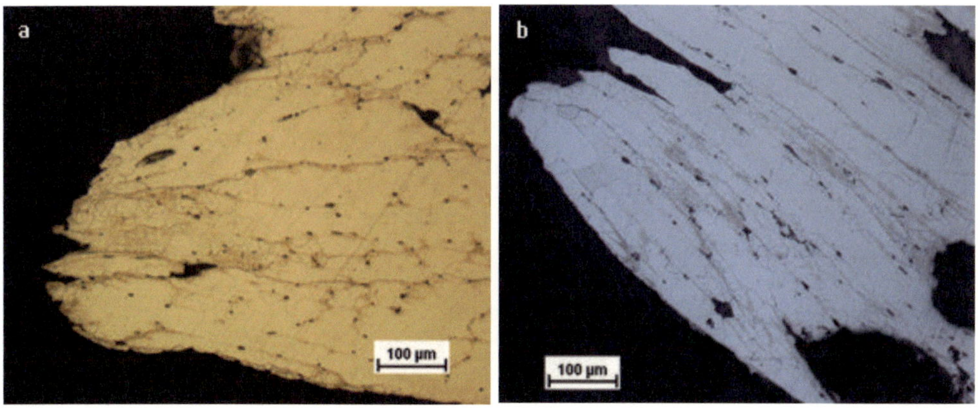

Figure 14. SRX close to fracture surface of hot deformed tensile test 7000 alloy at 400°C-$10^{-4}s^{-1}$ with Zr (a) and without (b)

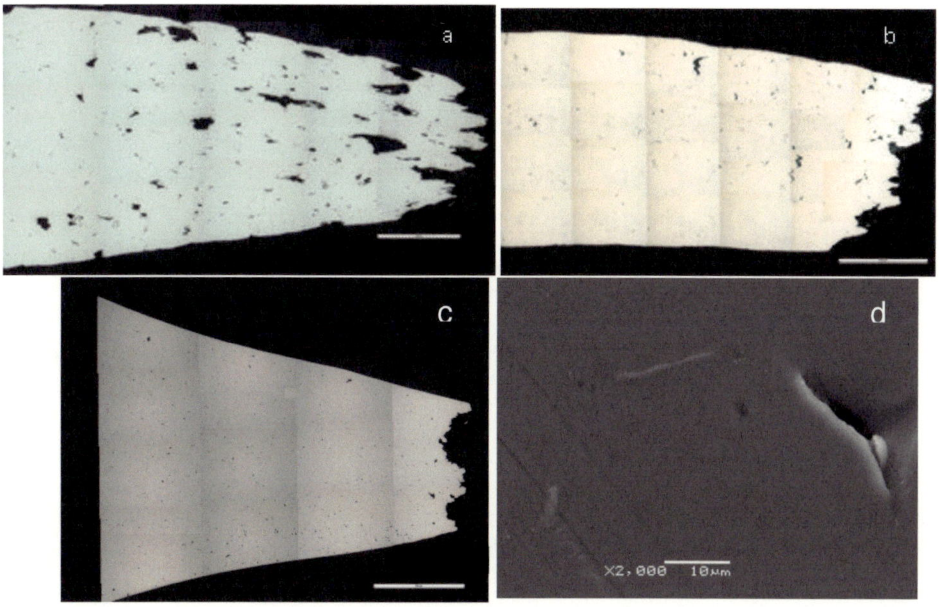

Figure 15. Optical micrographs of cavitation phenomena in as-cast 7000 (a), 7000Zr (b), solutionized 7000 alloys (c) deformed by tensile test at 350°C 10^{-4} s^{-1}. SEM micrographs of 7000 as-cast alloy (d) showing that cavities originate at brittle particles.

When cavity growth is controlled by plastic deformation, the simplest model for cavity growth assumes the following form [38]:

$$C_v = C_0 \exp(\eta \varepsilon) \tag{1}$$

where C_0 is the initial volume fraction of cavities and ε is the fracture strain. Following Lee and Huang [40], who based their analysis on the Stowell et al. relationship [41], one can express the cavity growth exponent as follows [42]:

$$\eta = \frac{3}{2}\left(\frac{m+1}{m}\right)\sinh\left[\frac{2(2-m)}{3(2+m)}\right] \tag{2}$$

Where m is the strain rate sensitivity of flow stress.

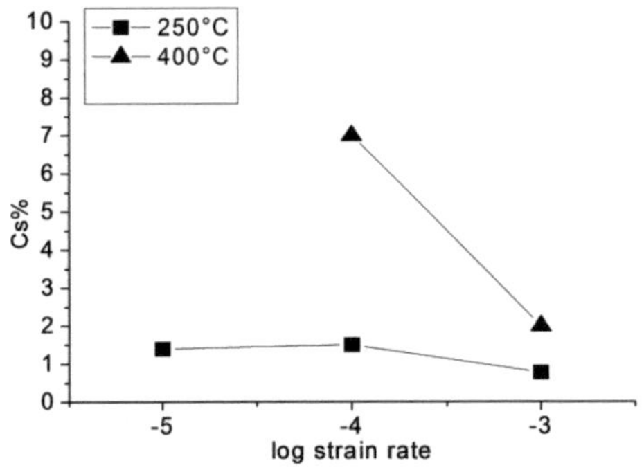

Figure 16. Cavitation (Cs%) versus log strain rate showing that the area of fissures increases with T and, at the highest T, with decreasing strain rate

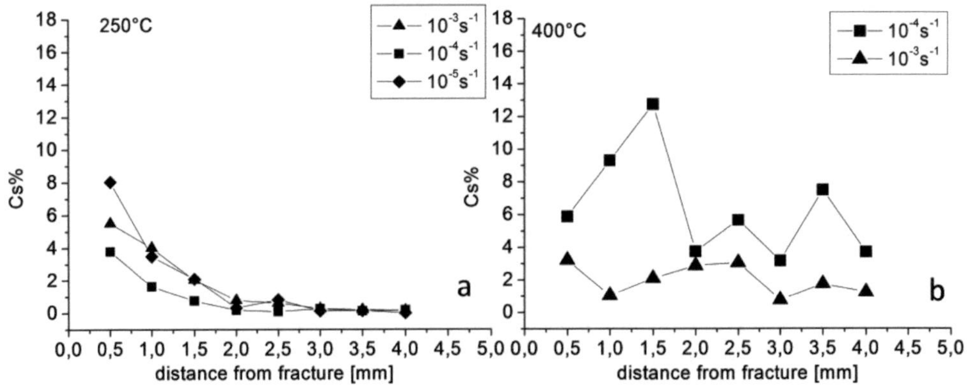

Figure 17. Cavitation (Cs%) versus distance from fracture at 250°C (a) and 400°C (b) showing that at the lowest T the area of fissures decrease as distance from fracture increases while at 400°C cavities are randomly distributed along the longitudinal area of the samples.

When the fracture strain is substituted into Eqn.1, an obvious approximation, since the fracture strain is affected by necking, the plot presented (with C_0 arbitrarily assumed to be

0.03%) in Fig. 18 are obtained. Although this calculation is based on very rough assumptions, the model of cavity growth controlled by plastic strain describes very well the observed cavitation trend. It can be thus reasonably concluded that, in the investigated range of experimental conditions, the cavity growth is mainly driven by plastic straining. GBS substantially contributes to enhance the fraction of cavities at the highest T and justifies the observed difference between the calculated value and experimental one (Fig.18).

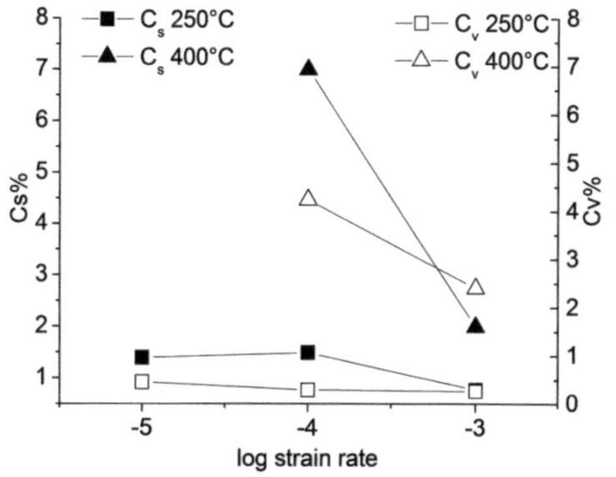

Figure 18. Comparison of measured Cs% and calculated Cv% versus log strain rate at 250°C and 400°C

4. Conclusions

The main conclusions are summarized in the following :

- In the as received state the grain size of 7000 Zr alloy is larger than that of 7000 alloy due to reaction between Zr and Ti that reduce the nucleation power of both elements and the microstructure of both alloys is characterized by dendritic microsegregation. The solution heat treatment at 490°C-2h leads to a strong reduction of segregation and to a complete dissolution of hardening Al-Zn-Mg particles as shown by EDS analysis. As consequence the hardness decreases. For the larger grain size and higher amount of alloying the electrical conductivity of 7000Zr is always lower than that of 7000 alloy.
- The response to heat treatment for both the as cast and solutionized samples is better at the higher temperature (190°C and 210°C) for the alloy containing Zr, while it is similar for both alloys at the lower temperature of treatment. This behaviour could be due to Al_3Zr compounds that don't solutionize at high temperature of treatment and moreover, harden the alloy and can act as nucleation sites for hardening precipitate η′ phase. Aging treatment of the as-cast alloys for the range of imposed times is ineffective in terms of increasing hardness.

- Concerning RT tensile tests, the SHR is higher for the as cast sample comparing to the solutionized. As consequence the recrystallization rate of as cast sample is faster too. Due to Zr effect it is incomplete and not homogeneous in the Zr modified alloy.
- During hot tensile test the as cast alloys exhibit high flow stress at low temperature due to reduced DRV. At higher T, both improved DRV and overaging of particles lead to reduced peak stress. The phenomena is much more evident during torsion because of T involved. For each fixed temperature T, the ductility decreases as $\dot{\varepsilon}$ increases for 7000 alloy deformed by torsion and tension while rises with $\dot{\varepsilon}$ for 7000 Zr tensile samples.
- The microstructure of both torsion and tensile samples hot deformed at the highest temperatures exhibit some SRX, more evident in the alloy without Zr.
- Cavitation in 7000 hot tensioned samples increases with T and with decreasing strain rate. At the lowest T, 7000 as cast alloy cavitation decreases as distance from fracture surface increases while at the highest T cavities are randomly distributed along longitudinal surface of samples, suggesting a more active role of GBS on cavitation nucleation and growth. Theoretical calculation have shown that cavity growth is mainly driven by plastic straining but GBS substantially contributes to enhance the fraction of cavities and justify the observed difference between the calculated and experimental cavitation values at the highest T and lowest strain rate. Cavitation is reduced if the alloy is solutionized before deformation.

Author details

P. Leo* and E. Cerri

Università del Salento, via per Arnesano, Lecce, Italy

References

[1] H.J. McQueen, P. Leo and E. Cerri; Aluminum Alloys: Fabrication, Characterization and Applications, : Weimin Yin, Subodh Das, Zhengdong Long Eds: TMS USA 2009 pp.37-44

[2] J.E. Hatch,ed. Aluminum, Properties, Physical Metallurgy, ASM, Metals Park, Oh, 1984

[3] H.J.McQueen, O.C. Celliers, Canadian Metal. Quart. 36 (1997), 73-86

[4] K. Laue, H.Stenger, Extrusion: Process Machinery, Tooling, American Society for Metals, Metals Park, OH (1981), 1-62; 124-152

[5] Y.Baba, H.Yoshido, Proc. 2nd Intnl.Al. Exrusion Tech. Sem., Aluminium Association, Washington, (1997) Vol. 1, 301-306

[6] T. Engdahl, V. Hansen, P.J. Warren, K. Stiller , Mater. Sci. Eng. A 327, (2002), 59

[7] K. Stiller, P.J.Warren, V. Hansen, J. Angenete, J. Gjonnes, Mater. Sci. Eng. A270, (1999),55

[8] S.K. Maloney, K. Hono, I.J. Polmear, S.P. Ringer, Scripta Mater. 41, 10, (1999), 1031

[9] H.Z.Li, V. Hansen, J. Jonnes, L. R. Wallenberg, Acta Mater. 47, 9. (1999), 2651

[10] J.C. Werwnskiold, A. Deschemps, Y. Bréchet, Mater.Sci. Eng. A293, (2000), 267

[11] M.J. Jones , F.J. Humphreys, Acta Mater. 51 (2003) 2149–2159

[12] H.J. McQueen, N.Owen, Materials in Automotive Industry E.Essadigi et al., Eds., Mat.Soc.Cim, Montreal (2001), pp.189-205

[13] W. Blum, H.J. McQueen, Aluminum Alloys, Physical Mechanical Properties, ICAA5 Mater. Sci. Forum 217-222,(1996), pp. 31-42

[14] H.J. McQueen, W. Blum. Aluminium Alloys, Phisical and Mechanical Properties ICAA6,T.Sato, ed Japan Inst. Metals (1998), pp. 99-112

[15] R.D.Doherty et al. Mat. Sci. Eng. A 238, 2 (1997) pp.219-274

[16] M. El Mehtedi, S. Spigarelli, E Evangelista, La metallurgia italiana, (September 2007) pp. 21-28

[17] H.J. McQueen, in Advanced Materials for 21st Century, J.R. Weertman Symp, Y.P. Chung Ed, TMS-AIME, Warrendale, PA, (1999), pp. 159-168

[18] S. Fulop , K.C. Cadien, M.J. Luton, H.J. McQueen, J. Test Eval. 5 (1977) pp.419-426

[19] N.D. Ryan, H.J. McQueen, J. Mech. Working Tech. 12 (1986) pp. 279-296

[20] C.M. Sellars, W.J. McG. Tegart, Int. Metall. Rev. 17 (1972) pp. 1-24

[21] H.J. McQueen in Hot deformation of Aluminum alloys, T.G. Langdon, H.D.Merchant, J.G.Morris and M.A.Zaidi Eds, Warrendale, PA, 1990, pp. 31-54.

[22] H.J. McQueen and J.J. Jonas in C.Q. Chen et al (eds.), Aluminum alloys 90 (Icaa2) Beijing, 1990, pp.727-742

[23] H.J. McQueen and K.Conrod in Microstructural control in Aluminum alloys, E.H. Chia and H.J. McQueen (Eds), TMS-AIME, Warrendale, PA, 1986, pp.197-219

[24] H.J. McQueen, E. Evangelista, J. Bowles and G. Crawford, Met Sci. 18, (1984), p. 395-402

[25] E. Evangelista, H.J. McQueen and E. Bonetti in Deformation of Multi-phase and particles containing Materials, J.B. Bildensoren et al (eds), Riso National Laboratory, Roskilde, 1983, pp.243-250

[26] H.J. McQueen in Hot deformation of Aluminum alloys, T.G. Langdon, H.D.Merchant, J.G.Morris and M.A.Zaidi (eds), Warrendale, PA, 1990, pp. 105-120

[27] E.Evangelista, A.Forcellese, F. Gabrielli and P.Mengucci in Hot deformation of Aluminum alloys, T.G. Langdon, H.D.Merchant, J.G.Morris and M.A.Zaidi (eds),Warrendale, PA, 1990, pp. 121-139

[28] B. Verlinden, P.Wouters, H.J. McQueen, E. Aernoudt, L.Delaey and S. Cauwemberg, Mat.Sci.Eng.A, 123 (1990) pp.229-237

[29] P.Wouters, B. Verlinden, H.J. McQueen and E. Aernoudt, Mat.Sci.Eng.A, 123 (1990) pp.239-245

[30] E. Evangelista, E. Di Russo, H.J. McQueen and P. Mengucci in Homogenization and Annealing of Al and Cu Alloys, H.D. Merchant, J. Crane and E.H. Chia (eds.), TMS-AIME Warrendale, PA, 1988, p. 209

[31] H.J. McQueen, E. Evangelista, A.Forcellese, I.C. Smith and E. Di Russo, in Modelling the deformation of Crystalline Solids, T.C. Lowe, A.D. Rollet, P.S. Follanbee and G.S. Daehn (eds.), TMS-AIME, Warrendale, PA, 1991, p.281-292

[32] A. Espedal, H. Gjestland, N. Ryum and H.J. McQueen, Scand J. Met. 18 ,1989, pp. 131-136

[33] E. Evangelista, H.J. McQueen and E. Cerri, , Modelling of plastic deformation and its Engineering Applications, S.I. Andersen et al. eds, Riso National Laboratory, Roskilde, DK, 1992, pp. 265-270

[34] P.A.Friedman and W.B.Copple, Proceeding of TMS,2003,Hot deformation of aluminum alloy III 211-219

[35] D. Nicolau, S.L. Semiantin, Acta Mat. 51, (2003) 613-623

[36] E.M. Taleff, P.J. Nevland and P. Krajewski, Metall.Mater.Trans. 32A (2001) 1119-1130.

[37] M.F.Ashby, C. Gandhi and D.M.R. Taplin, Acta Metall. 27 (1979) 699-729.

[38] M.J. Stowell, Cavitation in Superplasticity, in "Superplastic forming of structural alloys", N.E. Paton and C.H. Hamilton Eds., TMS-AIME, Warrendale, PA, 1982, 321-326

[39] H.J.McQueen, Materials Science Forum.Vols. 604-605 (2009)

[40] X.-G. Jiang, J.C. Erathman and F.A.Mohamed, J.Mater.Sci. 29 (1994) 5499-5514.

[41] C.J.Lee and J.C.Huang, Acta Mater. 52 (2004) 3111-3122.

[42] M.J.Stowell, D.W.Livesey and N.Ridley, Acta Metall. 32 (1984) 35-42.

[43] J. Polmear Light alloys –Metallurgy of light metals, Butterworth Heinemann 1995, 115-118

[44] Bjorn Ronning, Thesis " Constitutive relationships for AlZnMg, AlZnMgCr and AlZnMgZr alloys" October 1998, University of Trondeim, Norway

http://dx.doi.org/10.5772/3354

[45] P. Leo, E. Cerri, H.J. McQueen and S. Chiozzi, Mater. Sci. Forum 604-605; 2009, pp. 67-76

[46] L.K. Berg, G. Gjonnes,V. Hansen et al. Acta Mater, 49, (2001),3493

[47] H. Loffer, I. Kovacs, J. Lendvai, J. Mater. Sci. 18, (1983), 2215

[48] Z. Katz, N.Ryum, Scripta Met. 15, (1981), 265

[49] J.D.Verhoeven, "Fundamental of Physical Metallurgy" J. Wiley e Sons, NY (1975), Chapter 6,

[50] P.Leo, E. Cerri, H.McQueen and A.Taurino, ICA11, in 'Aluminium Alloys – Their Physical and Mechanical Properties' J. Hirsch, B. Skrotzki and G. Gottstein eds,Wiley-VCH (DE) 2008, pp. 1868-1874.

[51] M.E. Kassner, M.T. Perez-Prado, Fondamental of creep in metals and alloys, Elsivier 2004, 1-15

[52] R.W. Evans, B. Wilshire, Introduction to creep, Oakdale Printing Company Ltd, 1993, 16-22

Section 4

Durability, Degradation and Recycling of Aluminium Alloys

Chapter 11

Mechanical and Metalurgical Properties of Friction Welded Aluminium Joints

Mumin Sahin and Cenk Misirli

Additional information is available at the end of the chapter

http://dx.doi.org/10.5772/51130

1. Introduction

Aluminium alloys are alloys in which aluminium (Al) is the predominant metal. The typical alloying elements are copper, magnesium, manganese, silicon and zinc. There are two principal classifications, namely casting alloys and wrought alloys, both of which are further subdivided into the categories heat-treatable and non-heat-treatable. About 85% of aluminium is used for wrought products, for example rolled plate, foils and extrusions. Cast aluminium alloys yield cost effective products due to the low melting point, although they generally have lower tensile strengths than wrought alloys. The most important cast aluminium alloy system is Al-Si, where the high levels of silicon (4.0% to 13%) contribute to give good casting characteristics. Aluminium alloys are widely used in engineering structures and components where light weight or corrosion resistance is required [1].

Light non-ferrous metals such as aluminium and magnesium alloys have drawn attention with regard to application due to their energy-saving character. Above all, aluminium alloys are used more due to their superior workability and less cost. However, they are not entirely replaced by stainless steel, stainless steel having superior strength and weldability in certain structures. Therefore, it is necessary to join stainless steel and aluminium materials. Then, copper - aluminium joints are inevitable for certain applications due to unique performances such as higher electric conductivity, heat conductivity, corrosion resistance and mechanical properties. Aluminium and copper are replacing steels in electricity supply systems due to higher electric conductivity.

Friction welding is used extensively in various industries. Heat in friction welding is generated by conversion of mechanical energy into thermal energy at the interface of work pieces during rotation under pressure. Various ferrous and non-ferrous alloys having circular or non-circular cross sections and that have different thermal and mechanical properties can easily be joined by friction welding method. Friction welding is classified as a solid-state welding process where metallic bonding is produced at temperatures lower than the melting point of the base metals. Friction time, friction pressure, forging time, forging pressure and rotation speed are the most important parameters in the friction welding method [2].

In practice, friction welding is classified in two ways; continuous drive friction welding and inertia friction welding [3, 4]. In the continuous drive friction method (Figure 1), one of the components is held stationary while the other is rotated at a constant speed (s). The two components are brought together under axial pressure (P_f) for a certain friction time (t_f). Then, the clutch is separated from the drive, and the rotary component is brought to stop within the braking time while the axial pressure on the stationary part is increased to a higher upset pressure (P_u) for a predetermined upset time (t_u). Parameters of the method are shown in Figure 2.

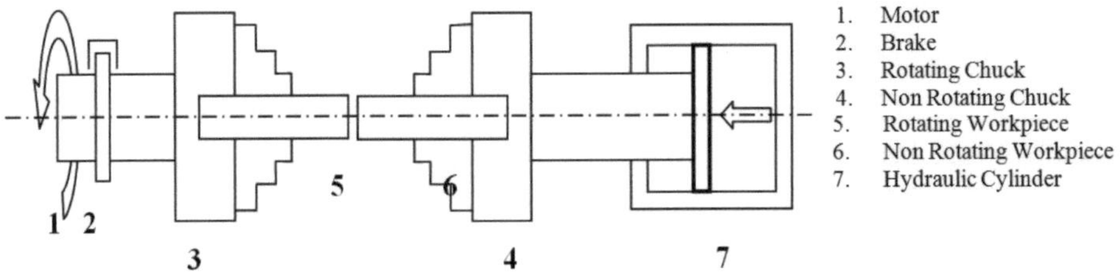

Figure 1. Layout of Continuous Drive Friction Welding

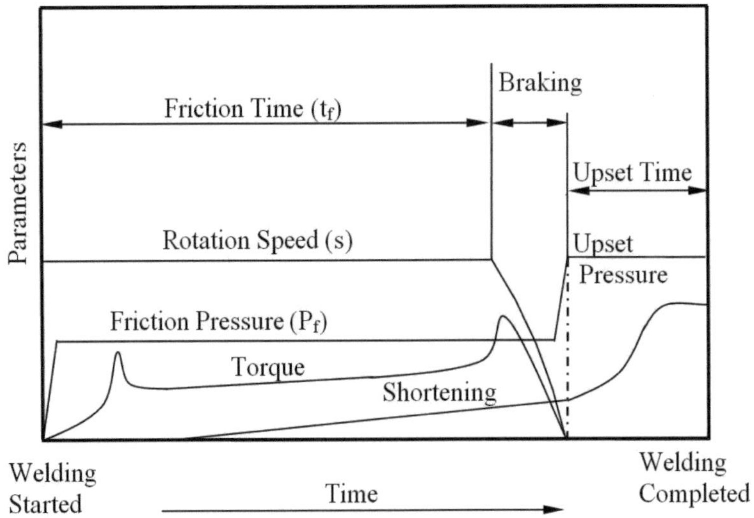

Figure 2. Parameters for Continuous Drive Friction Welding

http://dx.doi.org/10.5772/51130

In the inertia welding method, the second component is held stationary for welding, while one of the components is clamped in a spindle chuck, usually with attached fly wheels. The fly wheel and chuck assembly is rotated at a certain speed (s) to store a predetermined amount of energy. Then, the drive to the flywheel is declutched, and the two components are brought together under axial pressure (P_f) for welding. Friction between the parts decelerates the flywheel converting stored energy to frictional heat.

Vill, Kinley and Fomichev [2-4] studied the friction welding set-up and the strength of the joints. Murti et al. [5] directed a study about parameter optimisation in friction welding of dissimilar materials. Yılbas et al. [6] investigated the mechanical and metallurgical properties of friction welded steel-aluminium and aluminium-copper bars. Yılbas et al. [7] investigated the properties of friction-welded aluminium bars. Rhodes et al. [8] examined microstructure of 7075 aluminium using friction stir welding. Fukumoto et al. [9, 10] investigated amorphization process between aluminium alloy and stainless-steel by friction welding.

Then, Sahin and Akata [11] studied joining of plastically deformed steel (carburising steel) with friction welding. Sahin and Akata [12] carried out an experimental study on joining medium-carbon steel and austenitic-stainless steel with friction welding. Sahin [13, 14] studied joining austenitic-stainless steel with friction welding. Rhodes et al. [15] examined microstructure of 7075 aluminium using friction stir welding. Ouyang et al. [16] investigated microstructural evolution in friction stir welding of 6061 aluminium alloy (T6-temper condition) and copper. Maalekian M [17] performed a study on Friction Welding of dissimilar materials.

Surface cleanliness in terms of contaminants, especially grease, reduces the quality of joints. Furthermore, the cleanliness of the parts must be considered as important. Therefore, the ends of the parts were cleaned with acetone prior to the welding process to minimize the effect of organic contamination in the welding zone. However, the aim of this study is to investigate experimentally the microstructural and mechanical properties of friction welded aluminium-steel and aluminium-copper joints.

2. The experimental procedure

2.1. Material

In the experiments, AISI 304 austenitic-stainless steel and aluminium materials were used. The chemical composition and tensile strength of austenitic stainless steel is given in Tables 1. Chemical composition obtained by chemical analysis and tensile strength of aluminium and copper are given in Tables 2 and 3, respectively.

Material	% C	% P	% S	% Mn	% Si	% Cr	% Ni	Tensile Strength (MPa)
AISI 304 (X5CrNi1810)	0,07	0,045	0,030	2,0	1,0	17 - 19	8,5 – 10,5	825

Table 1. Chemical composition and tensile strength of austenitic-stainless steel [18].

Aluminium	%Sn	%Pb	%Zn	%Mn	%Fe	%Ni	%Si	%Mg	%Sb	%Cr	%Ti	%Cu	%Al	Tensile Strength (MPa)
	0,00500	0,03360	1,14000	0,11800	0,57400	0,01220	0,55400	0,17100	0,00300	0,02420	0,01340	0,59300	96,76000	200

Table 2. Chemical Compositions of Aluminium Used in the Experiments.

Copper	%Sn	%Pb	%Zn	%P	%Mn	%Fe	%Ni	%Si	%Mg	%Al	%Bi	%S	%Sb	%Cu	Tensile Strength (MPa)
	0,00222	<0,00200	<0,00100	0,00137	<0,00050	0,0381	<0,00100	0,00745	0,00376	0,00500	<0,00050	0,00251	<0,00200	99,93	300

Table 3. Chemical Compositions of Copper Used in the Experiments.

2.2. Geometry of Parts

Specimens were machined from materials according to geometry (Figure 3).

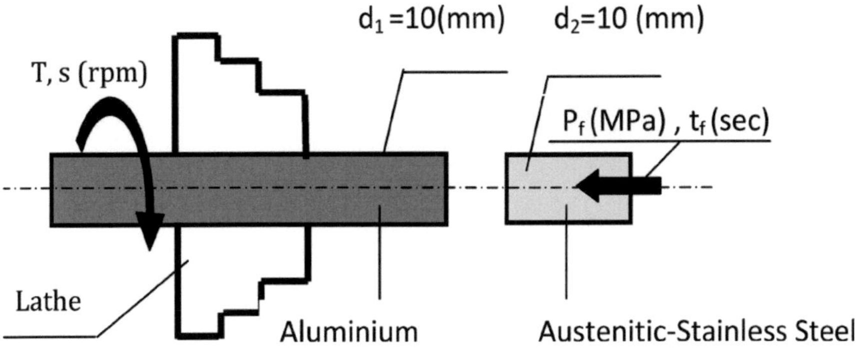

Figure 3. Equal Section Parts used in the experiments.

2.3. Experimental Set-up

An experimental set-up was designed and constructed as a continuous drive type. A solenoid valve and electrical control circuit was designed and constructed to control friction time and pressure in the set-up, thus allowing process control. The friction welding set-up is shown in Figure 4.

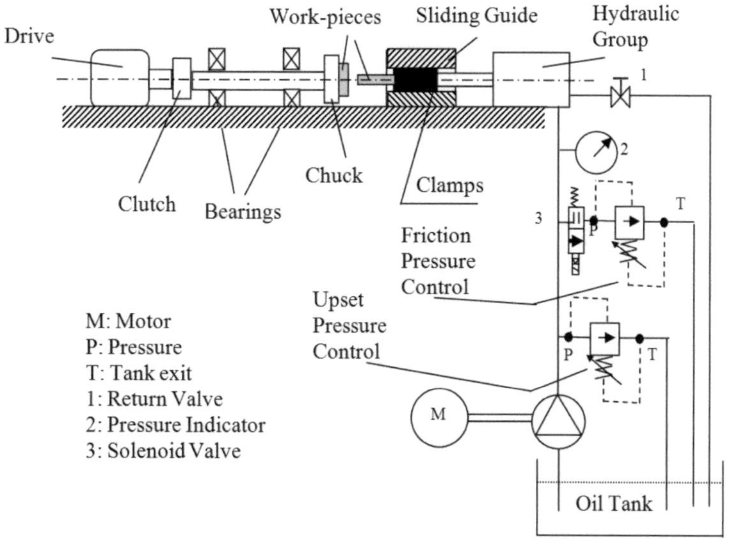

Figure 4. Continuous drive friction welding set-up.

The set-up was designed and constructed according to the principals of continuous drive welding machines. A drive motor with 4 kW power and 1410 rpm was selected as adequate for the torque capacity in friction welding of steel bars within 10 mm diameter taking into account the friction and the upset pressures. Friction and upset pressures can be seen on number2 pressure indicator, and the stages of the welding sequences are controlled by the number3 solenoid valve driven by an external timer.

Friction time, friction pressure and upset pressure have a direct effect on the tensile strength of joints. Therefore, linear statistical analysis was used in order to discover the effect of factors that have a significant role on the experimental results of previous studies [5, 6, 16].

3. Friction welded stainless steel and aluminium materials

Parameters having the least error by using the method of least squares were taken as the optimum welding parameters. Optimum parameters found in a previous different study [19] were used in the experiments (friction time= 4 sec., friction pressure= 30 MPa., upset time = 12 sec. and upset pressure = 60 MPa).

Subsequently, tensile tests, micro-hardness tests and metallurgical examinations were applied to the welded specimens.

3.1. Tensile Tests

Optimum parameters were found using statistical analysis for the welded parts. Later, many parts were machined and welded using the optimum parameters, and then these specimens were further tested. Effects of friction time and friction pressure on the strength of joints were examined in welding of equal diameter parts. Upset time was kept constant. The strength of joints was determined by tensile tests, and the results were compared with those of fully machined specimens. Tensile strength of the joints was estimated dividing the ultimate load by area of 10 mm diameter specimen. The relation obtained between tensile strength versus friction time and friction pressure is shown graphically in Figures 5 and 6.

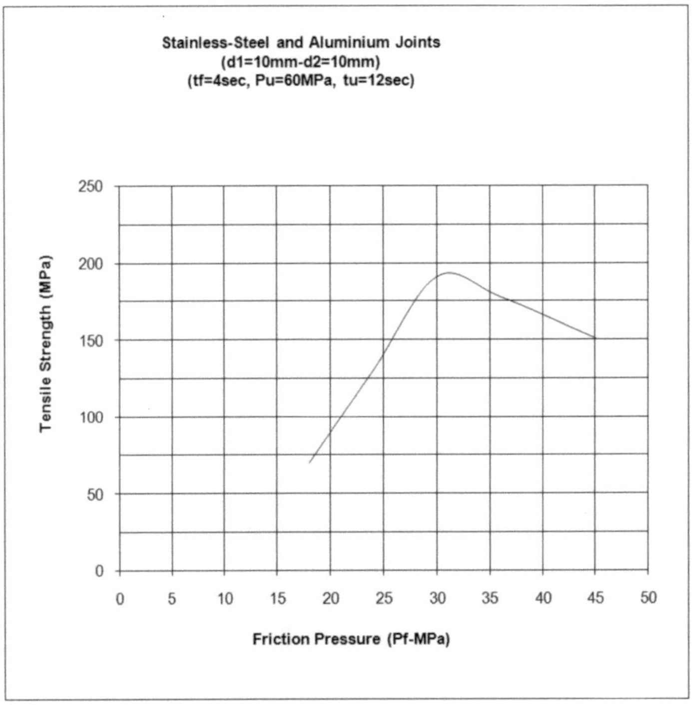

Figure 5. Relation between Tensile Strength versus Friction Pressure.

As friction time and friction pressure for the joints are increased, tensile strength of the joints increases (Figures 5 and 6). But, strength of the joints passes through a maximum, then, when friction time and friction pressure for the joints are increased, tensile strength of the joints decreases (Figures 5 and 6). Maximum strength obtained in the joints has about 94% that of aluminium parts having the weakest strength. Thus, it is shown that friction time and friction pressure have a direct effect on joint strength.

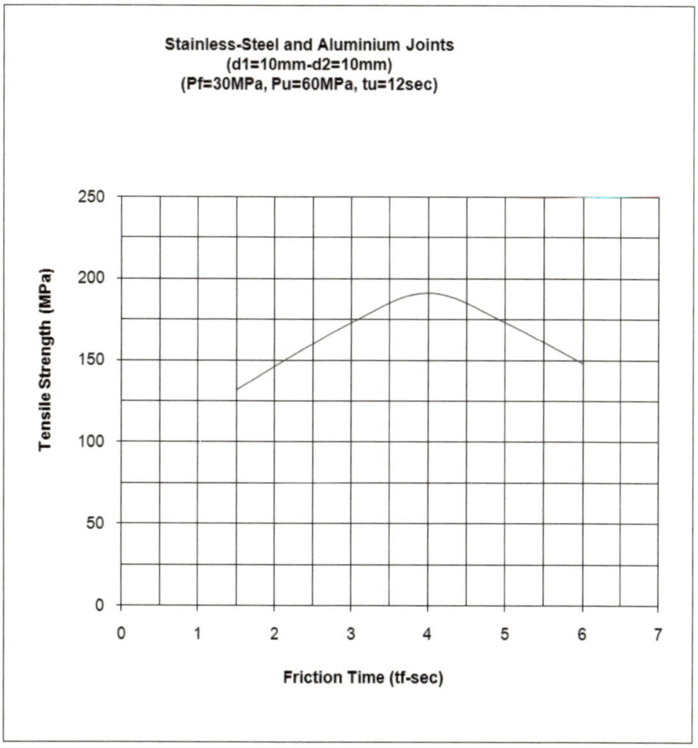

Figure 6. Relation between Tensile Strength versus Friction Time

3.2. Microstructure of Welded Parts

The photo and the macro-photo of the joint is shown in Figures 7 and 8,while the micro-structure-photos in the parent metals and interface region of the joints are shown in Figures 9, 10 and 11.

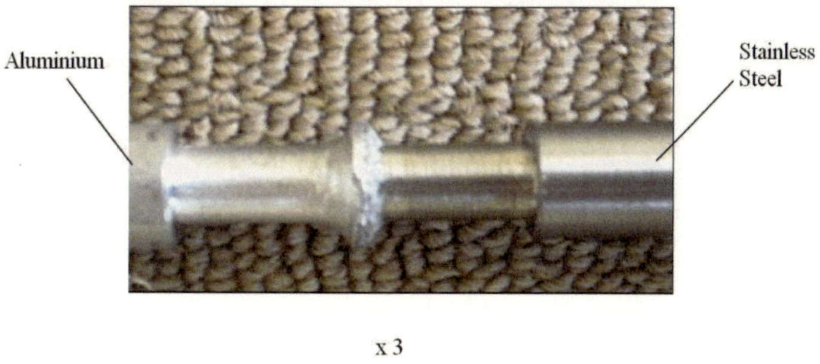

Figure 7. Photo of joint

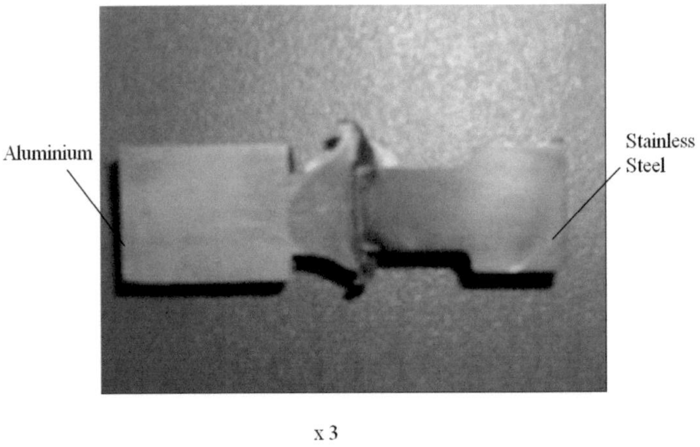

Figure 8. Macro-photo of joint

As shown in Figures 7 and 8, axial shortening in the aluminium side is much more than that of the stainless-steel side. However, the stainless steel was hardly ever deformed because the melting temperature of aluminium is lower than that of stainless-steel. Therefore, the weld flash consists of aluminium at the interface.

Figure 9. Micro-photo of stainless-steel

The microstructure of the base metal consists of austenitic grain structure.

x 1000

Figure 10. Micro-photo of aluminium

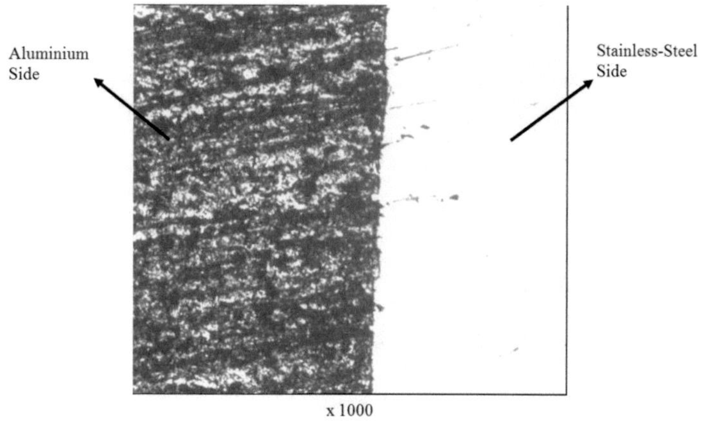

Figure 11. Micro-photo of interface region in joints

Micro-photographs (Figs. 9-11) show that aluminium was greatly deformed with grains elongated and refined near the weld interface. Stainless steel was slightly deformed and partly transformed at the faying surface from austenite to martensite owing to hard friction. Constituent elements of both materials had interdiffused through the weld interface, and intermetallic compounds such as FeAl and Fe_3Al, were formed at the weld interface.

3.3. EDX Analysis of Joints

Scanning electron microscopy (SEM) and energy dispersive X-ray (EDX) analysis were performed in order to investigate the phases that occur during welding at the welding interface. Observations were realized with a 25 kV field effect scanning electron microscope (SEM-JEOL JSM 5410 LV microscopy) associated to an EDS (energy dispersive X-ray spectroscopy) analysis. EDS point analysis was used in the examinations. The software allowed piloting of the beam, scanning along a surface or a line to obtain X-ray cartography or concentration profiles by elements, respectively. SEM microstructure of interface region in the friction welded steel-aluminium joint and EDX analysis results are given in Figure 12, while distribution of elements within the determined location are shown in Table 4. EDS analysis was carried out for various points of the SEM image.

(a) SEM microstructure of joint interface.

http://dx.doi.org/10.5772/51130

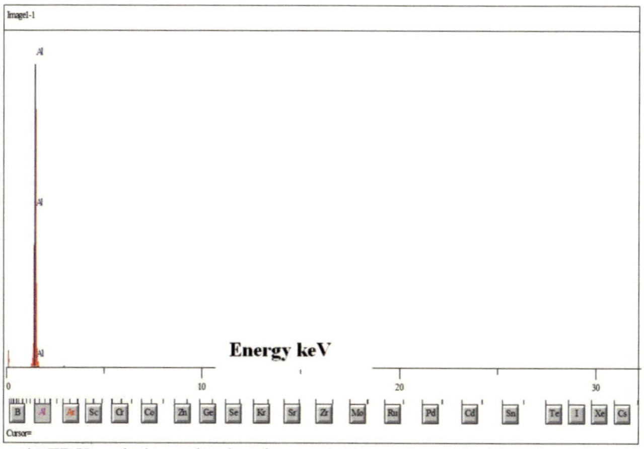

(b) EDX analysis result taken from point 1 represented to SEM image

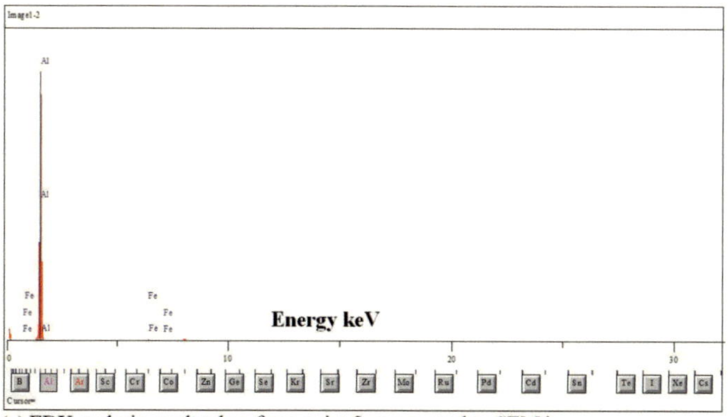

(c) EDX analysis result taken from point 2 represented to SEM image.

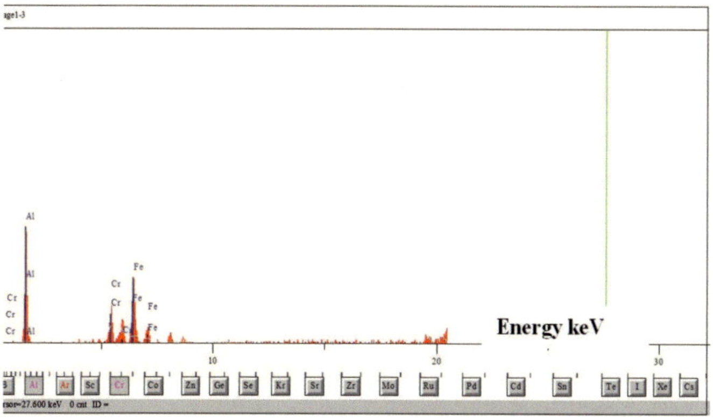

(d) EDX analysis result taken from point 3 represented to SEM image.

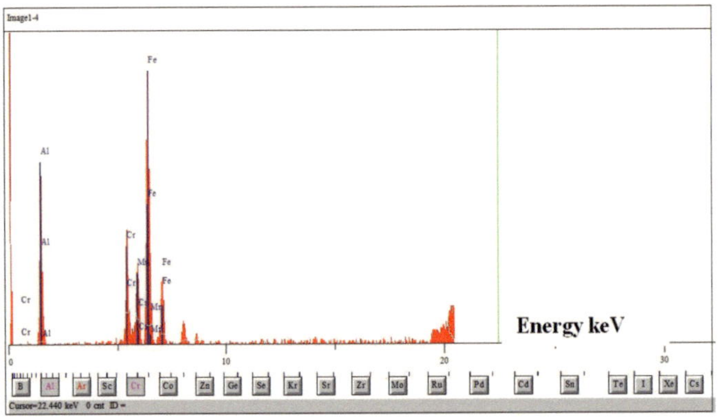

(e) EDX analysis result taken from point 4 represented to SEM image

Figure 12. SEM microstructure of interface region in the friction welded steel-aluminium joint and EDX analysis results.

Points	Elements	Line	Intensity (c/s)	Conclusion
1	Al	Ka	1928.84	100.000wt.%
				100.000wt.%Total
2	Al	Ka	1201.79	97.792wt.%
	Fe	Ka	10.30	2.208wt.%
				100.000wt.%Total
3	Al	Ka	57.68	36.742wt.%
	Cr	Ka	34.29	17.651wt.%
	Fe	Ka	60.68	45.607wt.%
				100.000wt.%Total
4	Al	Ka	97.01	21.117wt.%
	Cr	Ka	79.18	11.282wt.%
	Mn	Ka	76.10	12.472wt.%
	Fe	Ka	255.17	55.128wt.%
				100.000wt.%Total
5	Cr	Ka	370.48	18.189wt.%
	Fe	Ka	958.19	75.092wt.%
	Ni	Ka	58.40	6.719wt.%
				100.000wt.%Total

Table 4. EDS point analysis results according to SEM microstructure.

Fig. 12(a) shows EDX analysis points defined on the SEM microstructure in interface region of the friction welded St-Al joints. Fig. 12 (b), (c), (d) and (e) illustrate the EDX analysis results taken from the points 1,2,3 and 4 represented to St-Al joint, respectively. Then, Table 4 shows the EDS point analysis results represented to SEM. The EDS results confirm that St-Al joints contain some intermetallic compounds. Therefore, formation of brittle intermetallic compounds degrades the strength of the joints.

3.4. Hardness Variations of Welded Parts

Strength of the joints is related to hardness variation within the HAZ. Hardness variation was obtained under 500 g load by micro hardness (Vickers) testing, and measuring locations are shown in Figure 13. Hardness variations on horizontal and vertical distance from the centre in the welding zone of joints are shown in Figures 14 and 15.

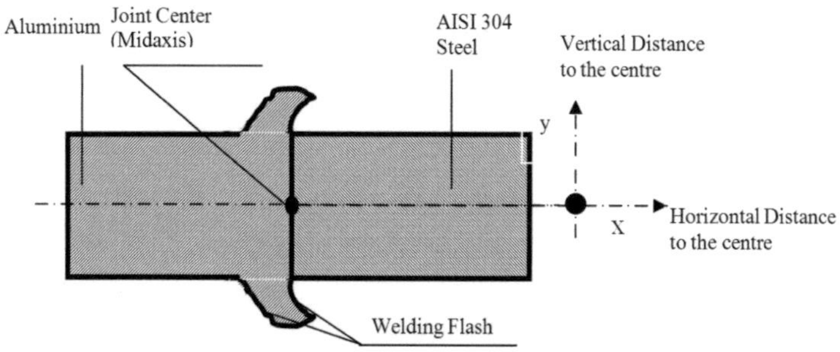

Figure 13. Hardness test orientation.

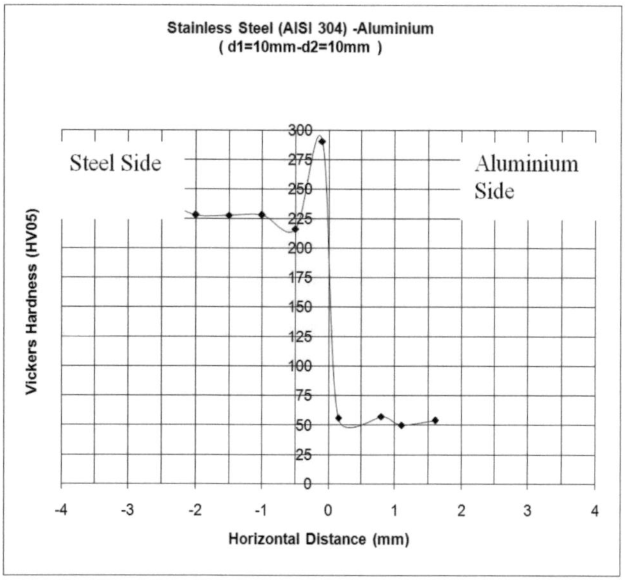

Figure 14. Hardness Distribution on the Horizontal Distance of Joints.

There are often significant differences between the tensile strength and hardness of the A heat affected zone (HAZ) the unaffected area of the welded component. The reduction in tensile strength of the HAZ under controlled conditions, particularly with the non-heat treatable alloys, can be somewhat predictable. The reduction in tensile strength of the HAZ for the heat treatable alloys is more susceptible to welding conditions and can be reduced below the required minimum requirement if excessive heating occurs during the welding operation.

Micro-hardness test results with respect to the horizontal distance from the center are shown in Fig.14. Increase in hardness corresponds to the steel side. HAZ with a small width was formed, resulting in softening of the aluminium alloy. As the aluminium used in the present study was a cold drawn bar, it was already work hardened before the friction welding procedure. The aluminium recovered and recrystallised as a result of friction heat and deformation, thus was slightly softened.

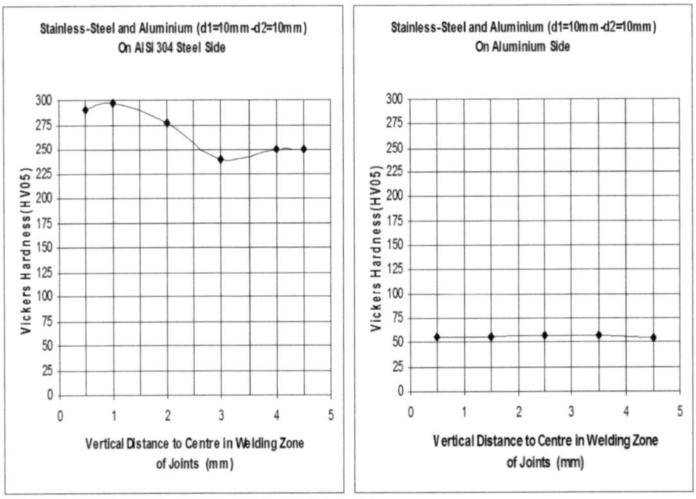

(a) (b)

Figure 15. Hardness Distribution on the Vertical Distance of Joints.

As shown in Fig. 15(a), the hardness on the stainless-steel side of the joints decreases as it is advanced towards the end of the parts. On the other hand, hardness on the aluminium side of the joints did not change significantly (Fig. 15(b)).

4. Friction welded aluminium and copper materials

Parameters having the least error by using the method of least squares were taken as the optimum welding parameters. Optimum parameters found in a previous different study [20] were found as; (60 MPa) for friction pressure, (120 MPa) for upset pressure, (12 sec) for upset time and (2,5 sec) for friction time.

Subsequently, tensile tests, micro-hardness tests and metallurgical examinations were applied to the welded specimens.

4.1. Tensile Tests

Optimum parameters for welded parts.were found using statistical analysis. Then, parts machined were welded using these optimum parameters. Effects of friction time and friction pressure on strength of the joints were examined welding parts with equal diameter. Upset time was kept constant. The strength of joints was determined by tensile tests, and the results were compared with those of fully machined specimens. Three specimens were tested at each condition and average of three specimens is presented. Tensile strength of the joints was estimated dividing the ultimate load by the area of the 10 mm diameter specimen. The relation obtained between tensile strength versus friction time and friction pressure is shown graphically in Figures 16 and 17.

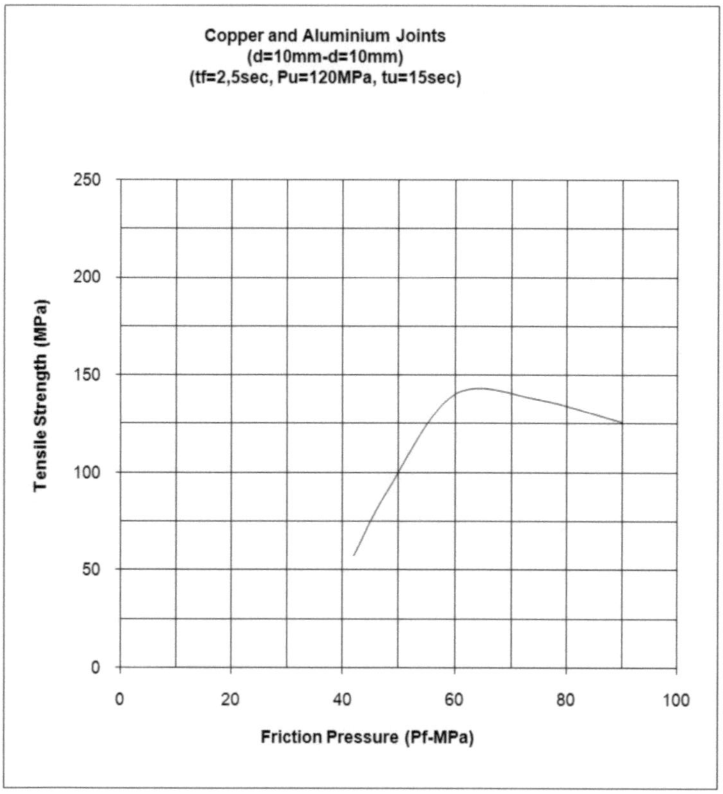

Figure 16. Relation between Tensile Strength versus Friction Pressure.

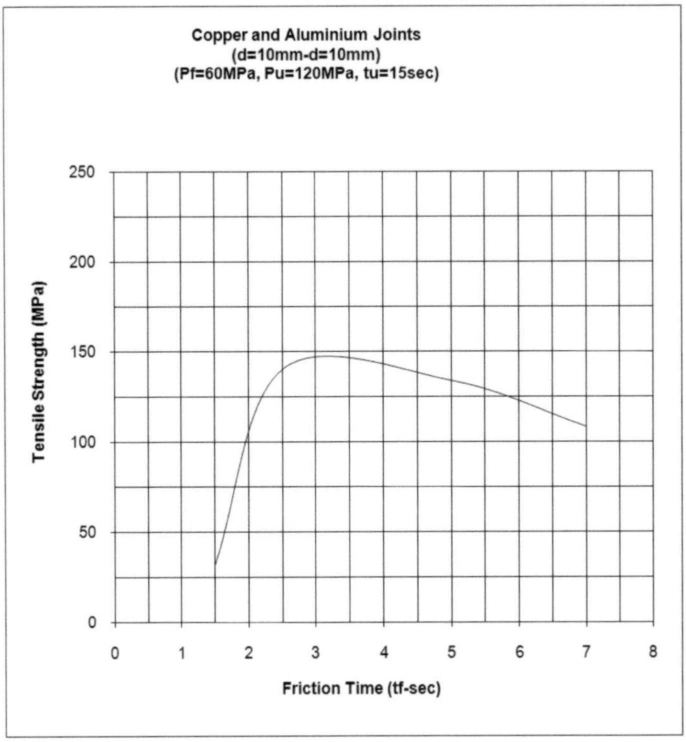

Figure 17. Relation between Tensile Strength versus Friction Time.

As the friction time and pressure for the joints is increased, tensile strength of the joints increases up to a peak strength then decreases with further increase in friction time and pressure (Figures 16 and 17). Peak strength corresponds to about 70% that of aluminium parts and 50% that of copper parts. A grey layer was observed at the fracture surfaces of welded parts. This layer results in a decrease in the strength of the joints.

4.2. Microstructure of Welded Parts

As regards joints, the photo and the macro-photo of the joint is shown in Figures 18 and 19. Then, the microstructure-photos in the parent metals and interface region of the joints are shown in Figures 20, 21, and 22.

It can be seen that the axial shortening on the aluminium side is more than that on copper side (Figures 18 and 19). Thus, the aluminium material has experienced weld flash at the interface. This is due to the fact that melting point of aluminium is lower than that of copper.

http://dx.doi.org/10.5772/51130

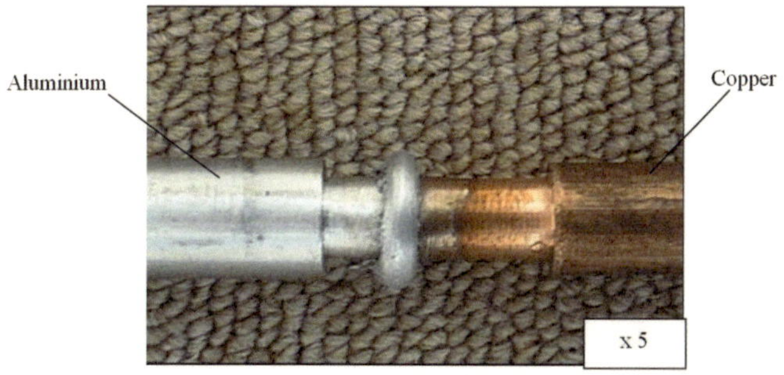

Figure 18. Macro-photo of Joint.

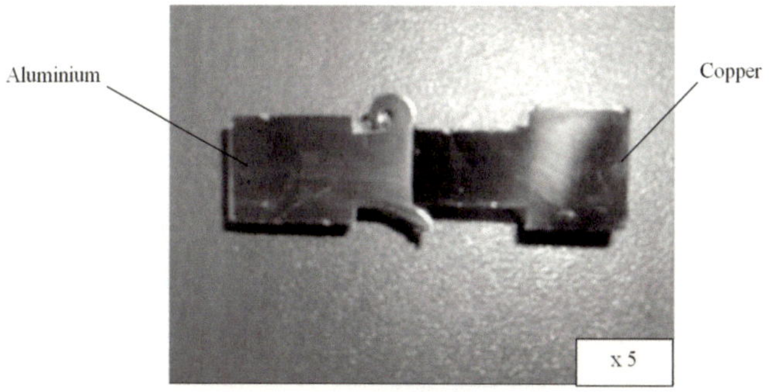

Figure 19. Macro-photo of Joint.

Figure 20. Micro-photo of Copper.

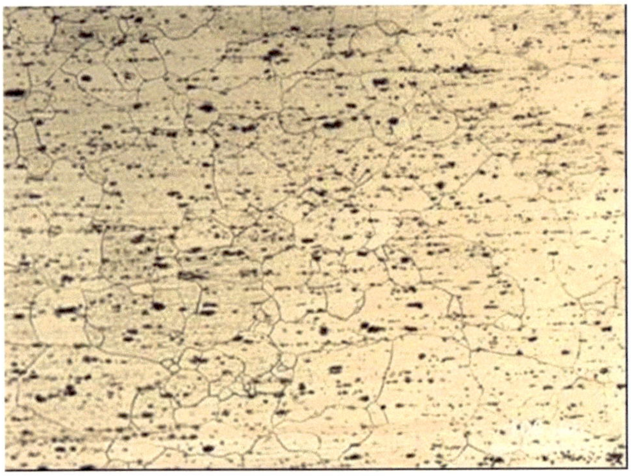

Figure 21. Micro-photo of Aluminium.

Figure 22. Micro-photo of Interface Region in Joints.

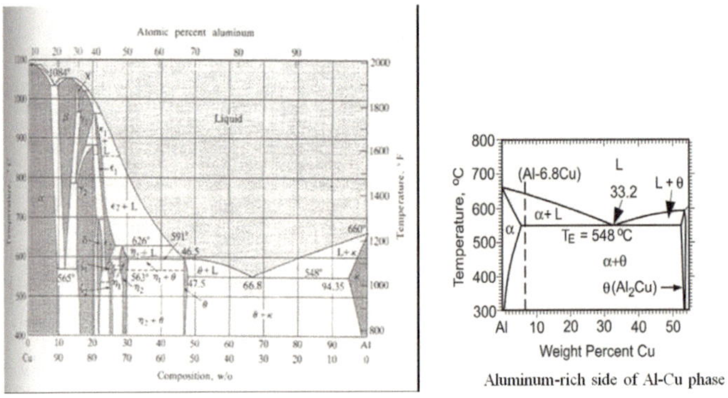

Figure 23. Al–Cu Binary Equilibrium Phase Diagram [17].

The copper substrate exhibits an irregular grain (Fig. 20). The grains of aluminium are elongated along the rolling direction (Fig. 21). The microphotograph of aluminium also contains insoluble particles of $FeAl_3$ (black). Relatively coarse $CuAl_2$ grains are clearly observed at the transition zone of the copper side [21, 22, 23 and 24].

Microstructural observations showed that a mixed layer of aluminium and copper that includes brittle intermetallic compounds such as $CuAl_2$, CuAl, and Cu_9Al_4 are formed in a dissimilar aluminium alloy/copper weld. The formation of intermetallic compounds can be understood by an analysis of the Al–Cu binary phase diagram (Fig. 11) [24].

4.3. EDX Analysis of Joints

Scanning electron microscopy (SEM) and energy dispersive X-ray (EDX) analysis were performed in order to investigate the phases that occur at the welding interface. Observations were realized with a 25 kV field effect scanning electron microscope (SEM- JEOL JSM 5410 LV microscopy) coupled to EDS (energy dispersive X-ray spectroscopy) analysis. EDS point analysis was used in the examinations. The software allowed piloting the beam to scan along a surface or a line so as to obtain X-ray cartography or concentration profiles by elements. SEM microstructure of the interface region in the friction welded copper-aluminium joint and EDX analysis results are given in Figure 24, while distribution of elements within the determined location are shown in Table 5. EDS analysis was carried out for various points of the SEM image.

Points	Elements	Line	Intensity (c/s)	Conclusion
1	Al	Ka	1054.36	89.113wt.%
	Fe	Ka	36.54	6.523 wt.%
	Cu	Ka	15.65	4.364 wt.%
				100.000wt.%Total
2	Al	Ka	1013.86	100.000wt.%
				100.000wt.%Total
3	Al	Ka	891.13	80.581wt.%
	V	Ka	0.96	0.115wt.%
	Fe	Ka	23.84	3.537wt.%
	Cu	Ka	65.64	15.767wt.%
				100.000wt.%Total
4	Cu	Ka	883.70	100.000wt.%
				100.000wt.%Total
5	Cu	Ka	790.24	100.000wt.%
				100.000wt.%Total

Table 5. EDS Point Analysis Results according to SEM Microstructure.

Fig. 24 shows EDS analysis points defined on the SEM microstructure in interface region. Table 5 illustrates the EDS analysis results taken from the points 1, 2, 3, 4 and 5, respectively represented by SEM.

The EDS results confirm that Cu-Al joints contain some intermetallic compounds. Formation of these brittle intermetallic compounds degrades the strength of the joints.

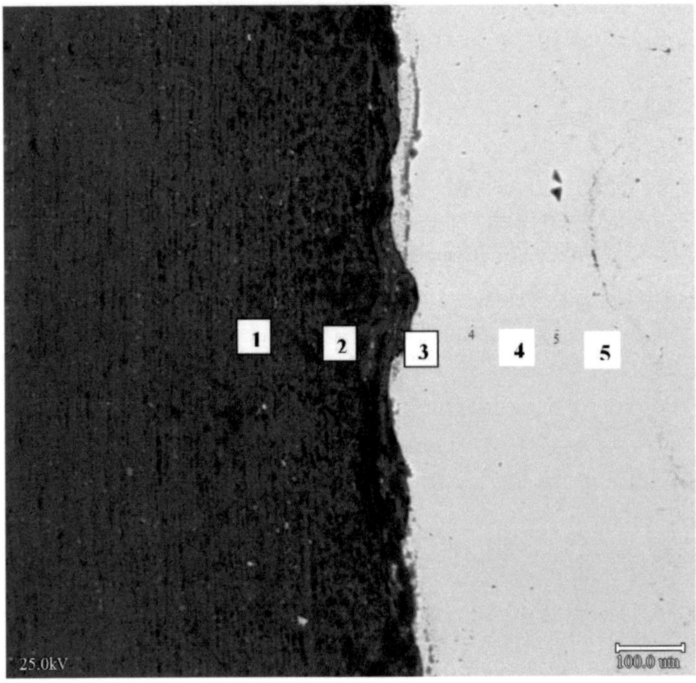

Figure 24. SEM Microstructure of Interface Region in the Friction Welded Copper-Aluminium Joint and EDX Analysis Markers

4.4. Hardness Variations of Welded Parts

Strength of the joints is related to hardness variation within the HAZ. Hardness variation was obtained under 200 g loads by micro hardness (Vickers) testing. Micro-hardness test results with respect to the horizontal distance from the centre are shown in Fig. 25.

As the aluminium used in the present study was a cold drawn bar, it was already work hardened before friction welding. Aluminium recovered and recrystallised as a result of frictional heat and deformation, thus was slightly softened. Hardness variations on the copper side are more than those on the aluminium side. This variation is due to comparatively high thermal conductivity of copper.

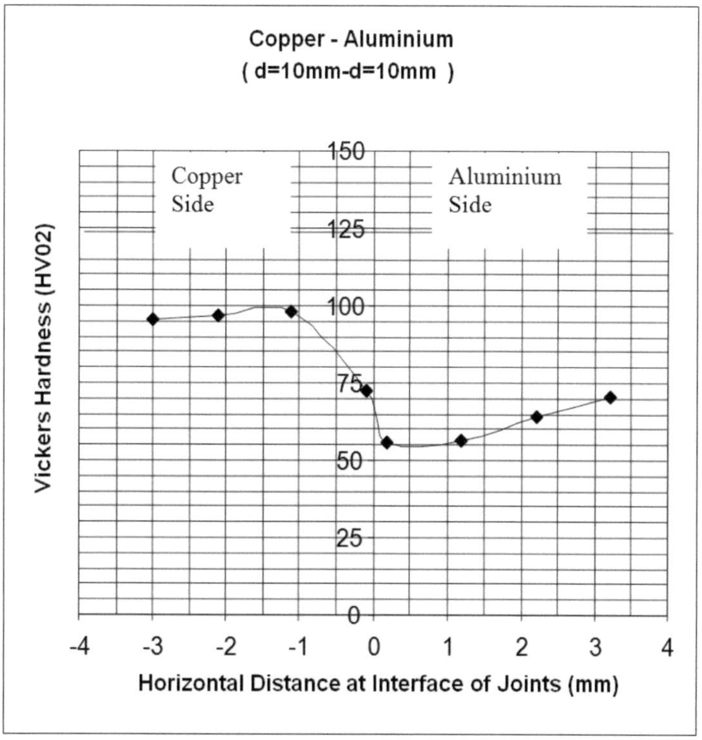

Figure 25. Hardness Results on Horizontal Distance at Interface of Joints

5. Conclusions

In the present study, austenitic-stainless steel (AISI 304)-aluminium and aluminium-copper materials were welded successfully. The welding process was investigated by tensile testing, microstructural observation, EDS measurements and hardness testing. As a result:

- Optimum welding parameters should be properly selected in the friction welding of parts.

- Tensile strengths for austenitic-stainless steel and aluminium parts yielded a positive result when compared to those of the base metals. The joint strength increased and then decreased after reaching a maximum value, with increasing friction time. Sufficient heat to obtain a strong joint could not be generated with a shorter friction time. A longer friction time caused the excess formation of an intermetallic layer. However, some of the welds showed poor strength depending on some accumulation of alloying elements at the interface, which are the result of a temperature rise and the existence of intermetallic layers such as FeAl.

- Although tensile strength for copper and aluminium joints were generally acceptable when compared with those of the base metals, some of the welds showed poor strength as a result of the accumulation of alloying elements at the interface. This was the result of temperature rise and the existence of a grey layer. This grey layer formed due to heat dissipation in friction welding and was found to contain a considerable amount of intermatallic compounds.

- The presence of contaminants at the interface of the metals reduces the joint quality. No significant effect was observed on welding properties with respect to the surface finish operations.

- In the microphotos, the broken up aluminium oxide film resulted in increased deformation at the interface. Formation of an oxide in the joints causes a barrier that prevents diffusion.

- The difference in weight of alloying elements can be clearly seen by analyzing spectrum of elements. EDX measurements clearly show that St-Al and Cu-Al joints consist of some intermetallic compounds. The intermetallic layer formed constituted mainly of FeAl, Fe_3Al, $CuAl_2$, CuAl, and Cu_9Al_4 together with some Al and Cu (saturated solid solution of Al in copper). Copper particles embedded in aluminium were observed. Then, it can be imparted in terms of galvanic effect that Fe_3Al particles are anodic to the matrix in St-Al joints. However, the copper band on either side of the grain boundary is dissolved while the grain boundary is cathodic due to the $CuAl_2$ and Cu_9Al_4 precipitates.

- Hardness of steel and aluminium materials in the vicinity of the weld interface was higher than that of the base metals. Then, in Al-Cu joints, hardness variations on the aluminium side were lower than those on the copper side as expected.

- Aluminium alloys are highly reflective and decorative. The high reflectivity is an inherent feature of aluminium; pure bulk aluminium can go up to 92 % total reflection. Alloying reduces this value slightly.

- The main value of this paper is to contribute and fulfil the detailed the Welded Aluminium Alloys that are being studied so far in the literature.

Acknowledgements

The author wishes to thank Hema Industry / Çerkezköy, the Mech. Eng. Dept. of Trakya University, Edirne and Metall. and Mater. Eng. Dept. of Yildiz University, Istanbul-Turkey for their help in the experimental and microstructure studies.

Author details

Mumin Sahin* and Cenk Misirli

*Address all correspondence to: mumins@trakya.edu.tr

Dept. of Mechanical Eng., Trakya University, Turkey

References

[1] Polmear, . J. (1995). *Light Alloys*, Arnold.

[2] Vill, V. I. (1962). Friction Welding of Metals. AWS, New York.

[3] Kinley, W. (1979, Oct). Inertia Welding: Simple in Principle and Application. *Welding and Metal Fabrication*, 585-589.

[4] Fomichev, N. I. (1980). The Friction Welding of New High Speed Tool Steels to Structural Steels. *Welding Production*, 35-38.

[5] Murti, K. G. K., & Sundaresan, S. (1983, June). Parameter Optimisation in Friction Welding Dissimilar Materials. *Metal Construction*, 331-335.

[6] Yılbaş, B. S., Şahin, A. Z., Kahraman, N., & Al-Garni, A. Z. (1995, Feb 15). Friction Welding of St- Al and Al- Cu Materials. *Journal of Materials Processing Tecnology*, 49(3-4), 431-443.

[7] Yılbaş, B. S., Şahin, A. Z., Çoban, A., & Abdul, Aleem. B. J. (1995). Investigation into the Properties of Friction-Welded Aluminium Bars. *Journal of Materials Processing Tecnology*, 54, 76-81.

[8] Rhodes, C. G., Mahoney, M. W., Bingel, W. H., Spurling, R. A., & Bampton, C. C. (1997). Effects of Friction Stir Welding on Microstructure of 7075 Aluminium. *Scripta Materialia*, 36(1), 69-75.

[9] Fukumoto, S., Tsubakino, H., Okita, K., Aritoshi, M., & Tomita, T. (2000). Amorphization by Friction Welding between 5052 Aluminum Alloy and 304 Stainless Steel. *Scripta Materialia*, 42, 807-812.

[10] Fukumoto, S., Tsubakino, H., Okita, K., Aritoshi, M., & Tomita, T. (1999, September). Friction Welding Process of 5052 Aluminium Alloy to 304 Stainless Steel. *Materials Science and Technology*, 15, 1080-1086.

[11] Sahin, M., & Akata, H. E. (2003). Joining with Friction Welding of Plastically Deformed Steel. *Journal of Materials Processing Technology*, 142(1), 239-246.

[12] Sahin, M., & Akata, H. E. (2004). An Experimental Study on Friction Welding of Medium Carbon and Austenitic Stainless Steel Components. *Industial Lubrication & Tribology*, 56(2), 122-129.

[13] Sahin, M. (2005). An Investigation into Joining of Austenitic-Stainless Steels (AISI 304) with Friction Welding. *Assembly Automation*, 25(2), 140-145.

[14] Sahin, M. (2006). Evaluation of The Joint-Interface Properties of Austenitic-Stainless Steels (AISI 304) Joined by Friction Welding. Materials & Design, In Press, Corrected Proof, Available Online 28 July

[15] Rhodes, C. G., Mahoney, M. W., Bingel, W. H., Spurling, R. A., & Bampton, C. C. (1997). Effects of Friction Stir Welding on Microstructure of 7075 Aluminium. *Scripta Materialia*, 36(1), 69-75.

[16] Ouyang, J., Yarrapareddy, E., & Kovacevic, R. (2006). Microstructural Evolution In The Friction Stir Welded 6061 Aluminium Alloy (T6-Temper Condition) To Copper. *Journal of Materials Processing Technology*, 172, 110-122.

[17] Maalekian, M. (2007). Friction Welding- Critical Assessment of Literature. *Science and Technology of Welding & Joining*, 12(8), 738-759.

[18] Stahlschlüssel, (1995). Verlag Stahlschlüssel Wegst Gmbh

[19] Sahin, M. (2009). Joining of stainless-steel and aluminium materials by friction welding. *Int J Adv Manuf Technol*, 41, 487-497.

[20] Sahin, M. (2010). Joining of aluminium and copper materials with friction welding. *The International Journal of Advanced Manufacturing Technology*, 49(5-8), 527-534.

[21] Koberna, M., & Fiala, J. (1993). Intermetallic Phases Influencing The Behaviour of Al-Cu Joints. *Journal of Physics and Chemistry of Solids*, 54(5), 595-601.

[22] Braunovic, M., & Alexandrov, N. (1994). Intermetallic Compounds at Aluminium-to-Copper Electrical Interfaces: Effect of Temperature and Electric Current. *IEEE Transactions on Components, Packaging, and Manufacturing Technology-Part A,*, 17(1), 78-85.

[23] Lee, W. B., Bang, K. S., & Jung, S. B. (2005). Effects of Intermetallic Compound on The Electrical and Mechanical Properties of Friction Welded Cu/Al Bimetallic Joints during Annealing. *Journal of Alloys and Compounds*, 390(1-2), 212-219.

[24] ASM Handbooks, (2002). Alloy Phase Diagrams, ASM International, Materials Park, Ohio, 3

Chapter 12

Elaboration of Al-Mn Alloys by Aluminothermic Reduction of Mn_2O_3

A. Flores Valdés , J. Torres and R. Ochoa Palacios

Additional information is available at the end of the chapter

http://dx.doi.org/10.5772/50682

1. Introduction

Aluminium is the most abundant metallic element in earth, posing an excellent combination of chemical, mechanical and physical properties which made it suitable for many applications. Among its most important properties a low density, good corrosion resistance, and excellent thermal and electrical conductivities can be found. On the other hand, some aluminium alloys require higher mechanical properties, which are attainable by thermal or thermomechanical treatments. This is the case for extrusions or hot or cold rolled products [1].

Alloys of the Al-Mn-Mg series have been used as raw materials for the elaboration of beverage cans. In these alloys, the second phase particles such as β-$Al_6(Fe,Mn)$ or α-$Al_{12}(Fe,Mn)_3Si$ play an important role in the development of microstructure [2]. Precipitation characteristics in Al-Mn alloys are significantly affected by the presence of iron and silicon. It has also has been reported that both elements accelerate precipitation rate [3].

One of the best solutions to the problem of the accumulation of waste is recycling. In this way, considerably decreases the amount of debris without polluting the environment. Aluminium recycling reduces pollution by 90%, and contributes to lower use of electricity, compared to processing primary raw materials [1].

Everyday increases the importance of recycling, especially when it comes to pollutants such as waste batteries. When running out and throwit away, they go to landfills, releasing highly toxic metal sand materials, adding to the water cycle through a in or gas produced by burning. The manufacture of batteries consume up to 50 times the energy they generate.

Recycling is extremely convenient for the production of goods, especially in countries that do not have the raw materials for the production of primary alloys. Some metals as aluminium can be recycled many times, as is the case of beverage cans. Chemical contamination of this kind of scrap often occurs, so the elaboration of alloys is based on smelting and refining special techniques. Nevertheless, recycling impacts directly in energy consumption savings and lower disposal costs.

1.1. Recycling of aluminium cans

Aluminium recycling is acomplex process, involving a variety of stages. The aluminium used comes primarily from two sources: waste either domestic consumptionor industrial (i.e., power lines, lithographic plates, beverage cans, other containers and packaging.); on the other hand, cut sand shavings produced during the manufacture of aluminium [3].

There are many types of aluminium alloys available in the market recovery, but basically can be groupedas follows [1]:

a. Rolled products (sheet of construction, printing plates, foil, vehicle body parts).

b. Extruded (profiles for windows, vehicle parts).

c. Molten aluminium alloys in metal molds, either by gravity or by injection(engine parts, handles of the doors).

d. Wiring for the manufacture of cables and other uses.

The most important alloys that can be elaborated using aluminium scrap are the Al-Si and Al-Si-Cu intended for the automotive industry. Nevertheless, other important alloys that can be elaborated from scrap are the well-known Al-Zn-Mg and complex Al-Si-Cu-Mg-Zn alloys.

Al-Mn alloys are not produced from scrap, rather they are obtained using primary smelting with commercial purity elements. Nevertheless, as it will be presented and discussed later, a good alternative is the usage of Mn_2O_3 powders obtained from discharged alkaline batteries, using aluminothermic reduction.

Aluminium does not changeits characteristics during there cycling, in addition beverage reservoirs can be madeentirely from recycled metal. Due to the high value of aluminium in the market, most cans are recycled in Mexico (98%) and worldwide (66%), allowing the city's image does not affector impair (hard to find canslying on street) and landfills do not increase their volume. In addition, recycling of metal packaging reduces contamination [4,5].

Aluminium cansare very light when empty, in fact, in the last 15 years, the amount of aluminium used in the production of a can has been reduced by 35%. Aluminium recycling provides energy savings and costs, when aluminium is used to manufacture the cans recovered instead of virgin materials. As large as 85% savings are obtained when recycling.

In conclusion, with its high level of recycling and being prepared with raw materials abundant in earth, aluminium cans are the quintes sentialeco-friendly-packaging materials. The malleability and resistance are some of the mainqualities and advantages of the cans, which allow a wide range of packaging products with different characteristics: from liquid and solid, to gaseous and aerosol products. Moreover, they can be used for packaging of some delicate products such as certain medications and some beauty products. Also flammable products that require optimal safety conditions such as paints, solvents, insecticides, fats, oils for industrial use,etc.

World wide consumption of cansis very high. By recycling this material generates a significant social conscience and is easy to convey the message to the population. Today the aluminium can recycling is an industry with large employment generation. It also has reduced energy consumption in the production of any material. This implies not only benefits the country level but also globally, because of environmental issues related to power generation [6].

Currently, the recycling of aluminiumis very common, and that all people want to recycle and benefit mostly by selling cheap, but not really worry about these contaminated waste.

1.2. Recycling of alkaline batteries

Several studies have been developed for the management of toxic substances that are dumped into the ecosystem, in order to reduce and/or eliminate pollution of the planet. A clear exampleis the subject of alkaline batteries, which have a high degree of toxicity components. The batteries are a source of energy producing electricity from chemical reactions, but once discharged, are especially dangerous waste. When a battery loses its protective covering, being exposed to the weather, it releases toxic metals and materials known to beharmful effects to the ecosystemand health of human beings. These metals and toxic material shave harmful effects on human health ranging from serious injuryin the central nervous system lesions in the liver, kidney affections, gastrointestinal disorders, carcinomas or getting even death [7]. Alkaline batteries have duration of three or more times than other primary batteries.

Kind of battery	Components	Uses
Alkaine	*Zinc: 15 % (anode) *Manganese dioxide: 35% (cathode) *Carbon: 2% *Potassium hidroxide: 8% *Plastic and sheet: 40%	Electric and electronic devices, known as common batteries

Table 1. Components and applications of alkaline batteries.

Alkaline batteries using an electrochemical system in which the zinc and manganese dioxideare the anode and cathode respectively, the overall reaction in analkaline battery is represented by:

$$Zn + 2MnO_2 \rightarrow Mn_2O_3 + ZnO \qquad (1)$$

So that the products at the anode and cathode of analkaline battery discharged are ZnO and Mn_2O_3, respectively.

Some dangers identified during use of the batteries are accidental break age of the shield, with the corresponding stroke of the components. When the batteries are subjected to extreme environmental conditions than those experienced on the interior of a computer, the shield is exposed to corrosion processes ending and releasing destroying cell components. Multiplying the number of batteries used by each in habitant, the number of people highlights the way in how they are polluting the planet [8,9]. Alkaline batteries are considered hazardous waste dischargedand forbidden to be deposited a long with the regular trash in landfills.

Different technologies have been developed for the handling of this kind of waste, in order to establish the most appropriate provisions of the discharged electric batteries [2]. There are many alternatives that can be applied, among which are: collection, stabilization, confinement, glazing, ceramization and recycling of components.

The use of recycled materials such as aluminium cans and alkaline cathodes as the raw material for the manufacture of aluminium-manganese alloys by metallothermic reduction is highly suitable. Therefore, one of the objectives of this study was to analyze the relationship between the microstructure and mechanical properties (hardness,tensile strength, and elongation) of Al-Mn alloys produced by aluminothermic reduction both in the as-cast condition and after homogenization annealing.

1.3. Metallothermic reduction of oxides

Metallothermic process relates to the production of metals andalloys by reducing their oxides or halides with metals. Generally, these reactions can be expressed by the equation [10,11]:

$$AX + B \rightarrow A + BX \tag{2}$$

Where Xis oxygen, chlorine or fluorine, and Aand B represent the two metals.

The main characteristic of this process is that the metal being reducedis converted into a solid or liquid, which dissolves in the molten metal. Metallothermic processes are used when the reduction with carbon, hydrogenor by electrolys is of a solution is not possible. Among the metals used as reducing agents are: Al,CaandMg.

Figure 1 shows a schematic representation of the aluminothermic reduction, in which the particles of manganese oxide (Mn_2O_3) is reduced by molten aluminiumat the boundary layer.

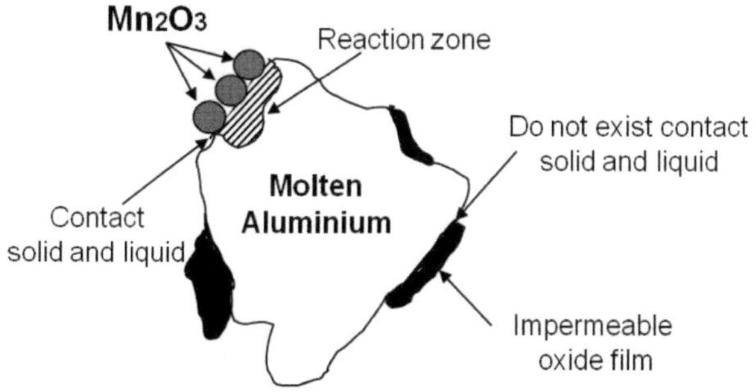

Figure 1. Scheme of the aluminothermic reduction of Mn_2O_3 [10,11].

1.3.1. Influenc surface tension in the metallothermic process

The nature of the interfaces of the species that are in contact is critical to achieving the wettability required for a reaction to occur between them, either dissolution or chemical reaction. For the alumin other mic reaction it is required very good contact between liquid and solid particles of the oxides to reduce. Therefore, without a good wettability, chemical reaction rates of mass transport and are greatly reduced. In the extreme case where the wettability between the solid oxideand the molten phase does not exist, the reaction will not occur [12].

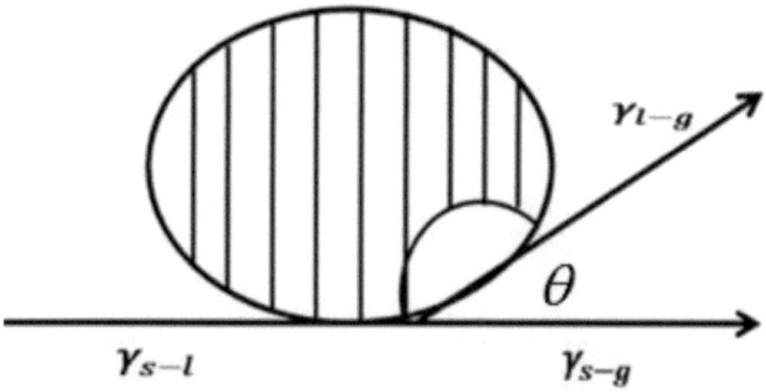

Figure 2. Scheme of the method of measurement of the contact angle between a solid and a molten phase [12].

Many solutes in liquid metals can act as surface active agents (such as soap in the water) which have a lower surface tension. Generally, measuring the contact angle,θ, is the most commonly applied to determine the wettability which occurs between the solid and liquid. Such measurements are performed in the liquid phase as shown in Figure 2. If the liquid phase takes hold on the solid surface with amorphology very close to that of a sphere, the contact angle will be large and will reach the limit of 90°, so that there will be no wettability. Otherwise, if the liquid phase tends to wet the solid phase, the value of angle θ is small and reaches the limit 0°.

The two limits represent the conditions for a complete and in complete wettability, respectively. Assuming that a liquid, L, lies on the solid surface, S, in contact with an inert gas G, in equilibrium, the liquid dropand the solid surface will be a tan angle θ as shown in the following equation:

$$\vartheta_{SL} = \vartheta_{SG} + \vartheta_{LG}\ coscos\varpi \quad (3)$$

Where γ_{SL}, γ_{SG} and γ_{LG} are the surface tension between the solid-liquid, solid-gas and liquid-gas, respectively.Therefore, the theoretical contact angle can be represented by the following equation:

$$Cos\,Cos\varpi = \left[\frac{\vartheta_{SG} - \vartheta_{SL}}{\vartheta_{LG}}\right] \quad (4)$$

Using this equation,the contact angle between the solid toliquid aluminium can be compared. Equation (4) maybe written also as follows:

$$Cos\,Cos\varpi = \left[\frac{\vartheta_{SG} - f(\vartheta_{SL})}{\vartheta_{LG}}\right] \quad (5)$$

Meaning that the surface tension, $\gamma_{SL,}$ varies proportionally with the surface tension γ_{LG}. Therefore, by increasing γ_{LG}, the contact angle, θ, is also increased. A lower contact angle means better wettability of the solid surface. Surface tensions of the liquids are high and the typical values of most representative metals are listed in Table 2.

Metal	V_{LG} (dyn/cm)	V_{SL} (dyn/cm)
Mg	559	56
Al	914	91
Si	865	86
Ca	361	36
Sr	303	30
Bi	378	38

Table 2. Surface tension for metals in the alumino thermic process [10].

The surface tension values for magnesium are minor compared to that of pure aluminium. Therefore, the magnesium improves the wettability between the solid and the liquid metal.

http://dx.doi.org/10.5772/50682

2.3.2. *Effect of magnesium in the metallothermic reduction*

In the metallothermic reduction process with aluminium, magnesium has attracted much attention due to their surfactant nature, since the value of the surface tension of this element is less compared with that of pure aluminium (γ_{Mg} = 559 dyn/cm., γ_{Al} = 914 dyn/cm) [10,13]. Therefore, the addition of magnesium reduces the surface tension of molten aluminium bath which leads to improve the wettability between the solid reactant and the liquid metal, thus increasing the kinetics of the reactions taking place at the solid-liquid interface.

From the thermo dynamic point of view, magnesium dissolved in theliquid has an affinity foroxygen, as seen by the Gibbs free energy values of the following reactions:

$$2Mg + O_2 = 2MgO \, \beta G°_{1073K} = -231 \, {}^{kcal}/_{mol} \tag{6}$$

$$^{4}/_{3Al} + O_2 = {}^{2}/_{3Al_2O_3} \beta G°_{1073K} = -319 \, {}^{kcal}/_{mol} \tag{7}$$

The thermo dynamic stability of magnesium oxide (MgO) suggests that the magnesium dissolved in the liquid aluminium in creases the reactivity of the process [10], and therefore the concentration of manganesein the molten metal could in crease.

For the case of the metallothermic reduction of manganese trioxide, reactions taking place are the following:

$$2Al + Mn_2O_3 = 2Mn + Al_2O_3 \beta G°_{1048K} = -156.940 \, {}^{kcal}/_{mol} \tag{8}$$

$$3Mg + Mn_2O_3 = 2Mn + 3MgO \, \beta G°_{1048K} = -185.331 \, {}^{kcal}/_{mol} \tag{9}$$

$$6Al + 3Mg + 4Mn_2O_3 = 8Mn + 3MgAl_2O_4 \beta G°_{1048K} = -656.151 \, {}^{kcal}/_{mol} \tag{10}$$

According to the above, it is possible to obtain alloys Al-Mn from recycled cathode alkaline discharge dusing the metallothermic reduction reaction with the molten metal obtained from the smelting of aluminium cans.

1.4. Al-Mn alloys

Aluminium alloys of the series 3000 (Al-Mn) are perhaps the most important of aluminium strain hardened alloys. Generally, they contain from 0.25-1.50 wt. %Mn, 0.05-0.2 wt. %Cu, 0.1 wt. % Zn and small percentages of elements such as Mg, Cr, Fe, Si, and Ni.Due to its excellent mechanical properties, they have several applications, such as food processing, textiles and automotive products, cookware, pressure vessels, construction elements, panels for refrigerators and heat exchangers.

According to the equilibrium diagram Al-Mn of Figure 3 [14], the aluminium forms aeutectic with the inter metallic Al_6Mn at a temperature of 658°C and a composition of 1% Mn. The presence of Fe, Mg and Siimpurities is inevitable, changing on set temperatures of reactions and solidification temperatures. First, the phase $Al_6Mn$is replaced by $Al_{15}(Mn,Fe)_3Si_2$appearing also particles of the Mg_2Si phase. The characteristic temperatures of solidificationare reduced some what by the presence of Si and Fe, starting with the formation of primary aluminium grain sat 652°C. The phase Al_6 (Fe,Mn) is formed around 644°C by an eutectic reaction, the temperature is at least 10°C lower than that of analloy of magnesium free (3003).

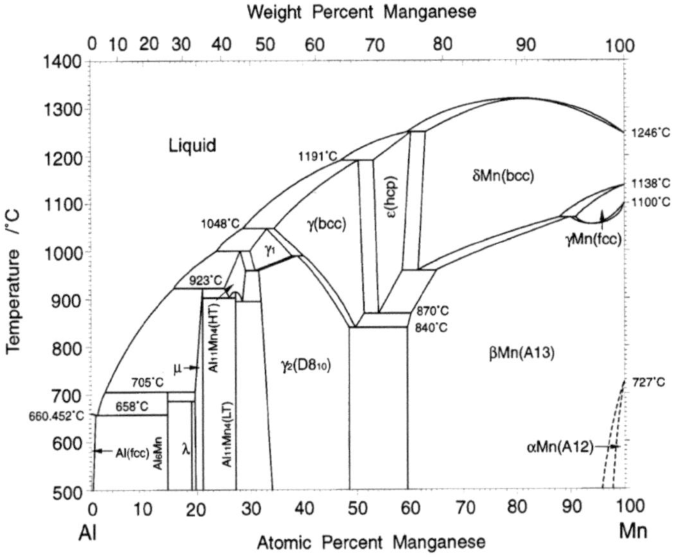

Figure 3. The Al-Mn phase diagram [14].

The inter metallic Al_6(Fe, Mn) reacts with the liquid phase through a peritectic reaction at 630°C to form $Al_{15}(Mn, Fe)_3Si_2$. An observation made by Bäckerud [10] is that, contrary to the usual trend, the peritectic reaction is complete at high cooling rates and partially at slower cooling rates. This explains why the particle size of β-Al_6(Fe,Mn) thicken at low cooling rates. Solidification ends around 580°C with the formation of Mg_2Si by an eutectic reaction. The particles of β-Al_6(Fe,Mn) remaining in the melt due to incomplete peritectic reaction, must be dissolved during annealing, resulting in the more stable phase $Al_{15}(Fe,Mn)_3Si_2$ [2].

Elemet	Si	Fe	Cu	Mn	Mg	Zn
wt. %	0.3	0.7	0.25	1.0-1.5	0.8-1.3	0.25

Table 3. Typical composition of a 3004 alloy, in wt.%[8].

http://dx.doi.org/10.5772/50682

These alloys have only a 20% higher hardness than pure aluminium. The chemical composition of alloy 3004 is presented in Table 3.

Min may be added effectively at only 1.5 wt. % (see Figure 3). Therefore, there are very feal-loys of this eries. However, the 3003, 3004 and 3005 alloys are used for making tools that need medium hardness and good formability, for applications such as beverage cans, kitchen utensils, heat exchangers, furniture, signs and architectural applications.

The 3004 alloy also contains magnesium, improving the mechanical properties while enhancing resistance to corrosion. Due to the high content of Mg in the molten phase coming from cans, it can be used to promote metallothermic reduction of manganese oxides, producing alloys of the 3000 series type, where the initial content of Mg could decrease below 0.3 in wt. %.

1. la
2. el
3. los
4. las
5. lo

2. Experimental procedure

This section describes both the experimental design and melting practices carried out for the production of Al-Mn alloys by aluminothermic reduction of Mn_2O_3.

2.1. Materials and Equipment

To obtain the Al-Mn alloy it is required:

- Alkaline Battery Cathodes (Mn_2O_3powders).
- Aluminium cans.

  Medium frequency induction furnace.

  Metal Molds

For characterization of the samples:

- Chemical analysis (Inductively Coupled Argon Plasma Spectrometry and Atomic Absorption)
- Scanning Electron Microscopy(EBSD).
- Quantitative Optical Microscopy(EDS).
- X-Ray diffraction.

Variables analyzed were target concentration of Mn in the alloy (1-1.5 wt. %), which caused that treatment time varied from 120 of up to 240 minutes, and temperature (1023, 1050 and 1073K). The powders used were of the constant size of 240 mesh, while agitation speed was maintained constant at 200 rpm.

It is worth mentioning that Mn_2O_3 powders were obtained in its purified form and of constant size, from discharged alkaline batteries, through a procedure reported elsewhere [8].

2.2. Methodology

2.2.1. Melting Practice

For Al-Mn alloys obtention using aluminothermic reduction of Mn_2O_3, the melting of aluminium cans was carried out using a medium frequency induction furnace. The cans were loaded to the furnace in charges of 10 kg. Once the temperature reached the value of interest, i.e. 1023 K, it was maintained constant up to completion of treatment time. During each experiment, samples from the melt were taken every 10 minutes. At the end of each experiment, fluxing and degassing was performed, in order to obtain solid enough samples for mechanical properties measurements. Table 4 shows the chemical composition of the alloy obtained from the fusion of aluminium cans.

Elemet	Si	Fe	Cu	Mn	Mg	Zn
wt. %	0.338	0.603	0.169	0.832	1.02	0.698

Table 4. Chemical composition of the starting alloys from recycled aluminium cans.

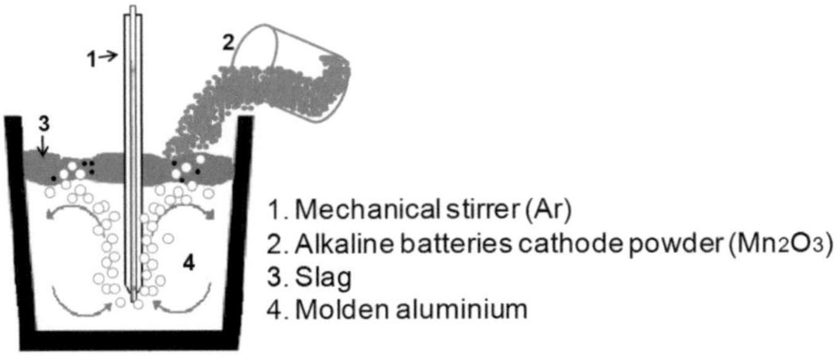

Figure 4. Scheme of the form of powders addition using a mechanical stirrer.

As reactive powders it were used those obtained from discharged alkaline cathodes (Mn_2O_3), previously separated from other components of the batteries and properly milled up to obtaining a -240 mesh size particles. The addition of Mg was done in the form of scrap, in order to reach for an initial concentration of 1.75 wt. % [10]. After finishing Mg adjustment, mechanical stirrer is placed, which is cylindrical in shape and made of graphite, using a constant stirring rate of 200 rpm. A scheme of the methodology of powders addition is shown in Figure 4. Mn_2O_3 is added depending on the target concentration of Mn needed. This additionis done inperiods of time of 20 minutes each. Treatment time varied from 120 minutes for the case of a concentration of 1 wt. % Mn, to 150 and 240 minutes for the higher Mn concentrations.

3. Results and discussion

This section presents and discusses the results obtained from the manufacture of Al-Mn alloys from recyclable materials (aluminium cans and alkaline battery cathodes discharged) by the aluminothermic reduction process.

3.1. Al-1 wt.% Mn alloys

Chemical analysis results show that the Mn contentin the alloy in creases with in creasing treatment time, suggesting that the aluminothermic reduction of Mn_2O_3 takes place with in the molten metal. Figure 5 shows the variation of the concentration of both Mg and Mn as a function of Mn_2O_3 powders addition time. It is observed that the Mg content decreases from 1.75% to 1.25 wt. %,so the metallothermic reduction reactionis carried out by Al and Mg from the molten alloy, while Mn increased from 0.83 to1.08 wt. %.

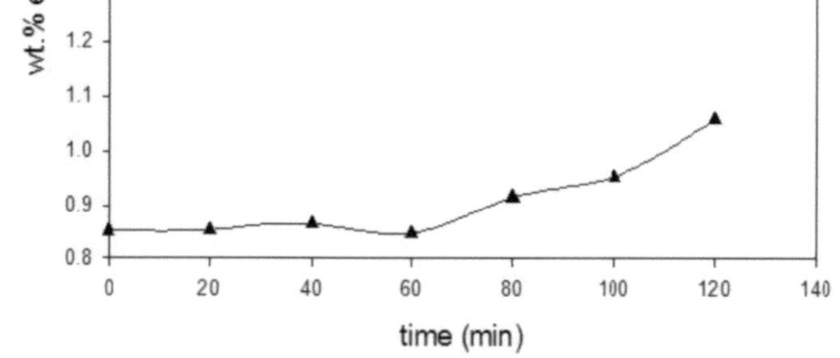

Figure 5. Variation of the concentrations of Mn and Mg,with respect to treatment time, during the metallothermic reduction of Mn_2O_3, for a target Mn concentration of 1 wt. % at constant temperature.

According to the equilibrium diagram Al-Mn-Fe, it is observed that manganese concentrations achieved in these experiments form on solidification the β-Al_6 (Fe,Mn) phase. When the concentration of Mn increases in the alloys, the percentage in area of the Mn rich phase increases. Furthermore, there is a stepped behavior, which can be attributed to the formation of reaction products around the particles acting as a barrier in the metallothermic reduction of Mn_2O_3. According to reactions (8) and (10), the stable compounds that are formed are Al_2O_3 and $MgAl_2O_4$, which surrounds the particles of manganese oxide, causing the metallothermic reduction becomes relatively slow. As this stage is the slowest one during the alumino thermic reduction process, it can be established that it control the diffusion of species such as Al, Mg and Mn through the layers formed. It probably requires more processing time to increase the concentration of Mn in the liquidor an increase in the amount of Mn_2O_3 for the metallothermic reaction.

Figure 6 shows a typical x-ray diffraction pattern of a slag sampled at the end of the experiment described above, where it is evident the formation of such mineralogical species as Mn_2O_3, Al_2O_3 and $MgAl_2O_4$. From this pattern, the most important features are the appearing of spinel and Al_2O_3, corroborating that these are the reaction products according to reactions (8) and (10) given above.

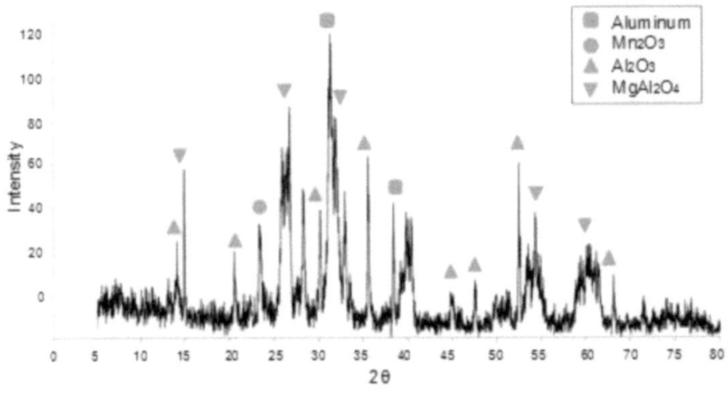

Figure 6. X-ray diffraction pattern of a slag sampled at the end of the experiment where an alloy Al-1Mn was obtained.

3.2. Al-1.25 wt. % Mn alloys

Figure 7 shows the variation of the concentration of both Mg and Mn as a function of Mn_2O_3 powders addition time, when the target concentration of Mn was 1.25 wt. %. It is observed that the Mg content decreased from 1.78 wt. % to 1.35 wt. %, while Mn concentration increased from 0.825 to 1.27 wt. %, as of the aluminothermic reduction reaction. Treatment time was 150 minutes, showing a slightly linear behavior of manganese concentration increase until 110 minutes, noting a sharp increase there after. This confirmed that the aluminothermic reduction of Mn_2O_3 requires more processing time to increase the concentration of Mn in the molten metal.

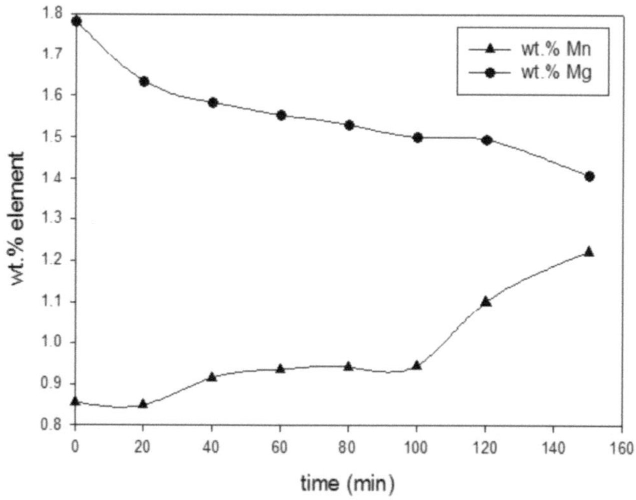

Figure 7. Variation of the concentrations of Mn and Mg,with respect to treatment time, during the metallothermic reduction of Mn_2O_3, for a target Mn concentration of 1.25 wt. % at constant temperature.

Also the reaction product obtained were Al_2O_3 and $MgAl_2O_4$, which are formed around the particles of manganese trioxide, causing that the aluminothermic reduction becomes relatively slow, controlling the diffusion ofthe chemical species such as Al, Mg and Mn through the formed layers.

3.3. Al-1.50 wt. %Mn alloys

Figure 8 shows the variation of the concentration of both Mg and Mn as a function of treatment time, when the target Mn concentration was 1.5 wt. %. From this graph, it is observed that the Mg content decreased from 1.76 wt. % to 1.20 wt. %, while Mn concentration increased from 0.83 to 1.50 wt. % after 240 minutes. It should be mentioned that according to the literature, it can only be added up to 1.5 wt. % Mn, no matterif the Mg concentration is increased to levels around 2 wt. %. In this case, it is established that the maximum concentration of Mn is 1.5%, even when treatment time increased.

Also the reaction products obtained were the stable compounds Al_2O_3 and $MgAl_2O_4$, which are formed around the particles of manganese oxide, having a similar behavior to that presented in the paragraphs previously discussed.

Figure 9 shows the trend in the variation of Mn concentration as of the aluminothermic reduction of Mn_2O_3, where the effect of temperature on reaction rate is evident. As it can be seen from this graph, the higher the temperature, the higher the final Mn concentration reached in molten aluminium. This is attributed to the fact that the reaction depends strongly on diffusion of chemical species such as Al, Mg and Mn to the boundary layer, where chemical reactions are taking place.

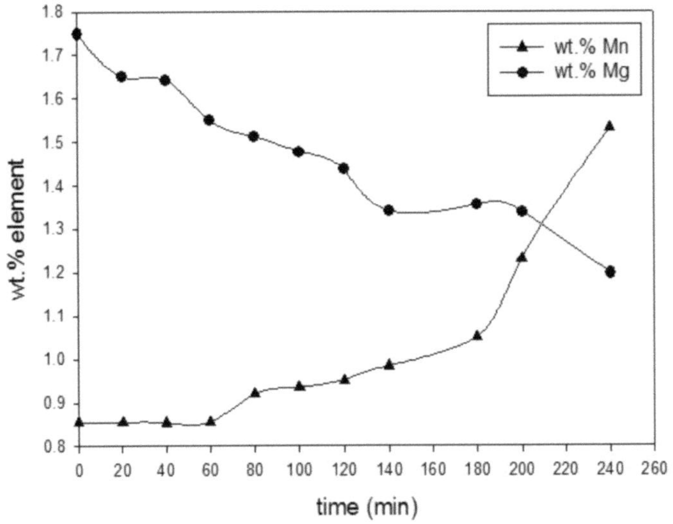

Figure 8. Variation of the concentrations of Mn and Mg, with respect to treatment time, during the metallothermic reduction of Mn_2O_3, for a target Mn concentration of 1.50 wt. % at constant temperature.

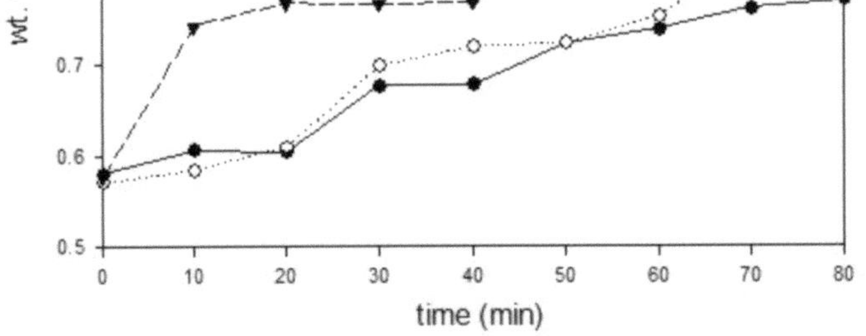

Figure 9. Trend in the variation of Mn concentration as a function of treatment time, showing the effect of temperature.

3.4. Reaction kinetics of the aluminothermic reduction process of Mn_2O_3

The results of Mn concentration increase in molten aluminium as of a function of treatment time for the aluminothermic reduction of Mn_2O_3 in liquid aluminium shown in Figure 9 could be used to determine some values of kinetic parameters of interest. After using the differential method, the reaction orderfor the aluminothermic reduction of $Mn_2O_3$is zero, being represented by the next equation:

$$\frac{d[x]}{dt} = -k(Ordenzero) \tag{11}$$

From the slopes of the linear relationship between Ln (dc_{Mn}/dt) versus Ln (C_{Mn}) shown in Figure 10, the values of the order of reaction were determined.

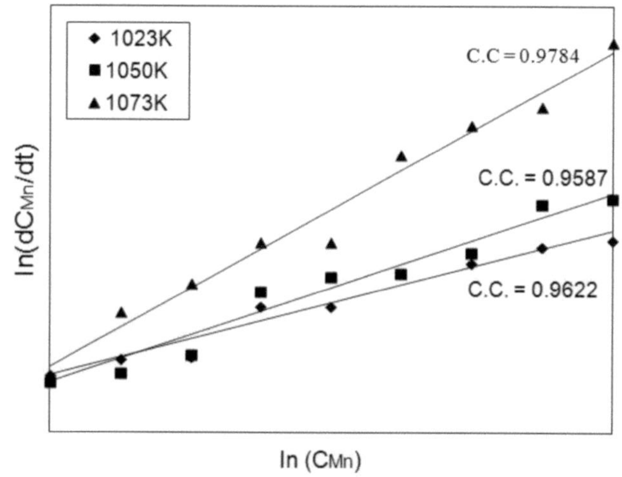

Figure 10. Linear plotof Mncontractionversusreaction time,for determining theorder ofreaction.

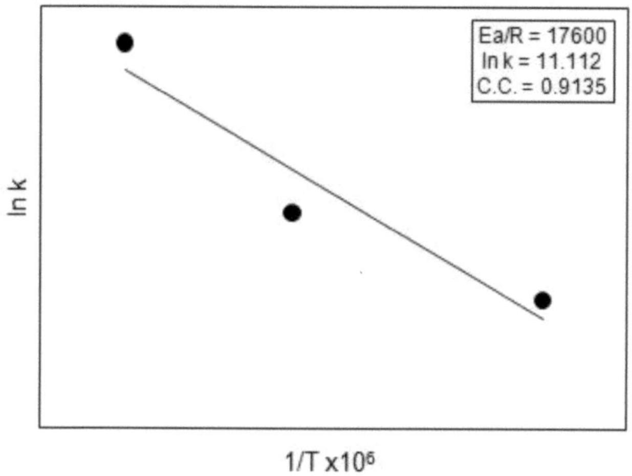

Figure 11. Linear plot of ln k vs 1/ T,for determining the activation energy in the aluminothermic reduction of Mn_2O_3.

Average value of reaction order was 1.0. Furthermore,from this graph, the values of the rate constant(k) could be determined. The rate constant values were also obtained using the differential method.

From the values of reaction rate(k) for each temperature, the activation energy(Ea) using the Arrhenius formula could be determined. Figure 11 shows the linear relationship of ln k versus ln 1/T, resulting in a value of Ea=146kJ/mol.

According to the value of activation energy for the aluminothermic reduction of Mn_2O_3, it can be established that the reactionis governed by diffusive processes [1].

3.5. Micro structures obtained

The incorporation of Mn from the chemically reduced metaloxide to the molten metal was corroborated by optical microscope. Figure 12 shows photo micrographs of alloys with the final manganese concentrations indicated.

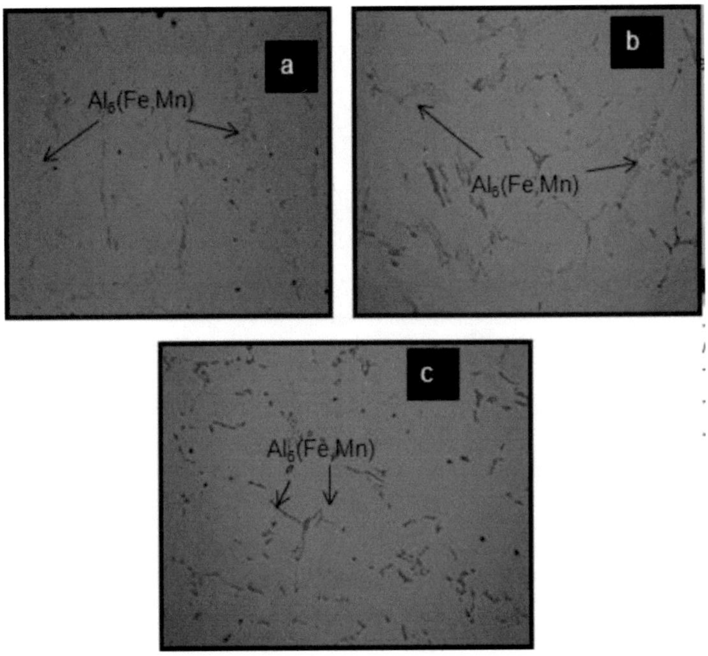

Figure 12. Photo micrographs showing the microstructure of: a) alloy with 1 wt. %Mn; b) alloy with 1.25 wt. %Mn; c) alloy with 1.5 wt. %Mn.

In these cases, the particles shown correspond to the formula β-Al_6(Fe,Mn), that were determined by microprobe analysis (EDS) in the scanning electron microscope (SEM). On solidification, Mn and Fe react with Al to form the β-Al_6(Fe,Mn), acquiring the morphology shown in the micrographs for all three compositions. Figure 13 shows a SEM micrograph corresponding to an alloy with a final Mn concentration of 1 wt. %, including a EDS pattern indicating the elements composing the particles.

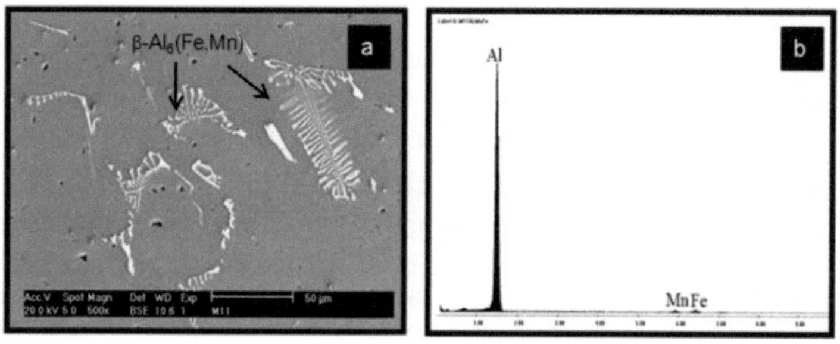

Figure 13. SEM micrograph showing the morphology of: a) β-Al_6(Fe,Mn): EDS pattern of the Mn rich particles, in an alloy with a final concentration of 1 wt. %Mn.

In Figure 13 it is observed the presenc particles of the β-Al_6(Fe,Mn) phase, as they can be found in the three different alloys. This phase in the Al-Fe-Mn system is formed around 644°C by an eutectic reaction. The particles of β-Al_6(Fe,Mn) remaining in the melt due to incomplete reaction, are dissolved during the homogenization treatment, resulting in the more stable phase $Al_{15}(Mn,Fe)_3Si_2$ on cooling. For this, silicon plays an important role in precipitation kinetics.

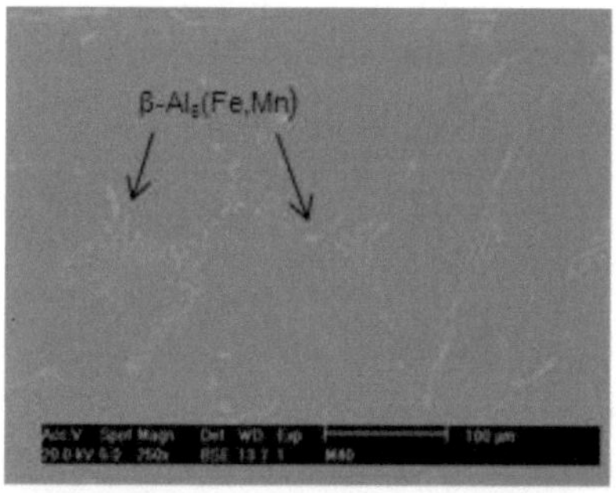

Element	Al	Fe	Mn
wt.%	59.3	20.1	20.5

Figure 14. Micrograph obtained by SEM showing the morphology of the particles of the β-Al_6 (Fe,Mn) phase, in analloy Al-1 wt. % Mn in the as cast condition.

The microstructure of Al-Mn alloys obtained by aluminothermic reduction process of $Mn_2O_3$was analyzed by the technique of scanning electron microscopy, in order to observe the morphology of the microconstituents in the solidified alloy, and estimating the semi-quantitative composition of the phases present. Figure 14 shows a micrograph of the alloy with 1% Mn, where it is observed that the second phase particles are β-Al_6(Fe,Mn), homogeneously distributed in the interdendritic spaces. It includes the semi-quantitative chemical analys is results determined by EDS in the SEM.

3.6. Heattreatment of Al-Mn alloys

The diversity of the demandson mechanical properties of the semi-finished parts made of aluminium alloys predetermines the choice of heat treatment. In some cases, they require high values of mechanical strengt hand high ductility. In other cases, high ductility is required to ensure good cold deformation to produce specific parts. The correct choice of the types of heat treatment allows obtaining products whose mechanical properties need to be optimized. In the as-cast condition the dendritic structure is not uniform, with precipitation of inter metallic phases in the interdendritic spaces and between the grain boundaries, giving very poor mechanical properties. According to the literature [15,16], the precipitates formed are mainly of the β-Al_6 (Fe, Mn) type, to a lesser extentof the α-Al (Fe,Mn)Si phase, and even some compounds of the Al-FeMnSiCu type.

The homogenization treatment for Al-Mn alloys obtained by the aluminothermic reduction process of Mn_2O_3, was carried out in aThermolyne brand Model 4500 furnace at 550°C for a holding time of 15 hours.

The homogenization annealing treatment eliminates chemical segregation and large solidification crystals produced by a faulty practice. Moreover, the homogenizationan nealingis used often to achieve a more uniform precipitation of Mn and Fe rich particles, which influence the behavior of the material during recrystallization and hot forming. It is performed at elevated temperatures, near the melting temperature of the alloy, and is mainly applied to alloys with susceptibility to segregation [16].

In Al-Mnalloys, casting structureis characterized by a heterogeneous solution, with primary particles of the -Al_6(Fe,Mn) phase and minor amounts of α-Al (Mn,Fe)Si particles. A homogenization treatment before hot rolling is necessary to reduce the concentration of manganese and control the size, density and distribution of particles, as they affect recrystallization, texture and mechanical properties of the alloy informing stages. The addition of alloying elements in the alloys Al-Mn has a strong influence on the behavior of the precipitation. Fe and Sigreatly decreasethe solubility of Mn in solid solution, accelerating the velocity of precipitation. Fe favors the precipitation of (Fe, Mn) rich particles, while Si favors precipitation of the phase α-Al_{12}(Fe,Mn)$_3$Si in the alloy [2].

The α-phase particles have an influence on the formability of the alloy,and has been found that during homogenization treatment, particles of the β-Al_6(Mn, Fe) phase are transformed into α-Al_{12}(Fe,Mn)$_3$Si [15]. According to Luigi [17], the α-Al_{12}(Fe,Mn)$_3$Si particles starts to precipitateon cooling at about 400°C. The disappearance of the β phase precipitates occurs because of silicon diffusion into the matrix, nucleating subsequently the precipitates of the α-Al_{12}(Fe,Mn)$_3$Si phase [18,19]. On heating, this transformation has a peak at the temperature of 560 °C, temperature at which the β-Al_6(Fe,Mn) particles starts its dissolution [2].

Figure 15 shows SEM micrographs oft he microstructure of analloy Al-1% Mn after being subjected to homogenization treatment. In these micrographs it is observed that the bright portion corresponds to the phase β-Al_6(Fe,Mn), which by effect of heat treatment are transformedin to α-Al_{12}(Fe,Mn)$_3$Si phase particles. However, complete transformation was not attained, because the treatment temperature was 10°C below the the oretical temperature of transformation and, therefore, it requires even more time to reachall of the transformation. Besides the micrographs, it also is included EDS patterns indicating the semi-quantitative chemical composition of these phases, after homogenization treatment.

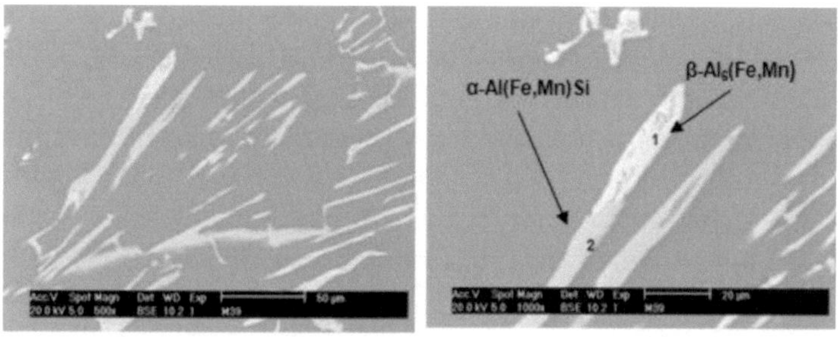

Phase	Al (wt.%)	Mn(wt.%)	Fe (wt.%)	Si (wt.%)
1	59.3	20.1	20.5	---
2	47.2	23.9	24.3	4.53

Figure 15. SEM micrograph of an Al-1 wt. %Mn alloy showing the morphologyof the β-Al_6(Fe,Mn) and α-Al_{12}(Fe,Mn)$_3$Si phases particles after a homogenizing treatmentat 550 °Cfor 15 hours.

The presence of elements such as ironand manganese in the supersaturated solid solution, or affinely dispersed precipitates of small size, difficult subsequent recrystallization processes by inhibiting the movement of atoms and dislocations. Furthermore, the presence of iron or higher contents of silicon, decrease the amount of supersaturated manganese in solid solution, facilitating its separation by forming different compounds [20].

Simple	σ_0(Mpa)	UTS (Mpa)	El (%)
As-cast	113	181	4
Homogenized	78	155	11

Table 5. Mechanical properties values the alloy Al-1% Mn obtained by metallothermic reduction of Mn_2O_3, in the as-cast condition and after homogenization heat treatment.

Additionally, Table 5 shows the values of yield stress (σ_0), ultimate tensile stress (UTS), and % elongation, in order to establish the effects of homogenization treatmenton the mechanical properties of the as-cast alloys. The transformation of the β phase to α phase cause dare duction in the values of σ 0 and UTS, and a significant increase in elongation affinely percentage,indicating that the material will have good mechanical behavior during forming.

From the results shown above, it can be stated that the Al-Mn alloy sobtained by aluminothermic reduction of Mn_2O_3 must be submitted to a homogenization treatment before its laminationor other deformation process.The phases obtained after homogenization treatmentand the values of the parameters measured indicate that these alloys could have the same behavior in mechanical forming than those obtained using pure elements.

4. Conclusions

It is possible to obtain Al-Mn alloys, with concentrations of manganese ranging from 1to 1.5 wt.%, using the aluminothermic reduction of Mn_2O_3 powders produced from alkaline batteries cathodes. Molten metal is obtained using aluminium cans which act as Mn_2O_3 reducer. Aluminothermic reductionis carried out effectively at 800 °C, allowing incorporating up to 1.5 wt. % of Mn in the solidified alloy. The cost effectiveness of the process is of beneficial interest, as 75% of savings in the purchase of raw materials for producing alloys from pure materials could be achieved. The Mg contentin aluminium cans play an important role during aluminothermic reduction, as it has direct influence on the wettability of the particles and the chemical reaction rate. The presence of elements such as Fe and Si, influence the behavior of the alloy on solidification, as well as in the homogenization heat treatment. The as-cast structure consists primarily of particles of the β-Al_6(Fe,Mn) phase, which during the homogenization treatment is transformed into the soft phase α-$Al_{12}(Fe,Mn)_3Si$, causing that the alloy can be deformed appropriately. The typ alloys obtained in this work must be submitted to a homogenization heat treatment before its laminationor other process of deformation. The phases obtained after homogenization treatmentand the values of the mechanical properties measured indicate that the Al-Mn alloy obtained by metallothermic reduction of Mn_2O_3 could have the same behavior in mechanical forming than those obtained from pure elements.

Acknowledgements

The authors wish to thank the National Council of Sciences from México (CONACYT), for the funds provided, project CB-2007-01-81251 and to Fundición J.V. S.A. de C.V. from Ramos Arizpe, Coahuila, México, for the technical facilities.

Author details

A. Flores Valdés *, J. Torres and R. Ochoa Palacios

*Address all correspondence to: alfredo.flores@cinvestav.edu.mx

Centro de Investigación y de Estudios Avanzados del IPN, Unidad Saltillo,, México

References

[1] Schlesinger, Mark E. (2007). Aluminium Recycling. Boca Raton, Fl, CRC. Press, Taylor and Francis Group.

[2] Bahadur, Aruna. (1988). Intermetallic phases in Al-Mn alloys. *Journal of Materials Science*, 23, 48-54 .

[3] Li, Y. J., & Arnberg, L. (2003). Quantitative study on the precipitation behavior of dispersoids in DC-cast AA3003 alloys during heating and homogenization. *Acta Materialia*, 51-3415.

[4] Carceaga, J. A. (1993). Manejo y reciclaje de residuos de envases. SEDESOL 4 www.sedesol.gob.mx (accessed1993).

[5] Moreno, C. (2008). Las latas, los envases ecológicos por excelencia. *Revista Mexicana del Empaque, Embalaje y Plásticos*, 313.

[6] Vásquez, R. O. (2011). Últimas noticias del aluminio en México y el mundo. *Revista Aluminia*, 16, 4-12.

[7] Castro, J., Díaz, M. L., & Díaz, Castro. (2004). Contaminación por pilas y baterías en México. *SEMARNAT. Instituto Nacional de Ecología.*, www.ine.org.mx, 5-73.

[8] Torres, Cabral. L. G. (2008). Recuperación de zinc y manganeso de baterías alcalinas descargadas utilizando lixiviación. *Master´sthesis.*, CINVESTAV-IPN, Unidad Saltillo, México.

[9] Cortinas, C. (2008). Base para Integrar Planes de Manejo de Pilas y Baterías Eléctricas a Base de Mercurio o de Níquel-Cadmio. *Editada por la Red Queretana de Manejo Ambiental de los Residuos*, 1-12.

[10] Langlais, Harris R. (1991). Strontium Extraction by Aluminothermic Reduction. *Master´s thesis*, McGill University Montreal, Canada.

[11] Levenspiel, O. (2002). Ingeniería de las Reacciones Químicas, 2da Edición, Barcelona España

[12] Neff, D. V. (1992). Paper presented at The Benefits of Flux Injection Combined With Rotor Dispersion for Producing Clean Aluminium Casting Alloys: proceedings of 3rd International Conference on Molten Aluminium processing, Orlando-Florida, USA.

[13] Davis, J. R. (1990). Metals Handbook. 9th Edition. Nonferrous Molten Metal Process. ASM U.S.A. , 15

[14] King, F. (1987). Aluminium and its Alloys. Chichester, Ellis Horwood Limided.

[15] Salvador, M. D., Amigo, V., Reig, L., Bloem, C., Carsí, M., & Caruna, G. (2007). Evolución microestructural y mecánica de una aleación de aluminio de la serie 3xxx obtenida por el proceso Hazelett. *Revista de metalurgia,*, 43(6), 424-433.

[16] Qiang, Du., Poole, Warren J., Wells, Mary. A., & Parson, Nick. (2011). Microstructural modeling of the homogenization heat treatment for AA3XXX alloys. *JOM*, 63(7), 35-39.

[17] Luiggi, A., & Ney, J. (2010). A Preliminary Study of the Phase Transformations in Rolled Al-Fe-Si Alloy. *Metallurgical and Materials Trasactions A*, 41(13), 3271-3275.

[18] Kamat, R. G., & Saimoto, S. (1994). Microestructure and Texture evolution in continuously cooled AA 3004 hot rolled sheet. *Materials Science Technology.*, 10(3), 215-221.

[19] Langlais, Alexander D.T.L. (2011). Solid-state intermetallic phase tranformations in 3XXX aluminium alloys. *Acta Materialia*, 50, 2571 -2583.

[20] Li, Y. J., Mueggerud, A. M. F., Olsen, A., & Furu, T. (2012). Precipitation of partially coherent α-Al(Fe,Mn)Si dispersoids and their strengthening affect in AA 3003 alloy. *Acta Materialia*, 60, 1004-1014.

Section 5

Application of Aluminium Alloys in Solar Power

Chapter 13

Aluminium Alloys in Solar Power – Benefits and Limitations

Amir Farzaneh, Maysam Mohammadi,
Zaki Ahmad and Intesar Ahmad

Additional information is available at the end of the chapter

http://dx.doi.org/10.5772/54721

1. Introduction

It has been widely accepted that fossil fuels have played significance roles in human's daily life and industrial developments. However, fossil fuels are associated with some problems like air pollution. Moreover, fossil fuels have limited sources in the world that will be finished in near future if the consuming rate of fossil fuels does not decrease. So, scientists have been encouraged to find suitable sources of energy as replacements for fossil fuels.

Sunlight, wind, see waves and hydrogen are examples of renewable sources of energy that have been subjects of researches for many years.

Among all of he mentioned renewable sources of energy, the sun is the most abundant and most renewable source that provides the energy for human much more than they need. If we can collect 0.01% of the sun's energy reaching the earth, it would be more than energy that all human use today.

Amount of the energy that reaches from the sun to a specific location on the surface of the earth can be 1000 W/m^2 [1]. Solar energy is had been received great world wide attention during the last decades as the most ideal renewable source of energy, which is mainly due to the points that this energy is safe, clean, free and unlimited [2]. Basic of fusion reaction in the sun and solar energy production have been discussed in many monographs [3-6].

The sun sends out huge quantity of electromagnetic radiations to the earth, which transfer approximately 4000 trillion kWh of energy to the earth surface each day [7]. This amount of energy is much higher than human's demand and also more than any other source of energy on the earth like nuclear power or fossil fuel burning. Solar energy is associated with some economic and environmental advantages, and has been considered as replacement source of energy for traditional sources of energy [8].

History of employing solar energy goes back to 212 BC, when Archimedes. The Greek scientist used metallic mirror to burn a Roman fleet [9]. Cooking food, heating water and home, and drying grains were the first applications of the sun's energy [10-12]. In 18th century, solar furnaces were constructed by glass lenses, mirrors and polished iron, which were capable to melt metals like iron and copper. During 19th century, solar energy was used to operate steam engine and convert solar to electrical energy [13].

Nowadays, solar energy are used in wide range of industrial, business and residential applications such as electricity generation, water heating, industrial processes, daylighting, heating and cooling [1]. As a consequence of rapid development of the solar power technologies, it is expected that solar systems will provide 12% to 25% of global electricity by 2050 [8].

Growing demand for heat and electricity generation in developed and developing countries causes rapid development in solar power generation systems. Quality and physical properties of materials that are used in solar systems determine the efficiency of each solar system. Different materials are used in various kinds of solar power systems such as glass, silver, steel, stainless steel and aluminium.

Among all of the mentioned materials, aluminium has special properties that make it an interesting material for many solar power companies. Light weight, high strength, proper corrosion properties, high surface reflectivity, excellent electrical and thermal conductivities, as well as special optic properties of its anodic coating are such as interesting properties of aluminium that make it inseparable part of solar power systems.

To sum up, aluminium plays an important role in various kinds of solar power systems include concentrating solar power (CSP), photovoltaic solar power (PV) and solar thermal collections. The application of aluminium and its alloys in these solar systems are explained in this chapter. Besides, its economical effect and future market are explored here.

2. Aluminium applications in solar power systems

In order to find the role of aluminium and its alloys in solar power systems, it is necessary to review different types of solar power plants, their properties, requirements and applications. Generally, solar power systems are divided into three widely used categories, which called concentrating solar power (CSP), solar thermal absorbers and photovoltaic solar cells (PV). Aluminium alloys have became a significant and inseparable part of each of the mentioned group of solar power systems, mainly due to special properties of aluminium and its alloys. Properties and applications of each kind of the mentioned solar power systems as well as the role of aluminium alloys in each of them will be discussed separately.

2.1. Concentrating solar power (CSP)

Concentrating solar power systems include reflector materials that concentrate heat energy .of the sun to a point or line to generate steam in a boiler, drive steam turbine and produce electricity [14-36]. Generation electricity, however, is not the only application of the CSP systems. Concentrated solar energy is suitable source of energy that can be used in wide range of materials processing such as producing metallic foams [37], synthesis of nanomaterials like carbon nanotubes and ceramic nanoparticles [38], fast heating of ceramic materials close to thermal shock [39], breaking down metal oxides to its metal counterpart [40-42], surface treatment [38] and metal sintering [43].

Cost of energy generation by CSP is much lower than that of PV, which is mainly due to higher average efficiency of CSP (42% compared with 15% for PV) and requires smaller field to produce certain amount of energy [8,44]. The cost of energy generation by solar collection devices is also lower than thermal solar collecting systems[45].

CSP has facilitated a system for production of energy which is neither noisy nor toxic. Therefore, this system can be used in cities without any safety problem [2,44].

Based on the focus geometry and receiving technology, CSP has been divided into four types.

2.1.1. Parabolic trough

In this kind of CSP, a series of curved mirrors concentrate sunlight on to a tube that located in the trough's focal line. This tube contains oil and is used as heat transfer medium. Temperature of the oil in the tube can reach as high as 400°C [46]. Figure1 shows a real and a schematic of a parabolic trough CSP system. This system was used in 1912 in Egypt for the first time [8]. The parabolic trough system is able to concentrate sunlight by 70-100 times and transfer solar to electrical energy with efficiency of 15% [47].

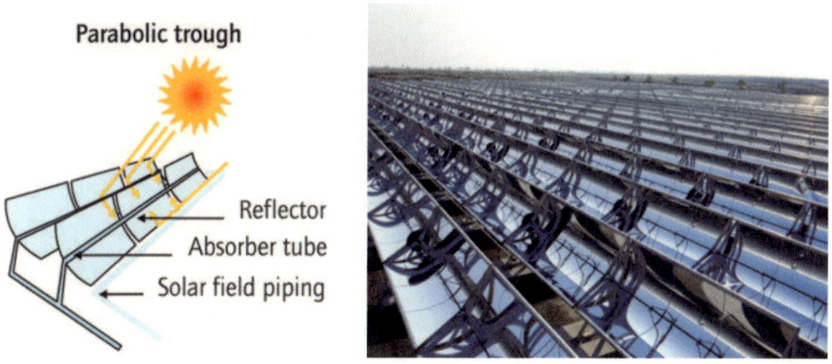

Figure 1. Parabolic trough concentrating solar power plant [47]

2.1.2. Solar tower

Solar tower plants use many reflectors called heliostats to collect sunlight reaching a field to a certain point on top of a tower, where a collector is located (Figure 2). The concentrated energy can generate electric energy with efficiency of 20-35% [8,46].

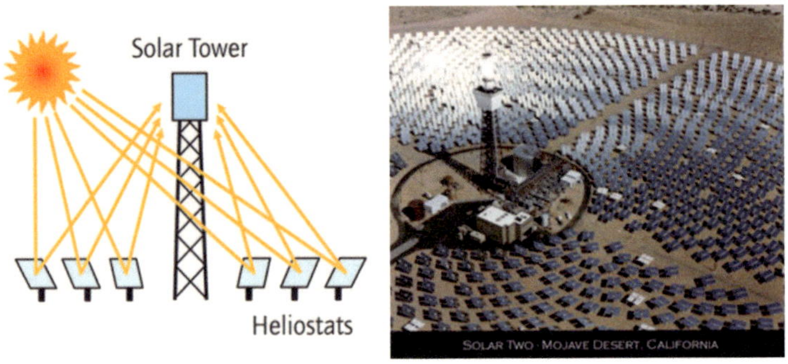

Figure 2. Solar tower concentrating solar power plant [47].

2.1.3. Parabolic dish

Parabolic dish concentrating solar system uses a reflector dish and concentrate sunlight onto its focal point. Figure 3 shows the schematic of this solar system. A receiver that located in the focal point of the dish can increase the temperature of the gas or fluid up to 750°C in a sunny day. Capacities of parabolic dish plants and their efficiency in changing solar to electrical energy are in the range of 0.01-0.4 MW and 25-30%, respectively. Due to its design, its optical efficiency is significantly higher than two other mentioned categories of CSP plants. Parabolic dishes are currently used in some simple application like cooking oven [8,13,46].

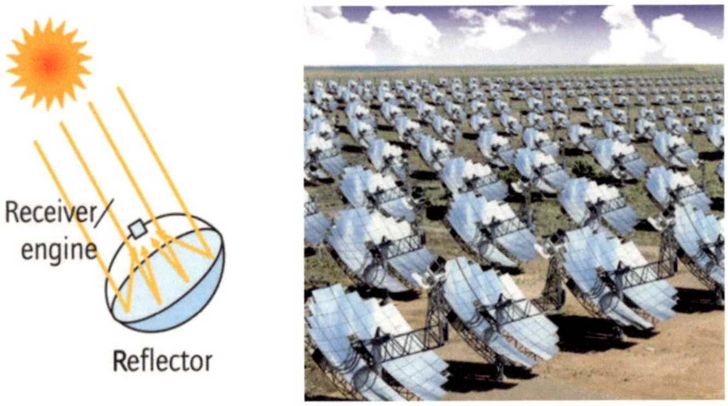

Figure 3. Parabolic dish concentrating solar power plant [47].

http://dx.doi.org/10.5772/54721

2.1.4. Linear fresnel

Linear Fresnel is a line-focus collector that consists of series of Fresnel lens behaving mirrors (Figure 4). Capacity of this solar concentrator is 10-200Mw and its efficiency to generate electric energy is between 8 and 10% [8].

Table 1 compares various properties of the mentioned CSP plants.

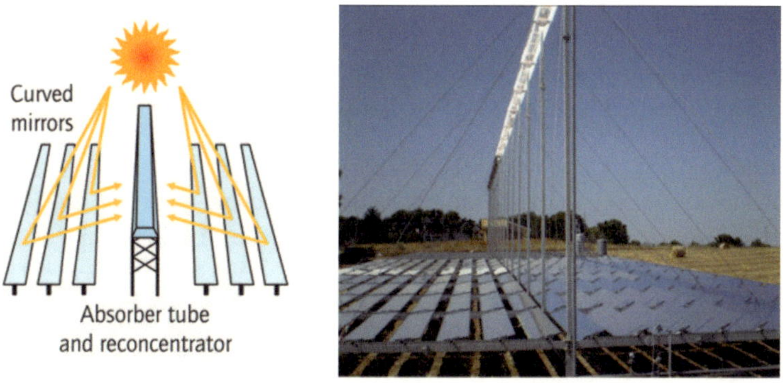

Figure 4. Linear Fresnel concentrating solar power plant [47].

	Capacity (MW)	Concentration	Annual solar to electric conversion efficiency	Cost (Dollars /ft^2 field)
Parabolic trough	10-200	70-80	15%	40
Solar tower	10-200	300-1000	20-35%	45
Parabolic dish	0.01-0.4	1000-3000	25-30%	
Linear Fresnel	10-200	25-100	8-10%	20

Table 1. Properties of CSP technologies [8,46,47]

Currently, capacity of installed concentrated solar power plants is 516MW in the world. 90.2% of this capacity belongs to parabolic trough systems, followed by 8.5% for solar tower, 0.8% for Fresnel trough systems and 0.24% for solar dishes [48].

Reflectors of concentrated solar power should have some properties, such as high reflectance ratio and low absorption, good wetability and low cost. Aluminium and silver are the most common reflectors for CSP systems due to their high reflecting efficiency in solar wavelength range [45].

Reflectors of concentrated solar power should have some properties, such as high reflectance ratio and low absorption, good wetability and low cost. Aluminium and silver are the most common reflectors for CSP systems because of their high reflecting efficiency in solar wavelength range [45].

Aluminium has some special properties that make it a useful mirror in various applications of solar cells, lasers and astronomer's instruments [2].

For example, aluminium can be deformed easily to have the best shape of reflectors and achieve the highest concentrating efficiency. Unlike glass mirrors, aluminium reflectors can not broken easily, which is a favourite property for outdoor applications [49].

Aluminium mirrors not only have better surface reflectivity than glass mirrors, they are much lighter. Compared to glass mirrors that have average weight of 11kg/m^2, aluminium reflectors have only weight of 7 kg/m^2.

Due to mechanical properties of aluminium and its low cost compared with silvered glass mirrors, aluminized reflectors found applicability to high temperature solar concentrating technologies [50].

Rolled aluminium also can be suitable for certain solar energy applications since it is cheaper than other reflector materials and can be cost-effective material in this application [45].

Thermal evaporation is one of the most practical methods to prepare aluminium reflector in order to use in concentrated solar power systems [2].

Ling et al. [2] studied performances of aluminium reflectors produced by thermal evaporation method on different substrates include galvanized iron, acrylonitrile butadiene styrene (ABS) and aluminium alloy. Experimental results clarified that reflection of thermally evaporated aluminium on ABS is comparable with that of silver mirror of ultra-white glass. It was also found that smoothness and roughness of the substrate have important effects on optical properties of the aluminium reflectors.

2.2. Solar thermal collectors

Solar thermal collector is a kind of solar power system that transforms solar energy from the sun rays into thermal energy. This solar system is widely used for generation of hot water, home heating and electricity generation [1,48].

Based on the kind of used collectors, solar thermal collecting systems divided into three types [1,13,48]:

2.2.1. Unglazed plastic collectors

This technology of solar thermal collectors provides a low cost heat and is used only for public bath heating and hay dying. Aluminium and its alloys have approximately no special application in unglazed plastic collectors.

2.2.2. Flat-plate collectors

Lower radiation and convection heat transfer losses in this solar thermal collectors compared to unglazed plastic collectors enables it to warm up to higher temperature. This system is widely used for space heating and hot water generation.

2.2.3. *Evacuated tube collectors*

Due to special design of this kind of solar thermal collectors, its conductive and convective losses are very low. So, it is capable to warm up the heat carry fluid up to 150°C.

According to figures recently released in Solar Heat World-wide report, the amount of energy that was generated by solar thermal collectors in the world in 2007 was approximately 147 GW. Fabrication of this amount of energy by solar thermal collectors requires approximately 210 square kilometres. Solar heating capacity of these solar systems had increased by 15% in the year of 2008, and became double compared to 2004. Capacity of solar thermal collectors does not equally divided among the mentioned solar thermal categories. According to published figures, approximately 50% of total amount of energy resulted from solar thermal collectors has been generated Evacuated tube collectors. Contributing proportions of Flat plate and Unglazed plastic collectors are 33% and 17%, respectively.

As mentioned before, Unglazed plastic collectors do not provide significant opportunity for aluminium usage. However, both of the other groups use aluminium and its alloys in different parts.

Figure 5 shows a Flat-plate collector and introduces its various parts; casing, absorber and frame. Aluminium, copper and steel are the materials that are used for absorbers. Casing and frames are usually made of aluminium and steel. However, aluminium is predominantly used because of lower weight of this alloy in that that of steel. Moreover, using aluminium as absorbers is growing. Special optic properties of anodic layer of aluminium and some aluminium alloys make aluminium a useful material for solar absorption. These qualities will be explained later. Today, approximately 35% of the solar absorbers are made of aluminium [1,13,48].

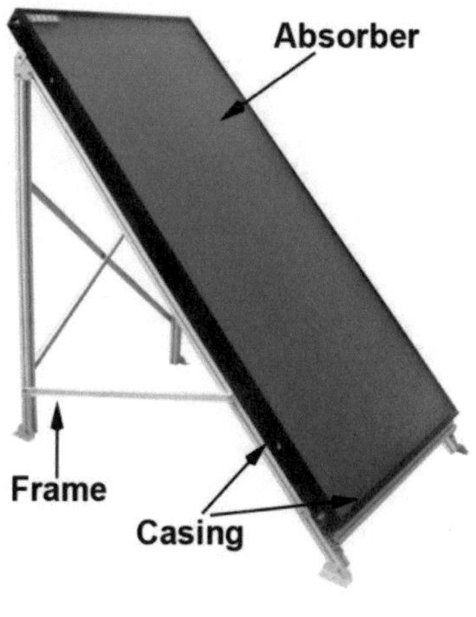

Figure 5. Different parts of a Flat-plate collector

Figure 6 shows main components of Evacuated tube solar thermal collectors, which are absorber, frame, heat pipes, header pipe and casing. Like what was mentioned for flat-plate collectors, using aluminium as absorber is growing. Low density of aluminium satisfies solar companies to use aluminium alloys for frames instead of stainless steel. Aluminium is also widely used in casing and header pipes [1,13,48].

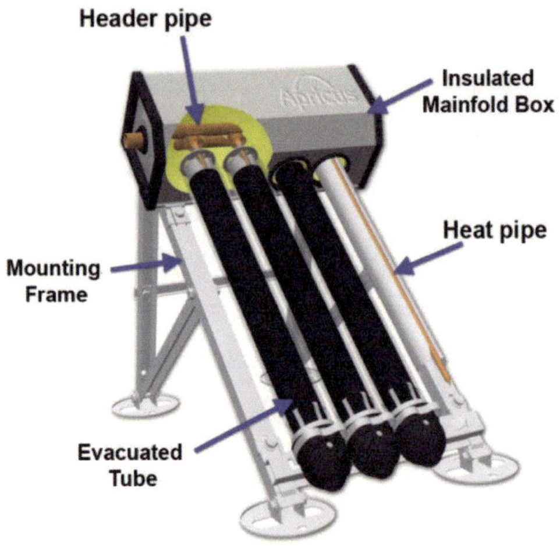

Figure 6. Different parts of a Evacuated tube solar thermal collectors

As mentioned before, aluminium is one of the most important materials utilized in solar absorbing, which is mainly due to special structure of the anodic layer that can fabricated on aluminium surface by anodizing process. This anodic layer is porous; with various pore sizes depend on the kind of anodizing electrolyte and anodizing process conditions [51-56].

It has been shown that these pores can be filled with metals by electrodeposition to have a coloured surface layer composed of metal particles in alumina dielectric matrix [57,57,58]. Using some special metal particles or ions, like nickel, provides suitable optical properties in anodized layer of aluminium for solar absorption application [57,59]. It was found that surface reflection of coloured aluminium anodized layer, as an undesirable property of solar absorbers, is a function of colouring time. It also was shown that reflectance decreases with electrolytic coloration time [57].

The effect of coloring time on solar absorption of aluminium anodized layer is more obvious in thicker anodic layers [53].

The proposed model of coloured and sealed anodized aluminium layer is shown in Figure 7. This model shows that the aluminium anodic layer composed of three parts, which have different properties. The top region (part 1) is protective layer that composes of Al_2O_3 and aluminium hydroxide; the middle region (part 2) is optical absorption layer that filled with metallic pigments; and the lowest region (part 3) is barrier layer that is a compact layer and has the highest density among all of the mentioned parts [57].

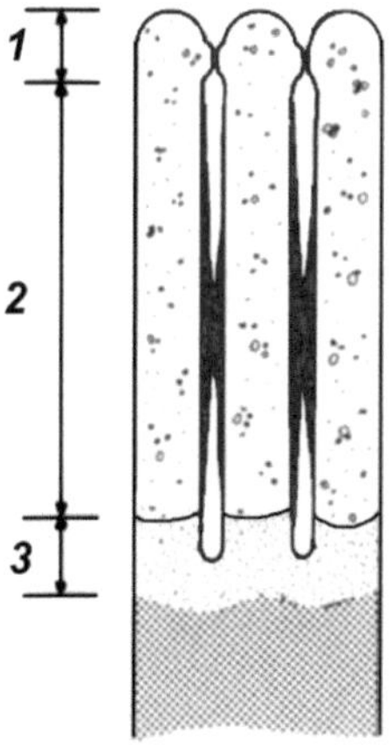

Figure 7. Proposed model of coloured and sealed anodized aluminium layer [57].

Absorption mechanism of solar radiation in colored anodic aluminium layer has been described by Granqvist [58].

For a photothermal conversion, the highest efficiency will be provided when the thermal losses by surface radiation are low enough, and solar absorption is high.

Microstructure of an alloy plays an important role on its efficiency as an absorber. As shown by Cody and Stephens [60], surfaces that its dielectric constant change gradually from the material/air interface to that of solid materials have low reflections.

Eutectic binary aluminium alloys such as Al-6wt.% Ni, Al-33wt.% Cu and Al-7.5wt% Ca have such microstructure and have acceptable optic properties to be used as absorber; i.e. low reflection high absorption [61,62].

Mechanical and thermal stability of the mentioned aluminium alloys, as well as possibility of regeneration of their surface by retching make them useful materials in this application [61].

2.3. Photovltaic solar system

Photovoltaic cells directly convert solar to electrical energy using semiconductor materials. Semiconductors can generate free electrons using energy of sunlight [63]. Photovoltaic property of materials had been discoveres by Becquerel in 1830, when he found this effect in Selenium [13]. Various aspects of photovoltaic solar systems have been reported in different books and references [64-72].

Apace was the first application of photovoltaic solar cells because sun is the only source of energy in space [13]. Photovoltaic solar cells have been used in wide range of applications include water pumping, solar home systems, remote building, solar cars and airplanes, satellites and space vehicles. Such a vast variety of applications is the main cause for increasing demand for photovoltaic solar cells [63].

Photovoltaic solar power system has experienced great development rate. The capacity of this solar system had increased from 0.1 GW in 1992 to 2.8 GW in 2007 and 5.95 GW in 2008. Table 2 compares the market demand of PV in some countries in 2007 and 2008 [1].

Countries	USA	Japan	Germany	Spain	Rest of Europe
Demand in 2007 (MW)	226	226	1328	650	170
Demand in 2007 (MW)	360	230	1860	2460	310

Table 2. Photovoltaic solar system market in different regions of the world [1]

Various advantages of photovoltaic solar systems make it a favourite technology in many industries. It has been observed that PV is one of the fastest growing industries in the world. This rapid growth requires new developments with respect of applicable structural materials [73].

Construction and structure of photovoltaic solar systems are the main part of this system that can be made of aluminium. Steel and aluminium are the most common materials that are used in construction of solar power systems.

However, the advantages of aluminium alloys over steel, other aluminium alloys and composite materials make it the core material in building of large scale solar generation fields. Significant proportion of the cost of solar generation system is related to supporting materials and frames. For instance, approximately 25-30% of the budget of CSP plants should be allocated to frames and support materials.

Some of the aluminium alloys can perform same as steel, but its weight is one-third of steel. Although aluminium alloys are more expensive than steel, using aluminium instead of steel can be economical. Lower density of aluminium gives the opportunity of easier, faster and cheaper transportation. Moreover, construction of a solar field would be faster if extruded aluminium is used instead of steel because it does not need crane and permanent joining process like welding. These properties of aluminium enable engineers to design and produce complex, efficient and stable structures.

6061 aluminium alloy that contains magnesium and silicon alloying elements is an example of useful aluminium alloys for structure of solar plants. This aluminium alloy is widely used in solar fields because of its high strength and machinability[74].

Another advantage of aluminium over steel is its higher corrosion resistance in outdoor environments, even if steel is galvanized. Even though aluminium is more chemically and electrochemically active than steel, a thin oxide layer that naturally formed on the aluminium surface in the air provides suitable protection for aluminium and enables it to have good performance for a long time. To illustration, consider the aluminium cap used in Washington monument was corroded only 0.13 mm after 73 years performance. The mentioned oxide layer is stable in pH range between 4.5 and 8.5. So, it may not provide enough protection if the aluminium is located in soil. However, most of solar collectors are mounted on concrete pads, which make the aluminium performance independent of soil conditions [74].

Since using solar power for electricity generation has became a serious competition among various companies, designing and using the most possible available, effective and efficient materials became very important. Extruded aluminium can be considered as one of these effective materials as it enables companies to create next generations of solar power plants with long life time and very low negative environmental effects.

PV inverter, which changes direct current to alternative current, and panel frame are the other components of a photovoltaic solar system that can be made of aluminium

Approximately 72% of aluminium input in photovoltaic solar systems is used in construction, while the proportion of aluminium used in panel frames and inverters are 22% and 6%, respectively [48].

2.4. Perspective of aluminium applications in solar power systems

Currently, CSP systems use approximately 55000 kilograms of aluminium per one megawatt generated energy, while used aluminium for photovoltaic cells is 45000 kg/MW.

CSP provides over 1000 MW of worldwide electricity, which looks to reach to 15000 MW in near future regarding to new solar projects in US, Spain, China, Morcco and India. Building these solar fields with the mentioned total capacity by aluminium frames requires 1080 million pounds of extruded aluminium [75].

If it can be assumed that the proportion of extruded aluminium that is used in construction of CSP plants remains 34%, the amount of the required aluminium for CSP plants will be approximately 635,000 tons in the next 20years [75].

Today, the amount of energy that is generated by CSP plants is approximately 0.5GW. Predictions reveal that capacity of solar collecting power plants will be 30GW in 2020, 140GW in 2030, and 800GW in 2050, which show a very rapid growth.

Based on these predictions and estimation of average use of aluminium, total amount of used aluminium in CSP plants will be 1.1 and 8 million tons in 2020 and 2030. The average amount of aluminium used in CSP plants in 2050 will be 51 million tons, which has the potential to be double. As a result, approximately 0.3% and 1.9% of annual aluminium production will be used in CSP plants in the decades of 2010-2020 and 2020-2030, respectively. This proportion will be 5.7% in average for the period of 2031-2050 [48].

Today, extruded aluminium used in photovoltaic solar plants is approximately 12% of total amount of aluminium that are used in this kind of solar power plants. If, like what mentioned in future market of CSP plants, it is assumed that this proportion remain constant in future, approximately 1,500,000 tons extruded aluminium will be used for these systems in the next two decades [75].

As mentioned before, these calculations are based on the assumption that the proportion of used aluminium in solar systems will not increase. Considering the growth of aluminium usage in solar systems during the last years, however, clarifies that the solar industries prefer to use extruded aluminium instead of steel frames. Consequently, demands for aluminium related to steel will increase in the course of time.

According to the report of International Energy Agency published in 2010 [47], about 14GW energy are produced by photovoltaic solar system in the world. It also is predicted that the average capacity of this solar system will be 87GW in 2020, 225GW in 225 and 597GW in 2050. Based on this prediction, total amount of aluminium used in photovoltaic solar system will be 3, 7 and 19 million tons in 2020, 2030 and 2050, respectively. Consequently, 0.64% of total annual aluminium production will be used in PV systems in decade 2010-2020, which will reach to 1.21% in decade 2020-2030 and 1.63% in period of 2030-2050.

Temperature is another important factor in efficiency of the photovoltaic solar systems. Researches have revealed that increasing temperature reduces the efficiency of PV solar cells. So, it is important to set a cooling system for PV cells, which provide an excellent opportunity for aluminium to extend its role in solar cells in near future [75].

Another advantage of aluminium over other kinds of materials that encourages many companies in different fields, especially in solar power systems, to use this metal and its alloys is successful and cheap recycling process for aluminium.

Recycling is an important industrial process that can reduce energy usage, air and water pollution and save money. So, recycling should be considered in any kinds of materials that will be employed in industrial or home applications. Recycling is not successful for all materials. Recycling of plastics, for example, is very problematic, but glass can successfully be recycled. Aluminium is the most successful material in recycling. Its recycling is much cost saving process as producing aluminium goods from recycled aluminium cost less than 10% of producing from bauxite ore. Low melting point is the main cause for cost-saving property of aluminium recycling [49]

2.5. Photovoltaic modules and corrosion

The performance of solar fields and their power sale revenue are directly dependent on the reflectivity (for solar thermal) and solar panel absorptivity (for solar PV). Unfortunately the mirrors and panels are subjected to exposure to dust moisture and aerosols resulting in a significant drop in system efficiency [76].

Field data has shown that without any external cleaning, reflectivity dropped off with time at a rate of 0.45 percentage points per day and a loss of 16 percentage points in a month (figure 8). Currently the mirrors are cleaned using demineralized water by high pressure spray, deluge type cleaning methods, and fixing of piping spray systems. These methods, however, require demineralized water in the order of 0.23gallons/m^2 of the area [77].

http://dx.doi.org/10.5772/54721

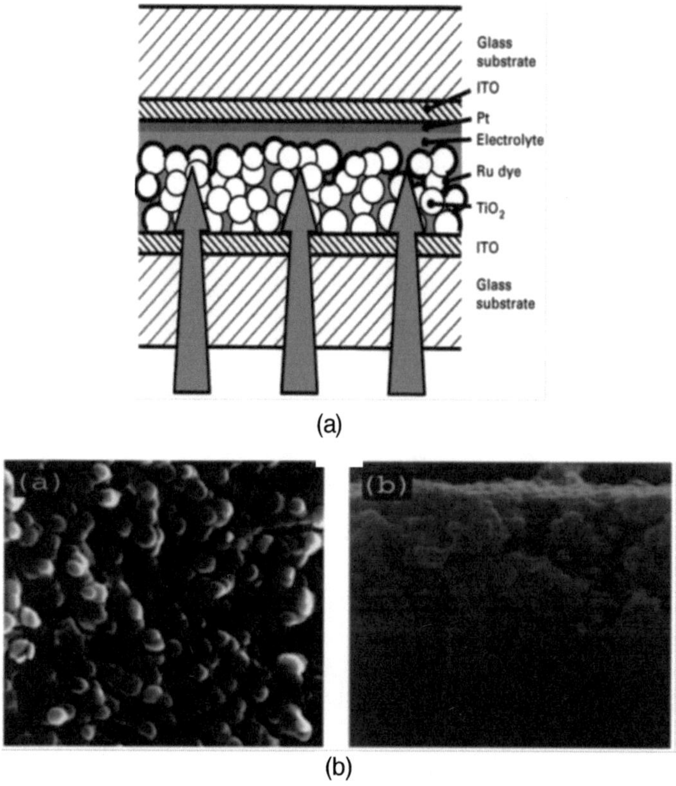

Figure 8. a): Schematic sample setup of the photoelectrochemical cells using nanocrystalline TiO_2. The arrows indicate the direction of incident light [79]. (b): SEM micrographs of a sol-gel TiO2 layer with a thickness of 5 μm on an ITO glass substrate. (a) surface (magnification 15000). (b) cross section (magnification 9000) [79].

Due to high accumulation rate of dust large quantities of water are require. The washes unfortunately recovered only 1-3% of the loss in reflectivity. Photovoltaic systems use semiconductor technology in dust and moisture laden region such as the east coast of Saudi Arabia. There is insufficient experience in dust laden and moisture laden region and the problem is very serious. It has been stated that the accumulation of dust on a PV panel causes a loss of 0.5% in PV efficiency. Currently commercially available films have a barrier property of water vapor transmission rate of 10^{-3}g/m^2 per day or one thousand gram square meter per day at 90%. The barrier oxide film therefore develops cracks and pinholes. The life of PV cell is drastically reduced in desert with expended span of humidity and suspended particles (SO_2, NO_X, Cl) [78].

The harsh environment has put a serious demand for fabrication of moisture resistant, corrosion resistant, water and dust repellent surface to prevent PV cells and module. This problem needs three approaches,

a. A dry approach

b. A wet approach

2.5.1. The dry approach

This approach is based on surface modification without using water. In certain cases the surface chemistry may be modified by applying low pressure compressed air and installing the system against the direction of winds, and creating conditions which do not allow the dust particles to settle down. Nanostructured surface could be created with ultra-fine grain size which may provide self-cleaning.

2.5.2. The Wet approach

The wet approach would depend on fabricating hydrophobic films. One good example is provided by Titanium dioxide films for photovoltaic cells derived from a sol-gel process.[79]

Figure 8 describes the set-up of photo electrochemical cells using nano crystalline TiO_2. A sol-gel process is used for spin coating. After being sensitized with a type the films are for transport of electrons in the photo electrochemical cells.

The particles having a size of ~ 100nm. In this process the heating process and application of TiO_2 coating by spin coating technique needs to be controlled to produce high efficiency. The demand to produce dust and water repellent coatings is increasing because of its application in MEMS/NEMS and photovoltaic cells. However a considerable progress has been made in producing hydrophobic and dust repellent coating. The best symbol of hydrophobicity is lotus leaf which exhibits water repellent and self-cleaning characteristic. A detail study of lotus leave shows two levels of structures a nano like hair structures with wax covering the surface and micro scale mounds protruding from the surface figure 9 [80].

Attempts have been made to fabricate nano coatings with inherent self-cleaning properties. The lotus coating is microscopically thin and optically transparent. The coating sheds dust particles utilizing anti- contamination and self-cleaning properties. It is designed to preserve optimum long time performance. The coating can be applied to metal glass surface and epoxy composite substrates. In a desert there may be no rain, however, this technology needs to be complemented by dust management technologies which include, electro dynamic dust shields, thermal control systems and surface power systems. Tribo electric charging has been conducted under Mars surface conditions. The charging showed that large amounts of dust could be accumulated [81].

The super hydrophobicity or wettability is characterized by contact angle formed a liquid droplet and solid surface. A surface is consider hydrophobic if the water contact angle (WCA) is greater than 90° and super hydrophobic if the WCA is between 150° and 180°. The contact angle hysteresis is the difference between the advancing angle (ϕ_{adv}) receding (ϕ_{rec}) contact angles. Roughening a surface enhances its water repellent property. The effect of surface roughness has been described by different theories [82].

$$Cos\phi_w = \gamma Cos\phi_y$$

Where ϕ_w and ϕ_y are contact angle on the rough and smooth surface and γ is the roughness factor define as the ratio of solid-liquid area A_{SL} to the projection on a flat plane A_f.

Figure 9. a) Lotus leaves, which exhibit extraordinary water repellency on their upper side. (b) Scanning eletron microscopy(SEM) image of the upper leaf side prepared by glycerol substitution shows the hierarchical surface structure consisting of papillae,wax clusters and wax tubules. (c) Wax tubules on the upper leaf side. (d) Upper leaf side after critical point (CP) drying. The wax tubules are dissolved, thus the stomata are more visible. Tilt angle 15^0. (e) Leaf underside(CP dried) shows convex cells without stomata [80].

$$\gamma = \frac{A_{SL}}{A_f}$$

According to Wenzel, the space between protrusions is filled by liquid. Cassie and Baxter model [83] assumed that air trapped by asperities of hills on the surface and a composite interface is formed. The composite interface may form air pockets trapped in the cavity. The relationship between apparent contact angles C_β and ideal contact angle is given by

$$Cos C_\beta = r_f fCos\phi_f + (f-1)$$

Where r_f indicates roughness factor of the wetted area and f is the area fraction of the projected wet area. The product r_f is called solid fraction of limit f reached to zero the macroscopic contact angle C_β approaches to 180° leading to a super hydrophobic behavior. If ϕ_r is high, the drop would be in a Cassie and Baxter and low it would be Wenzel region. Both regions are shown in figure 10 [84].

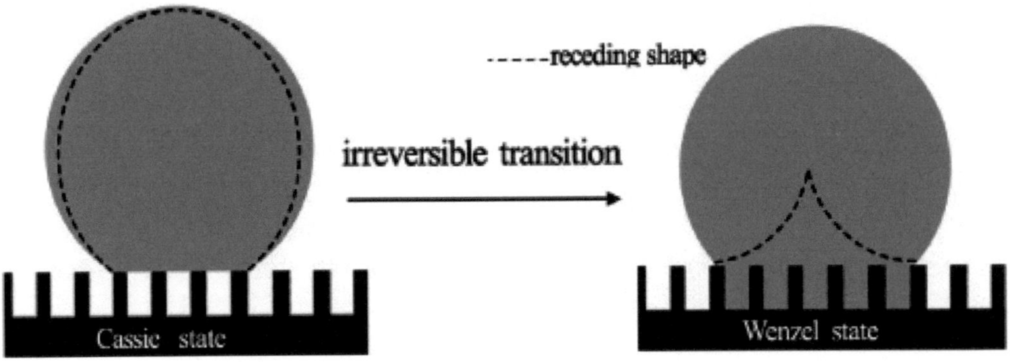

Figure 10. Irreversible Transition from the Cassie State to the Wenzel State by the Wicking Test [84].

Two methods are in used for fabrication of super hydrophobic surface; a) making a surface with low energy materials or making a rough surface and coating with low surface materials like tetrafluoroethylene (TEFLCN) or polydimethylsiloxanes (PDMS).

Modifications of a surface with low energy compounds the process such as plasma etching [85] laser etching [86] chemical etching [87] lithography [88] and sol get techniques have been used to produce super hydrophobic surfaces.

Wet coating techniques have been used with success to obtain super hydrophobic surfaces base the critical factor is the preparation of an optimal rough surfaces. Processes such as sand blasting and annealing have been used to obtain a rough surface with grain size in nano dimension [88].

In a study on making a super hydro surface on aluminum samples were mechanically polished to obtain a rough surface in the range of 0.05 to 1.5μm. The surface was modified by treating polydimethyl siloxane vinylterminated (PDMSVT) with 1 weight % Curing agent as cross linked perfluoroethylene (PFPE).

http://dx.doi.org/10.5772/54721

PFPE film was applied by spin coating (3000 rpm). Curing was done for C_9F_{20} at 80°C and PDMSVT and PFPE at 120°. Sol get process titanium coating have been produced and they are extensively reported in literature. Optically transparent super hydrophobic silica based films have been formed by control of hydrolysis and condensation reaction.

In a work under taken [89] three types of originally modified silica gels have prepared as shown in the Table 3.

Sols	Composition	Catalyst	pH	surface bonds after etching
A	TEOS,MPS	HCL	2	Si-O, Si-OH
B	TEOS,MPS	HCL	6	Si-O, Si-OH
C	TEOS	NH_4OH	10	$Si-O_2$, Si-OH

Table 3. Types of Modified Silica Gel

After mixing, the surface chemistry was modified with self assembles monolayers using chlorotrimethylsilane (CTMS) (CH3) (SiCl) and tetra fluorotetrahydroxydimethylchloro silane, hydro phobicity followed the order C, B and A.

Substrate of soda lime glass coated with a thin layer of tin oxide was doped with fluorine (SnO2:T). The transparent conductive oxide (TCO) form, the contact for the solar cell layers deposited on tin oxide. An ethylene vinyl acetate provides both electrical isolation between solar cells and binds the glass back sheet to the module. An extruded aluminum sheet provides structured mounting points and grounding points. Damp heat tests were applied to the modules [90]. A voltage of +600V was applied to the output of one modules and -600V to the other module. The modules work placed in an environmental chamber. Module biased at -600V (amorphous silicon) visible damage occurred. Modules subjected in positive bias showed no or little damaged. The dependence of corrosion damage to polarity suggests the reaction of Na^+ of the glass with fluorine. Humidity acts as an electrolyte for corrosion to take place. Virtually no damage was observed under positive biased polarity. the positive biase only reacts with contaminants outside the glass surface and does not need the mobility of ion.

on subjecting the modules to -600v and 85% RH at 85,72 and 60 °C/ leakage current were observed (fig 11). It was shown that water vapor enhance the leakage of current [89].

Electrochemical corrosion is the result of transport of metal ions between the cell and module frame. under the influence of large cell frame voltage(100-500-V) cell frame voltage present in applications with high system voltages above or below the ground potential [90].The problem is being solved with microencapsulation with low conductivity and control of ionic conductivity of encapsulated free surface and interface. The leakage current is highly dependent on ionic conductivity and moisture level. The transport of ionic elements is associated with corrosion of internal cell material by leakage of current from the cell string to the module exterior. Both processes electro migration and ionic migration need to be controlled [91].

Numerous experiments have been conducted on module, fabricated on the oxide coated glass and operated at voltages and elevated temperatures in a humid climate. The phenomena of electrochemical corrosion appeared to be in all cases related to humidity. Experiments conducted BP Solar a-Si/a- SiGe tendem module fabricated on tin-oxide coated glass and encapsulated with another sheet of glass with EVA, showed that sodium reacted with water to form sodium hydroxide and hydrogen and highly alkaline solution PH>9 such that sodium hydroxide will dissolves Silicate glass rapidly at temperature (≥ 100°C).

In addition, the generation of hydrogen bonds near the interface may lead to a weakening of interfacial bonds because of reduction of tin-oxide. The relative durability of Zinc Oxide appeared to be due to the fact that Zinc oxide is not reduced [92].

A common observation in all investigations shows that corrosion can be minimized by use of low alkali or high resistivity glass, by increasing the adhesion of the transparent conducting oxide to the glass surface or using zinc oxide rather than tin-oxide as a transparent conductive contact. The use of anti-reflection coated (ARC) glass is being used in an increasing percentage of PV modules due to expected high power energy output. The use of ARC glass declined because of the inability of the coating to maintain performance over long period of time. Recent progress made has given some confidence to the consumers to use it again. Several defect such as coating degradation, soiling and optical degradation have been observed. Recent progress in ARC glasses has been shown by ARC glass is developed by the sun power. It has been shown that a well-designed ARC coating protected the glass from humidity and sand blasting, whereas the uncoated glass showed chippings. More than three years of field data showed that the energy gain from ARC significantly exceeded by 3.5 to 5% over uncoated glass which is the consequence of the improved coating gains in diffuse and off angle lights, due to effect of refractive index and light scattering within the coating. The sun power modules are slowly emerging in the market.

Metal electro migration is a big concern in electronic industry along with corrosion, module and discoloration. The metal migration reduces the service life of module. SAFlex PS-41 is the first module to be produced to suppress electro migration by exploring the activity of embedded encapsulation which prevents electro migration when in contact with metals such as silver, copper and nickel. The first encapsulant Saflex PS41 is specifically designed to protect against metal diffusion from solar cell stacks, adhesive and bus ribbons [93].

Whereas enough evidence has been shown how the adverse effects of humidity vapor pressure and current leakages on the glass substrate the back sheet materials in ARC also play a pivotal role. In recent year high moisture barrier and high resistivity coatings on polyethylene terepthalate (PET) have been fabricated for application in PV module back sheet application. It is necessary for the back sheet completely insulating to prevent a conduction path from the back contact to the grounded metal frame. To prevent the penetration of moisture and create low water vapor transmission rates WVTR, g/m^2, d., cost effective coatings have been created on inexpensive polymers such as polyethylene terepthalate (PET) and biaxial oriented polypropylene (BOPP). Table 4 shows the thickness and transmission rates of various coatings applied PET [94].

Material	Thickness(mm)	WVTR(g/m²d)
Tedlar/ Al/Tedlar C	0.1	
NREL coated PET	0.18	0.1-0.2
Tedlar / PET/ EVA (TPE)	0.2	3.0
PET	0.1	3.4
EVA	0.4-0.5	27-33

Table 4. WVTR for Polymer laminates at 37.8° and 85% Relative humidity

These back sheets can be used as a substitute for glass if they resist the ingress of moisture and transport of current. The treated polymer has a dramatic improvement over the untreated polymer. The peel strength for different back sheets is shown in Table 5.

Material	t=0	t=1week	t=2weeks
Uncoated PET	1-1.7	—	~ 0.8
NREL coated PET	7-8	7.10	6.5-6.8
TPE	—	—	~4.0

Table 5. Peeling Strengths

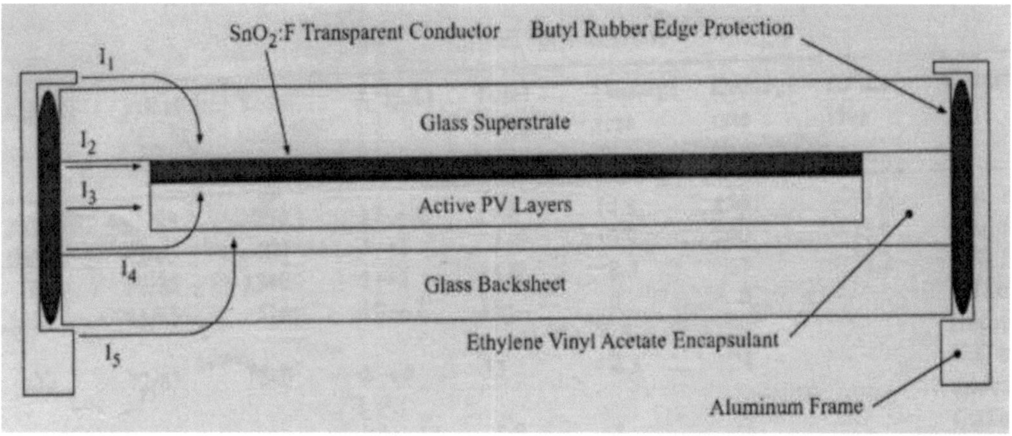

Figure 11. A cross section of a PV module constructed with an SnO_2:F transparent conducting oxide(TCO) layer deposited on a glass superstrate. The active semiconducting layers are deposited over the tin oxide, and the entire package is encapsulated with ethylene vinyl acetate (EVA) between another sheet of glass. Not shown are the laser scribes that form the individual solar cells connected in series. Five possible current paths between the frame and the TCO are illustrated (1) along the surface and through the bulk of the glass superstrate, (2) along the glass superstrate-EVA interface, (3) through the EVA bulk,(4) along the glass backsheet-EVA interface and through the EVA bulk, and (5) along the surface and through the bulk of the glass backsheet, and through the EVA bulk [95].

The above coating shows good adhesion, weather ability and low water vapor transmission rates. These coatings have a good adhesion to EVA after UV or damp heat exposure. The amount of water vapor that diffuses into EVA controls the stability of glass EVA surface. It may that glass/ EVA interface favors condensation reactions and the hydrolysis reactions are difficult to achieve. The exact mechanism is not known, however it is confirmed by research studies that humidity and leakage current play a very predominant role in degradation of PV modules. The important pathways are showing figure 11 [95].

In a study H-V induces leakage current from eight modules, a pair from each module type of C-Si, Pc-Si. (Bulk Si) and tendem junction and multi-junction a-Si were monitored for eight years. It was observed that the leakage currents from C-Si Pc-Si module were thermally activated and the activation energy varied with RH ranging from 0.86-1.0ev at high RH to 0.8ev as low RH. The leakage currents for a-si modules were much lower than bulk Si modules by a factor of 10 to100 times. After operation of HV tests for 24 hrs a day the loss rates were 0.0% year for pL –s, 0.1% year for c-s, (positive polarity) between 0% as 0.05% year for ja-si module. The bulk of the modules degraded compared to modules not biases at HV. For thin film modules the losses were insignificant. The detail of modules discussed as shown in Table 6 [94].

Module type	Structure front to back	Area (m^2)	Perimeter (m)	Mounls
C-Si	Glass/ C- Si cells/ Tedar	0.60	3.4	All edges frames
Pe-Si	Glass/ Pc- Si cells/ Tedar	0.52	3.1	All edges frames
2 Ja-Si	Glass/ TCO/a- Si /Al/ glass	0.76	3.7	Rear Brackets
3Ja- Si	Flouro polymer/TCO/a-Si/ SS	0.45	3.3	All edges frames

Table 6. Construction and size of PV Modules Tested

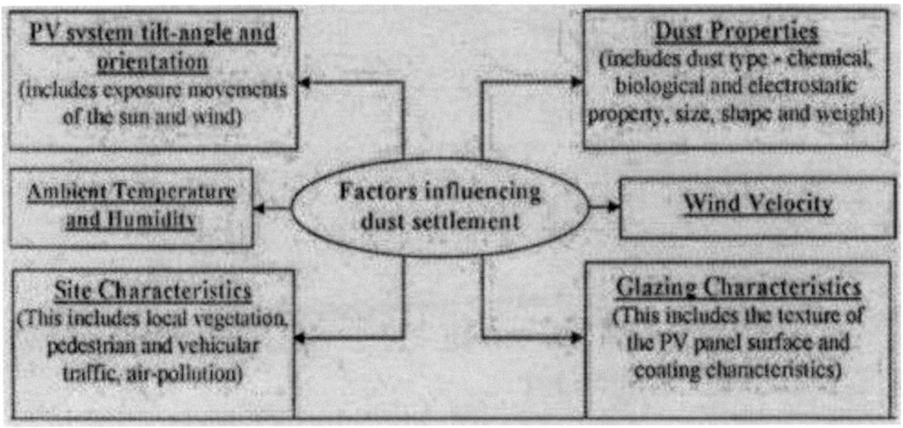

Figure 12. Factors influencing the dust settlement [96].

http://dx.doi.org/10.5772/54721

2.6. Importance of dust

Dust is a term which is applied to solid particle with less than 500μm. The main sources of dust are; dust laden winds, volcanic eruptions etc. It also includes micro pollens, microfiber, which are scatter with atmosphere. The factors influence the dust settlements are shown in figure 12 [96].

Dust promotes dust. It settles in region of low vapor pressure induced by the high pressure movements on inclined /vertical surface. The PV system is affected by several environmental factors as shown in figure 13 [96].

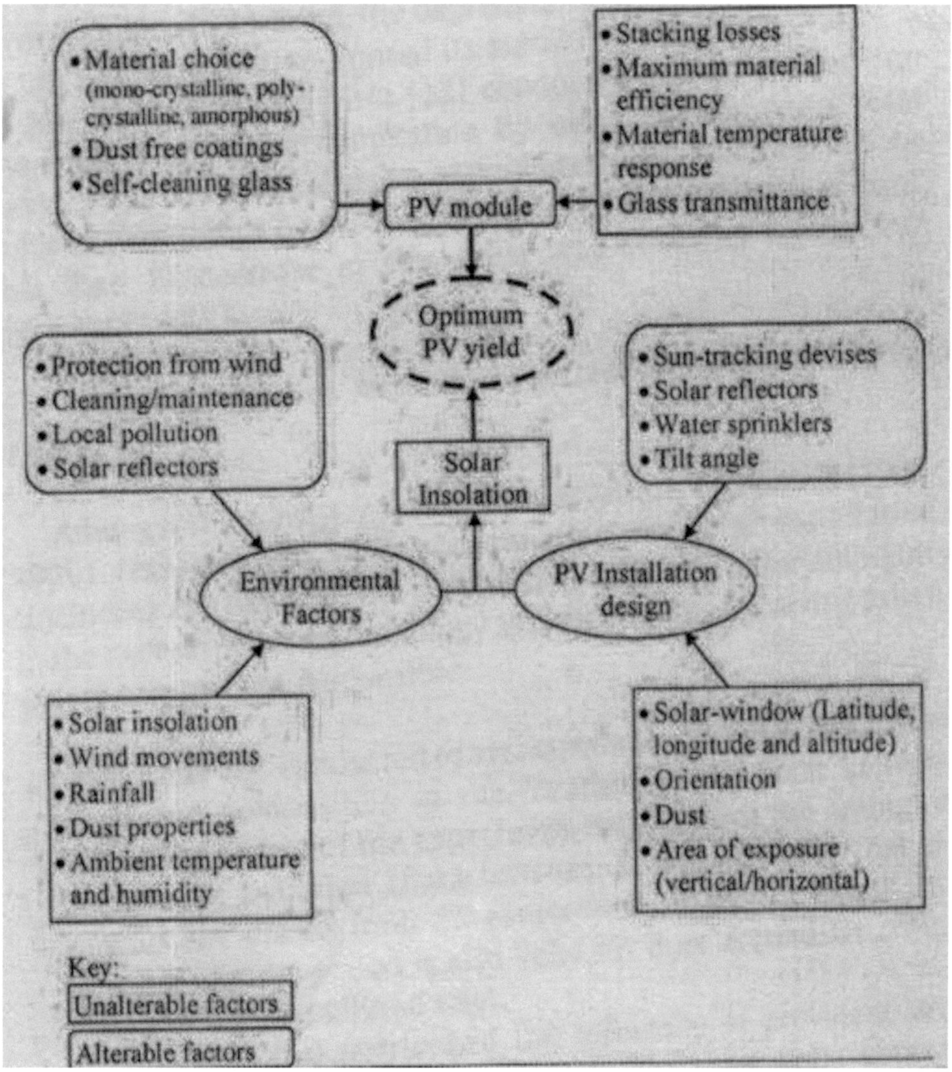

Figure 13. Alterable and unalterable factors determining maximum PV system yield [96].

Amongst all the environmental factors; removal of dust is the most complex factors in PV module. As reported in the literature degradation of 26-40% in the efficiency of thermal panels are photovoltaic cell was reported for installation in Saudi Arabia which is prevalent with dessert wind [97].

The impact of the dust on the reflection of glass in shown figure. Despite few developments in surface technology such as lotus surfaces of work in program with NASA, cleaning by wet method is still predominant.

Future exploration on moon would require mitigating the difficulties posed by Lunor regolith includes dust. NASA is tasked with the development of mitigation strategies of lunar dust. What has been achieved so far is the development of an Electrodynamics dust shield to minimize dust accumulation, a technique which also could be used to remove dust from PVC modules. The dust removal is achieved by applying a multiphase travelling electric field to the electrode that are embedded in the surface to lift and transport charged and uncharged particles off the surface. Following is a brief description of the electrodynamics dust shield technology being developed by NASA. A schematic of three phase Electrodynamics dust shield is shown in figure 14 [98].

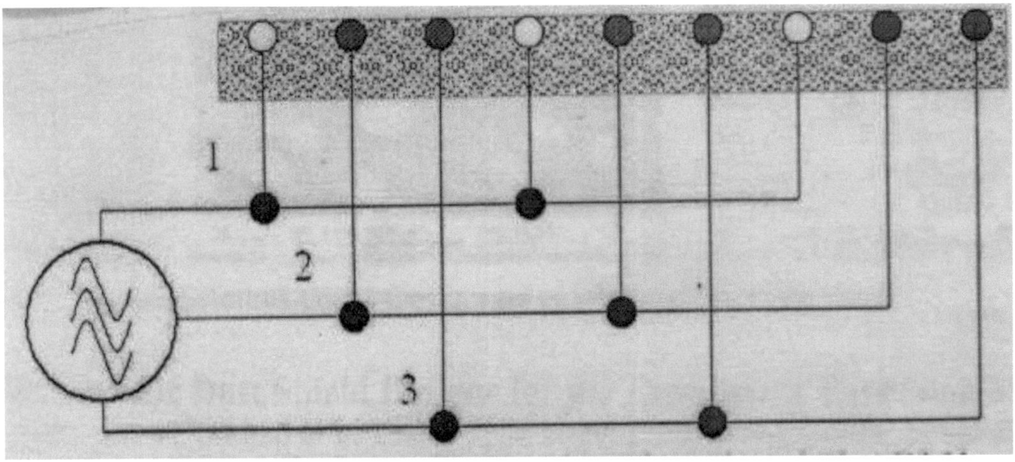

Figure 14. Schematic diagram of a three-phase Electrodynamic Dust Shield [98].

It consists of a series of a parallel electrodes connected to a multiphase AC source that generates a propagating electrodynamics wave. The wave transports the dust particle to a specified location. An electric field is generated by the signal output. The strength of electric field varies proportional to the potential difference between electrodes which is controlled by phase shift. The uniform field carries the charged particle [99].

NASA developed transparent 20cm diameter EDS Indium tin oxide (ITO) electrodes on a polyethylene (PET) film. For testing three electrodes of copper 20cm X 25cm for EDS were constructed and two of them were coated with a lotus film. A lotus coating is a two layer system containing microfills (protrusions) ~2.0 mm, and micro valleys containing epicuticular wax crystals and nano hairs with nano pores. The two layered hierarchical surface is covered by low energy compounds with very low surface energy like PDMS, fluorocarbons and other low energy compounds. Such a two level surface can be created by chemical etching or laser etching to make it a rough surface where the average grain size is in nano region. Sandblasting and short penning or cavitation shotless penning can also be used to make a two level surface. Work on stainless steel has shown the effectiveness of this process. Nanostructured films of TiO_2 were produced by mixing tetra-n-butyltitanate,ethylacetoacetate and ethanol by sol gel technique [100].

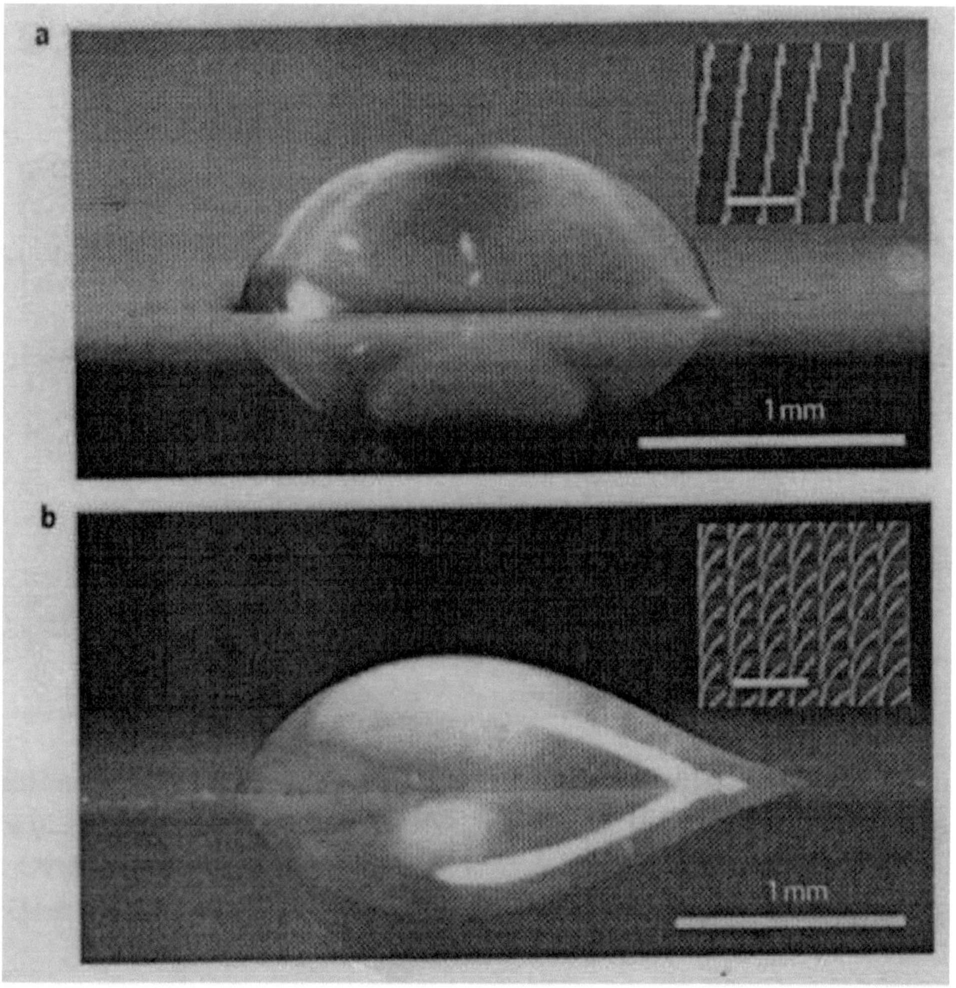

Figure 15. Comparison of wetting behavior on symmetric and asymmetric nanostructured surfaces. a, Axially symmetric liquid spreading of a 1 µl droplet of deionized water with 0.002% by volume of surfactants (Triton X-100) deposited on typical vertical nanopillars with diameters of 500 nm, spacings of 3.5 µm and heights of 10 µm(inset).b, Unidirectional liquid spreading of a droplet on the same dimension nanostructures as a, but with a 12⁰ deflection angle(inset). The images show the characteristics of a spreading droplet at one instant in time. The scale bars in the insets are 10 µm [102].

The secret lies in preparing a tailored morphology of the surface. The surface can be textured for hydrophobicity (water repelling). The surface exhibit micro convexity with clusters of nanoparticles (99,100). Such surface can trap a large amount of air which has the ability to induce large wet contact angles for hydrophobicity (170^0).Water drops on such surfaces become detached, rolls down and carries the dust with them. However in arid regions, there is hardly any rainfall for this phenomena to occur. Super hydrophobic surfaces can be prepared on metals, glasses and plastics. Similarly a hydrophobic surface can be prepared by depositing films of TiO_2 by hydrolysis of titanium aloxides and hydrolysis of TEOT (Triethylorthotitanate). These films are hydrophilic (water repelling) and also remove contaminants and microbes by the photocatalytic reaction induced by TiO_2 particles. Hydrolysis and condensation of titanium aloxide yield Ti-O based network.

The above surface obtained is hydrophilic. It is also possible to control the direction of flow of water (uni-directional) on asymmetric nanostructured surface by allowing the liquid to flow in one direction and pin on other direction which can be very helpful for various geometries of photovoltaic modules [101]. Figure 15 shows a comparison of wetting behavior on symmetric and asymmetric nanostructure surface [102].

The lotus coatings being worked out by NASA is developed on the principle described above.

The lotus surface developed is expected to mitigate dust without use of water as in the case of hydrophobic surface. The lotus coating two level would shed particles utilizing anti contamination and self cleaning properties which would minimize dust accumulation. Such coatings based on the structure of lotus flower such as hydrophobic coatings on glass and plastics would have the capability to repell dust. NASA is developing both hydrophobic and hydrophilic coatings which are next generation coatings to minimize dust on solar cells and thermal radiation. Self cleaning and anti-contamination systems being developed have also the capability to kill bacteria, chemical agents, pathogens and environmental pollutants. In future the super hydrophobic coatings would play a leading role not only in lunar environment but also in solar cells and most importantly in space exploration. The hybrid coating for photovoltaic solar arrays are shown in figure 16 [103].

In order to understand the working of EDS, it is important to understand the forces which are responsible for lifting the sands.Two types of forces are applied by the electrodynamic field; a: Electrostatic force and b): di electrophorelectric force.

Most airborne dust particles acquire an electrostatic charge during their detachment process. Each sand particle is subjected to a sinusoidal excitation voltage generated by the electric field. A charge particle experienced two forces of repulsion, one tangential and other normal to the contact angle. The lift force for the particle is provided by centrifugal force which is induced by the curvilinear motion of the particle. Another particle charged with –q will be subjected to repulsive force and it would levitate of the lifting force is larger than the adhesion force due to the cumulative effects of lifshitz –vander walls forces, electrostatic forces and capillary forces. There is hardly any capillary force in the desert region. On energizing of three phase voltage,the charged particles are lifted from the surface by the vertical component of the field and the travelling wave component as mentioned earlier carries the dust to the screen. Single phase excitation lifts the particles. This process becomes more effective when a three phase voltage is applied.

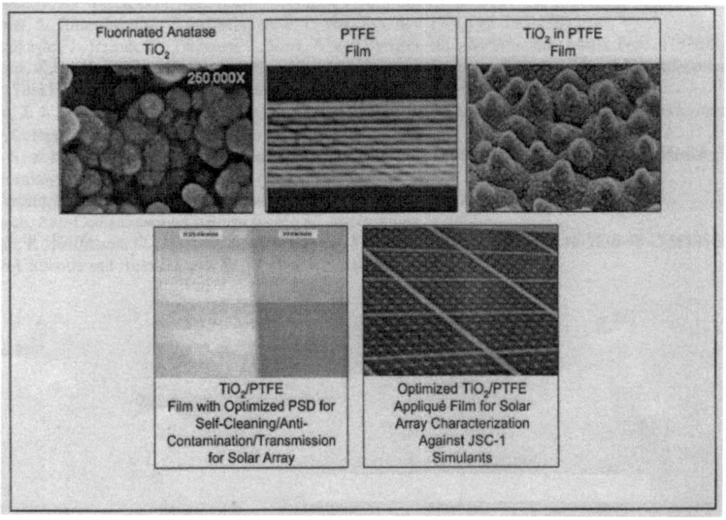

Figure 16. Hybrid coating for Photovoltaic solar arrays [103]

Another force to be considered is dielectrophoretic force. It is experienced by charged or un-charged particles in any AC or DC field (E). Because the particles +q and – q are charged and separated by a distance, a dipole moment (qd) is formed. Because of induced dipole mo-ment these particles experience dielectrophoretic force. The applied voltage creates a gradi-ent in the electric field. The divergence of electric field applies a dielectrophoretic force F_d and a torque T. This force causes the movement of neutral particles on the surface and indu-ces electrostatic charging by triboelectrification. This acquired charge would induce to the columbic force of repulsion to lift the particles. In summary, columbic and dielectrophoretic forces move the dust particles to the surface and hence the particles acquire charge. These charge particles are repelled by electrostatic forces.

This mechanism applies also to conducting particles deposited on the shield. The charge q is proportional to E^2. Particle is the vicinity of electrodes acquire electrostatic charge and they are repelled when the force of repulsion $F_{Repulsion}$ =9 E_0= E_0 r^2 is > force of adhesion ($F_{Adhesion}$).The particle would be lifted.

By applying a three phase voltage 90% of the dust is removed in about two minutes.

The energy requires for dust removal is only a small fracture of the energy output of the modulesc [104].

The electrode grid uses indium tin oxide or carbon nanotubes. Figure 11 shows transparent EDS coatings in glass.The role of aluminium has become very predominant in solar power system. The solar power system has been divided in four distinct groups, parabolic trough, parabolic dish, linear fersenel and solar tower. Aluminium is one of the most important material utilized in solar absorbing due to the capability of its anodic layer formed on its surface by the process of anodization. Eutectic binary aluminiumm alloys such as Al-0 wt% Ni, Al-33 wt% Cu and Al-7.5wt% Ca have been successfully used as absorber(low reflection and high absorption).The mechanical and thermal ability of aluminium alloys and regeneration of surface is etching enhances their properties in solar power system.

Aluminium extrusion provides a clear economic advantage in the product of solar application. White steel costs less than aluminum on a dollar per pound basis, the lower weight of aluminium (1/3rd of steel) allow far more material to be used at a lower cost.

Because of its recyclability,light weight, high strength and high corrosion resistance,it has become a preferred material. By using aluminium Alcoa is saving on costs of solar pond and transportation.

Hydro in Germany is making mirrors for concentrated solar power as well as absorber sheets for solar thermal application. This company launches the first tailored Hybridlife solar aluminium alloy for concentrated solar power and launched high select coatings for solar thermal systems. Hydro serves its customer in solar thermal concentrated solar power and photovoltaic for all area of solar energy product.

Ali Baba has manufactured 1 KW, 2 KW, 3 KW and 10 KW solar power systems,aluminium based with a life span of 25 years.(AliBaba.com). Light technology is a leading and provides standard aluminium profile for mounting systems for solar energy.

Pacific power management (PPM) has announced the installation of a 800 KW power plant for Sierra Aluminium Company. It contains 4,480 Mistzubish Electric,180 watts modules and 2 satcons 5oo Kw invertors. It generates 1.4m Kwh per year.

It would reduce Sierra carbon footprint by 48% and supply power to 28,000 homes over a 25 year period.

The use of solar mirrors could reduce the cost by 20%. For a 50 megawatt power plant, a saving of 20 million euros could be made.

2.7. Conclusion

Aluminum is playing a predominant role in solar power system because of its technical capability,ease of fabrication and ease of transport use, recyclability and resistant to corrosion. The promising future of aluminium in solar power is reflected by the projections on market growth from 210 mm^2 to 11 bmm^2. By 2050, the amount could reach 39 mtons from the existing 17 mtons. The major attributes are large energy area for collection, solar directed installation and dynamic development. However there are several technical problems associated with solar power such as the ingress of moisture causing corrosion and leakage of current causing deterioration of modules. The water vapours ingresses through the edges and increases the conductivity of the front glass surface and also the magnitude of leaking current. In the four types of modules a). C-Si, b). PC-Si, c). 2J a-Si (Glass/TCO/a-Si/Al/Glass) and 3 Ja-Si (Fluropolymer/TCO/a-Si/stainless steel), the two modules containing a-Si showed the

http://dx.doi.org/10.5772/54721

maximum resistance to HV operation. In HV operation, all modules degrade at rates higher than the modules not biased on HV. Films on PET showed promising properties as a back sheet replacement for glass. These coatings exhibit excellent moisture resistance properties and a good cohesion after exposure to damp heat. The corrosion effect can be minimized by increasing adhesion of transparent oxide by using Zinc oxide in place of Tin oxide and by using low acetate and high resistivity glass.

Dust is still haunting the scientists and engineers working on solar and space equipment. It is of vital importance to solar panels and equipment used in space exploration. A substantial amount of research has been done on electrodynamics system to remove dust. This is couples with creating a lotus surface (two level) hierarchical surface (nano/micro hybrid) to create self cleaning properties for removal of dust by mimicking the surface of a lotus flower. Various paints containing self cleaning agents have also been designed to remove dust. The wet chemistry route creating a superhydrophobic surface is an outstanding achievement but it cannot be applied in dessert conditions. Intensive work is undertaken by NASA to create dust shields. It appear that new techniques would be developed to mitigate the degradation of PV modules and the use of aluminum would continue to rise.

Acknowledgements

I am highly indebted to Ms. Zahra Khan (Comsats) and Ms. Tayyaba Abid (Comsats) for their dedicated help in preparing the manuscript of the chapter for the book. The above colleagues have put in a very dedicated work in all technical aspects related to the formatting processing and editing of this script.

Author details

Amir Farzaneh[3*], Maysam Mohammadi[4], Zaki Ahmad[1,2] and Intesar Ahmad[5]

*Address all correspondence to: amir.frz@gmail.com

1 KFUPM. Dhahran, Saudi Arabia

2 Dept Of Chemical Engineering, Comsats Lahore, Pakistan

3 Dept of Materials Engineering, University of Tabriz, Tabriz, Iran

4 Department of Materials Engineering, University of British Columbia, Vancuver, B.C, Canada

5 Department of Electrical Engineering, Lahore College For Women University, Lahore, Pakistan

References

[1] T.K. Ghosh MAP. Energy resource and systems, Volume 2: Renewable sources. 1st ed. : Springer; 2011.

[2] Xiaoming L., Duowang F., Fan Y. A study of the organization and performance of thermally evaporated aluminium reflector for solar energy system. 2nd Conference on Environmental Science and Information Application Technology, ESIAT 2010, July 17, 2010 - July 18; 2010; Wuhan, China: IEEE Computer Society; 2010.

[3] Mazria E. Passive solar energy book. Emmaus: Rodale Press; 1979.

[4] Stine WB HR. Solar energy fundamentals and design. New York: Wiley; 1985.

[5] Kreith F KJ. Principles of solar energy. Washington, DC: Hemisphere Publishing; 1978.

[6] Tiwari G,. Solar energy: fundamentals, design, modeling and applications. New York: CRC; 2002.

[7] Gupta PK. Renewable energy sources-a longway to go in India. Renewable Energy 1999;16(1-4):1216-9.

[8] Ummadisingu A, Soni MS. Concentrating solar power - Technology, potential and policy in India. Renewable and Sustainable Energy Reviews 2011;15(9):5169-75.

[9] Anderson B. Solar energy: fundamentals in building design. New York: McGraw-Hill; 1977.

[10] Malik MAS, Tiwari GN, Kumar A, Sodha MS. Solar distillation. New York: Pergamon Press; 1985.

[11] Kreider JF KF. Solar heating and cooling. New York: McGraw-Hill; 1977.

[12] Kalogirou S. Solar water heating in Cyprus: current status of technology and problems. Renewable Energy 1997;10(1):107-12.

[13] Kalogirou SA. Solar thermal collectors and applications. Progress in Energy and Combustion Science 2004;30(3):231-95.

[14] Bean JR., Diver RB. Performance of the CPG 7.5-KWe dish-stirling system. Proceedings of the 28th Intersociety Energy Conversion Engineering Conference, August 8, 1993 - August 13; 1993; Atlanta, GA, USA: Publ by SAE; 1993.

[15] Corrigan RD, Peterson TT, Ehresman DT. Update of the Solar Concentrator Advanced Development Project. Part 1 (of 6): Aerospace Power Systems and Power Conditioning, August 6, 1989 - August 11; 1989; Washington, DC, USA: Publ by IEEE; 1989.

[16] Dang A. Concentrators: A Review. Energy Conversion and Management 1986;26(1): 11-26.

[17] Fend T, Hoffschmidt B, Jorgensen G, Kuster H, Kruger D, Pitz-Paal R, et al. Comparative assessment of solar concentrator materials. Solar Energy 2003;74(2):149-55.

[18] Zhao J, Wang A, Blakers AW, Green MA. High efficiency prismatic cover silicon concentrator solar cells. Twentieth IEEE Photovoltaic Specialists Conference - 1988, September 26, 1988 - September 30; 1988; Las Vegas, NV, USA: Publ by IEEE; 1988.

[19] Gregory GG, Koshel RJ. Modeling the operating conditions of solar concentrator systems. Photonics for Solar Energy Systems, April 5, 2006 - April 6; 2006; Strasbourg, France: SPIE; 2006.

[20] Hasuike H, Yoshizawa Y, Suzuki A, Tamaura Y. Study on design of molten salt solar receivers for beam-down solar concentrator. Solar Energy 2006;80(10):1255-62.

[21] Heath J, A.R., Hoffman EL. Recent gains in solar concentrator technology. J.Spacecraft Rockets 1967;4(5):621-4.

[22] Hinsch A, Zastrow A, Wittwer V. Sol-gel glasses. A new material for solar fluorescent planar concentrators? Solar energy materials 1990;21(2-3):151-64.

[23] Irshid MI, Othman MO. V-Troughs with High Concentration Ratios for Photovoltaic Concentrator Cells. Solar Cells 1988;23(3-4):159-72.

[24] Isshiki N, Watanabe H, Shishido K, Ohtomo M, Watanabe K. Studies on solar-dish heated stirling engines TNT-3, NAS-2. Proceedings of the 28th Intersociety Energy Conversion Engineering Conference, August 8, 1993 - August 13; 1993; Atlanta, GA, USA: Publ by SAE; 1993.

[25] Kribus A, Huleihil M, Timinger A, Ben-Mair R. Performance of a rectangular secondary concentrator with an asymmetric heliostat field. Solar Energy 2000;69(2):139-51.

[26] Li M, Wang LL. Investigation of evacuated tube heated by solar trough concentrating system. Energy Conversion and Management 2006;47(20):3591-601.

[27] Maish AB. PV Concentrator Array Field Performance Measurement.Solar Cells 1986;18(3-4):363-71.

[28] Morel DE, Ayers SR, Gulino DA, Tennyson RC, Egger RA. Solar Concentartor Materials Development. 21st Intersociety Energy Conversion Engineering Conference: Advancing toward Technology Breakout in Energy Conversion. San Diego, CA, USA: ACS; 1986.

[29] Palavras I, Bakos GC. Development of a low-cost dish solar concentrator and its application in zeolite desorption. Renewable Energy 2006;31(15):2422-31.

[30] Rowan B, Mc Cormack S, Doran J, Norton B. Quantum dot solar concentrators: An investigation of various geometries. High and Low Concentration for Solar Electric Applications II, August 26, 2007 - August 28; 2007; San Diego, CA, United states: SPIE; 2007.

[31] Singh P, Liburdy JA. Solar concentrator design for uniform flux on a flat receiver. Energy Conversion and Management 1993;34(7):533-43.

[32] Tecpoyotl-Torres M, Campos-Alvarez J, Tellez-Alanis F, Escobedo-Alatorre J, Aguilar JQ, Sanchez-Mondragon J. RF control system of a parabolic solar concentrator. High and Low Concentration for Solar Electric Applications II, August 26, 2007 - August 28; 2007; San Diego, CA, United states: SPIE; 2007.

[33] Valade FH. Solar concentrator advanced development program update. Proceedings of the 23rd Intersociety Energy Conversion Engineering Conference, July 31, 1988 - August 5; 1988; Denver, CO, USA: Publ by IEEE; 1989.

[34] Wong WA, Geng SM, Castle CH, Macosko RP. Design, fabrication and test of a high efficiency refractive secondary concentrator for solar applications. 35th Intesociety Energy Conversion Engineering Conference, July 24, 2000 - July 28; 2000; Las Vegas, NA, USA: IEEE; 2000.

[35] Holbert KE, Haverkamp CJ. Impact of solar thermal power plants on water resources and electricity costs in the Southwest. 41st North American Power Symposium, NAPS 2009, October 4, 2009 - October 6; 2009; Starkville, MS, United states: IEEE Computer Society; 2009.

[36] Ortiz-Rivera E, Feliciano-Cruz L. Performance evaluation and simulation of a Solar Thermal Power Plant. 2009 IEEE Energy Conversion Congress and Exposition, ECCE 2009, September 20, 2009 - September 24; 2009; San Jose, CA, United states: IEEE Computer Society; 2009.

[37] Cambronero LEG, Canadas I, Martinez D, Ruiz-Roman J. Foaming of aluminium-silicon alloy using concentrated solar energy. Solar Energy 2010;84(6):879-87.

[38] Flamant G, Ferriere A, Laplaze D, Monty C. Solar processing of materials: opportunities and new frontiers. Solar Energy 1999;66(2):117-32.

[39] Martinez D, Rodriguez J. Materials surface treatments by concentrated solar light: A renewable energy option; Tratamiento superficial de materiales mediante luz solar concentrada: una opcion mediante energias renovables. Revista de Metalurgia (Madrid) 1998;34(2):104-8.

[40] Bakos J, Miyamoto HK. Solar hydrogen production - Renewable hydrogen production by dry fuel reforming. Solar Hydrogen and Nanotechnology, August 14, 2006 - August 17; 2006; San Diego, CA, United states: SPIE; 2006.

[41] Baykara SZ. Hydrogen production by direct solar thermal decomposition of water, possibilities for improvement of process efficiency. Int J Hydrogen Energy 2004;29(14):1451-8.

[42] Mori M, Kagawa H, Nagayama H, Saito Y. Current status of study on hydrogen production with space solar power systems (SSPS). 4th International Conference on Solar Power from Space, SPS '04 - Together with The 5th International Conference on

Wireless Power Transmission, WPT 5, June 30, 2004 - July 2; 2004; Granada, Spain: European Space Agency; 2004.

[43] Roman R, Canadas I, Rodriguez J, Hernandez MT, Gonzalez M. Solar sintering of alumina ceramics: Microstructural development. Solar Energy 2008;82(10):893-902.

[44] Brenna M, Foiadelli F, Roscia M, Zaninelli D. Evaluation of solar collector plant to contribute climate change mitigation. 2008 IEEE International Conference on Sustainable Energy Technologies, ICSET 2008, November 24, 2008 - November 27; 2008; Singapore, Singapore: Inst. of Elec. and Elec. Eng. Computer Society; 2008.

[45] Ronnelid M, Adsten M, Lindstrom T, Nostell P, Wackelgard E. Optical scattering from rough-rolled aluminium surfaces. Appl.Opt. 2001;40(13):2148-58.

[46] Machinda GT, Chowdhury S, Arscott R, Chowdhury SP, Kibaara S. Concentrating solar thermal power technologies: A review. 2011 Annual IEEE India Conference: Engineering Sustainable Solutions, INDICON-2011, December 16, 2011 - December 18; 2011; Hyderabad, India: IEEE Computer Society; 2011.

[47] IEA. Technology roadmap – concentrating solar power. 2010;.

[48] Bödeker JM, Bauer M, Pehnt M. Aluminium and Renewable Energy Systems – Prospects for the Sustainable Generation of Electricity and Heat. 2010;.

[49] Mok SC. Aluminium economy for sustainable development: Aluminium as core material for energy storage and energy saving products: Low cost, high performance, and easy processing in developing countries. 2011 IEEE Global Humanitarian Technology Conference, GHTC 2011, October 30, 2011 - November 1; 2011; Seattle, WA, United states: IEEE Computer Society; 2011.

[50] Fend T, Jorgensen G, Kuester H. Applicability of highly reflective aluminium coil for solar concentrators. Solar Energy 2000;68(4):361-70.

[51] Granqvist CG. Radiative Heating and Cooling with Spectrally Selective Surfaces. Appl.Opt. 1981;20(15):2606-15.

[52] Keller F, Hunter MS, Robinson DL. Structural features of oxide coatings on aluminium. Electrochemical Society -- Journal 1953;100(9):411-9.

[53] Henley V. Anodic Oxidation of Aluminium and its Alloys. : Pergamon Press; 1982.

[54] Woodman TP. Light Scattering in Porous Anodic Aluminium Oxide Films-1. Colour Effects. Thin Solid Films 1972;9(2):195-206.

[55] Woodman TP. Light Scattering in Porous Anodic Aluminium Oxide Films - 2. Polarization Effects. Thin Films 1972;9(3):389-94.

[56] Pavlovic T, Ignatiev A. Optical And Microstructural Properties of Anodically Oxidized Aluminium. Thin Solid Films 1986;138(1):97-109.

[57] Pavlovic T, Ignatiev A. Optical Properties of Spectrally-Selective, Anodically-Coated, Electrolytically-Colored Aluminium Surfaces. Solar energy materials 1987;16(4): 319-31.

[58] Granqvist C. Optical properties of integrally colored anodic oxide films on aluminium. Journal of Applied Physics 1980;51(6):3359-61.

[59] Kumar SN, Malhotra LK, Chopra KL. Nickel Pigmented Anodized Aluminium as Solar Selective Absorbers. Solar energy materials 1983;7(4):439-52.

[60] Cody GD, Stephens RB. Optical Properties of a Microscopically Textured Surface. 1978;40:225-39.

[61] Chang V, Bolsaitis P. Study of Two Binary Eutectic Aluminium Alloys as Selective Absorbers for Soalr Photothermal Conversion. Solar energy materials 1980;4(1): 89-100.

[62] Pellegrini G, Brughera P, Quazzo F. On the Properties of the Superplastic Aluminium-Calcium Alloy as Material for Solar Collectors. Solar energy materials 1982;7(3): 351-7.

[63] Parida B, Iniyan S, Goic R. A review of solar photovoltaic technologies. Renewable and Sustainable Energy Reviews 2011;15(3):1625-36.

[64] Wurfel P. Thermodynamic limitations to solar energy conversion. : Elsevier; 2002.

[65] Feltrin A, Freundlich A. Material considerations for terawatt level deployment of photovoltaics. Renewable Energy 2008;33(2):180-5.

[66] Duran Sahin A, Dincer I, Rosen MA. Thermodynamic analysis of solar photovoltaic cell systems. Solar Energy Mater.Solar Cells 2007;91(2-3):153-9.

[67] Brown GF, Wu J. Third generation photovoltaics. Laser and Photonics Reviews 2009;3(4):394-405.

[68] Miyashita N, Shimizu Y, Kobayashi N, Okada Y, Yamaguchi M. Fabrication of GaInNAs-based solar cells for application to multi-junction tandem solar cells. 2006 IEEE 4th World Conference on Photovoltaic Energy Conversion, WCPEC-4, May 7, 2006 - May 12; 2006; Waikoloa, HI, United states: Inst. of Elec. and Elec. Eng. Computer Society; 2007.

[69] Green MA. Thin-film solar cells: Review of materials, technologies and commercial status. Proceedings of the International Conference on Optical and Optoelectronic Properties of Materials and Applications (ICOOPMA 2006) 233 Springer Street, New York, 10013-1578, United States: Springer New York LLC; 2007.

[70] Rey-Stolle I, Garcia I, Galiana B, Algora C. Improvements in the MOVPE growth of multi-junction solar cells for very high concentration. J.Cryst.Growth 2007;298:762-6.

[71] Conibeer G, Li Y, Slaoui A, Tao M, Topic M. Advanced inorganic materials and concepts for photovoltaics. Energy Procedia 2011;10:v-.

[72] Bosi M, Pelosi C. The potential of III-V semiconductors as terrestrial photovoltaic devices. Prog Photovoltaics Res Appl 2007;15(1):51-68.

[73] Jager-Waldau A. European Photovoltaics in world wide comparison. J.Non Cryst.Solids 2006;352(9-20):1922-7.

[74] Tjotta S. Advantages of extruded aluminium in solar power generation systems processes. Advanced Materials and Processes 2011;169(1):28-9.

[75] Werner C. Aluminium extrusion in solar power applications. Light metal Age 2010(August):24-8.

[76] G.E. Cohen, D.W. Keatney and G.J. Kolb, Final report on "Operation and Maintenance Improvement Program for Concentrating Solar Power Plants" Sandia lab report CSPSAND,1998.

[77] Parabolic Trough Workshop "Cooling for Parabola Trough Power Plants, DOE presentation, NREL, PR-550-4002r,August 16-18,1999, Ontrio 2007.

[78] Zaki Ahmad and M. Ahsan, Nano-paints for Desert Environment, jr of xyz or ACE, Nov 2006

[79] Yongxiang Li, Jurgen Hagen, Winfried Schaffrath, Peter Ostrchik, Dielet Harry Solar Energy Materials and Solar Cells, 56(1995)[67-67]

[80] Hans J, Ensikat, Ditsche-Kuru P, Neinhuis C, Barthlott W. Superhydrophobicity in perfection: the outstanding properties of the lotus leaf. Beilstein Journal of Nanotechnology 2011; 2, 152–161.

[81] Mark J. Hyatt, Sharin A. Starka, Dust Management Project Advance Capabilities project office, 21000 brookport Rd, Glenn research Centre at lewis fields, Cleveland, OH 44135 USA,2006.

[82] R.N Wenz, Resistance to solid surface to wetting by wear, Ind Eng. Chem vol 28, p 546, 1944

[83] Yongxiang Li, Jurgen Hage, Winfried Schaffartg,Peter OOtschik,Dieter, Harer, Titanium Dioxide Films For Photovoltaic Cells Derived From Sol Gel Processes,Solar Energy Materials And Soalr Cells,56,1991,page no 166-174.

[84] Wu XD, Zheng LJ, Wu D. Fabrication of superhydrophobic surfaces from microstructured ZnO-based surfaces via a wet-chemical route. Langmuir 2005; 21, 2665-2667

[85] M.H. Jin, X.J Feng, J.M. Xi, J. Zhai, K.W. Chow, Feng etal; Super hydrophobic PDMS surface with low Adhesive Forces, Micro molecular rapid communication, vol. 26,2002 pp1805.

[86] X.Y. Song, J. Zhai, Y.L. Wang, L. Jiang; fabrication of super hydrophobic surface by self-assembly and their water adhesion property, J. Phy chem 13, vol. 109, 2005 pp4048-900.

[87] Z. Gaou, F.Zhau, J.Hao, W.Liu, stable Biomimetic superhydrophobic Engineering Materials J. Am. Chem. Soc,Vol 127,2005pp 15670-15070.

[88] Y.H.Yang, Z.Y.Lee, B.Wang. C.X.Wan, D.H.Chen,K.G.N.Yang, Mehanical ,Electochem and Tribological Properties of Nanocrystalline surfaces,306 stainless steel,J Phy.condens,mats, Vol 17 ,2005 pp 5441-5446.

[89] G.Z Cao, H. Tian ,Synthesis of highly porous inorganic Hybrids by Ambient Pressure, sol gel processing, Jr Sol Gel.Tech, Volume 13,1998p 315.

[90] C.R.Osterwall, T.J.Cueto, Electro chemical Corrosion of SnO_2 F. Transparent conducting layers in thin film photovoltaic module, Solar energy, Materials and solar cells, 79(2000),21-33(2003).

[91] RG Ross,Jr, G.R.Mon,L.C.Wen and R.S sigma, Measurement and characterization of voltage and current induced Degradation of thin film Photovoltaic modules, Solar Cells,27(1989) pp 289-298.

[92] Kumar S, Drevillon, A Real time Eilipsometry study of growth of Amorphous silica on transparent conducting oxide, Jr. App Phy 1989,65(8),pp 3023-3034.

[93] Solution Inc, 575, Maryvillee Centre Drive St Louis, Missouri,6314 USA,2012.

[94] G.D.Barbar,G.J.Jorgensen, K.Terwilliger,S.H.Glick.J.Pern, T.J McMohan,29TH IEE Specialist Conference , New Orlenas, Louisana, May 20-24,2002.

[95] Osterwald CR, McMahon TJ, del Cueto JA. Electrochemical corrosion of SnO2:F transparent conducting layers in thin-film photovoltaic modules. Solar Energy Materials and Solar Cells 2010; 79, 21–33.

[96] Mani M, Pillai R. Impact of dust on solar photovoltaic (PV) performance: Research status, challenges and recommendations. Renewable and Sustainable Energy Reviews 2010; 14, 3124–3131.

[97] Nimmo B, Seid Sam, Effect of dust on performance of Thermal and Photovoltaic plate collector in Saudia Arabia in Veziroglu, TN, Ed, Proc 2nd Miami Conf,1979,Page 33-35.

[98] Calle CI, Immer CD, Ferreira J, Hogue MD, Chen A, Csonka MW, Van Suetendael N, Snyder SJ. Integration of the Electrodynamic Dust Shield on a Lunar Habitat Demonstration Unit. ESA Annual Meeting on Electrostatics 2010; Paper D1.

[99] C.I.Calle, A.Chen, C.D.Immer, M.Csonoka, M.D.Hogue,S.J.Synder, M.Rodriguef and D.R.Margretta, ARSC.Aerospace Kennedy space Centre,3289. Dust removal Technology Demonstration for a lunar Habital,2010.

[100] L.Hu, T.Yoko,H.Kozuka, S.Sakka, Jr Thin solid Films(1992), 18,P 219.

[101] Kuang.Haczu, Rongxia and Evelyn. Wang, Unidirectional Liquid spreading on asymmetric nanostructure surface, Nanomaterial Letters,28th March,2010, DOI: 10-10.38/NM7276.

[102] Chu KH, Xiao R, Wang EN. Uni-directional liquid spreading on asymmetric nano-structured surfaces. NATURE MATERIALS 2010; 9(5), 413-417.

[103] Pirich R, Weir J, Leyble D. Self-Cleaning and Anti-Contamination Coatings for Space Exploration: An Overview. Conference on Optical System Contamination - Effects, Measurements, and Control 2008: Proceedings of the Society of Photo-optical Instrumentation ENGINEERS (SPIE), AUG 13-14, 2008, San Diego, CA; 2008.

[104] Malay Mazumdar, Adhesion society Meeting,Feb,14-16,2011, Boston, MA 002215.